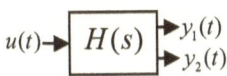

Control System Analysis & Design
in MATLAB and SIMULINK

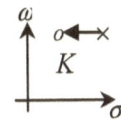

Mohammad Nuruzzaman

Electrical Engineering Department
King Fahd University of Petroleum & Minerals
Dhahran, Saudi Arabia

LULU Press, Inc.
Raleigh, North Carolina, USA
www.lulu.com

Dr. Mohammad Nuruzzaman
Electrical Engineering Department
King Fahd University of Petroleum and Minerals
KFUPM BOX 1286
Dhahran 31261, Saudi Arabia
Email: nzaman@kfupm.edu.sa, nzaman@ymail.com, mzamandr@gmail.com
Skype: nzaman1769
Web Link: http://faculty.kfupm.edu.sa/EE/NZAMAN/

© Copyright 2014, Mohammad Nuruzzaman
All rights reserved.
No part of this book may be reproduced, stored in a retrieval system, or transmitted by any means, electronic, mechanical, photocopying, recording, or otherwise, without the written permission from the author.

ISBN: 978-1-312-13951-0

Printed in the United States of America

This book is printed on acid-free paper.

To my parents
Mohammad Shamsul Haque & Nurbanu Begum

Other titles by the author:

1. M. Nuruzzaman, "*Finite Difference Fundamentals in MATLAB*", July, 2013, CreateSpace, South Carolina.
2. M. Nuruzzaman, "*Digital Image: Theories, Algorithms, and Applications*", June, 2012, CreateSpace, Washington.
3. M. Nuruzzaman, "*Digital Audio Fundamentals in MATLAB*", July, 2010, CreateSpace, California.
4. M. Nuruzzaman, "*Modern Approach to Solving Electromagnetics in MATLAB*", January, 2009, BookSurge Publishing, Charleston, South Carolina.
5. M. Nuruzzaman, "*Signal and System Fundamentals in MATLAB and SIMULINK*", July, 2008, BookSurge Publishing, Charleston, South Carolina.
6. M. Nuruzzaman, "*Electric Circuit Fundamentals in MATLAB and SIMULINK*", October, 2007, BookSurge Publishing, Charleston, South Carolina.
7. M. Nuruzzaman, "*Technical Computation and Visualization in MATLAB for Engineers and Scientists*", February, 2007, AuthorHouse, Bloomington, Indiana.
8. M. Nuruzzaman, "*Digital Image Fundamentals in MATLAB*", September, 2005, AuthorHouse, Bloomington, Indiana.
9. M. Nuruzzaman, "*Modeling and Simulation in SIMULINK for Engineers and Scientists*", January, 2005, AuthorHouse, Bloomington, Indiana.
10. M. Nuruzzaman, "*Tutorials on Mathematics to MATLAB*", March, 2003, AuthorHouse, Bloomington, Indiana.

Preface

"Control System Analysis & Design in MATLAB and SIMULINK" is blueprinted to solve undergraduate control systems and engineering problems in MATLAB platform. Control engineering is exciting and challenging in its very nature. Contemporary science and engineering are realizing many biologic and artistic facts, most of which are by dint of control engineering. The discipline is an established one and has a strong mathematical foundation. In today's computer centric world numerical analysis and design is an indispensable part in every branch of science and engineering, control system/engineering is no exception. Ultimate objective of this course is to enable the engineers to design specific input-output based controller, select appropriate parameter for desired time or frequency performance, or compensate certain criteria. Ever increasing computer processor speed assists control solution finding very fast which was time consuming in earlier days of the engineering. Never had the analysis and design of a control system with assumptions, limitations, or constraints been so swift as it is today. Verification of classically established control theories is just one click operation with the blessings of a computer. Besides unsolvable control problems in traditional computing can be tried with finite difference or element technique. It is always recommended to engage computer tools to obtain satisfactory results, our all approach and effort have been exercised in this regard.

MATLAB whose elaboration is matrix laboratory coincides fantastically in one aspect with control system. Modern control system theory at the heart is state space or matrix model based, computing element of MATLAB is array or matrix oriented too. As computing software, increasingly scientific and engineering communities are being attracted to it. Embedded built-in functions in MATLAB do not call for reprogramming an already solved problem, nor does clumsy compiling often encountered in base language such as in FORTRAN or C. SIMULINK is an add-on to MATLAB. Most study materials of control engineering assume input-output model, SIMULINK blocks or boxes also follow the same. MATLAB along with SIMULINK form an excellent instructional tool for the discipline. Self-learners will find the text helpful and interesting. Spectrum of MATLAB as well as SIMULINK on computing or graphing is extremely large frankly several volumes are required to get the full extent of both packages. We aim at presenting undergraduate control system analysis and design training through academic approach in MATLAB and in SIMULINK as well.

Some contemporary applications on control system related components or devices are in the following:
- Controllers are employed in automobile dashboards (odometer, tachometer, and fuel indicator),
- Automatic door opening or closing in shopping malls/hospitals/public offices/private offices uses controllers,
- Recent development on drones certainly are the hi-tech applications of controllers,
- Every airport landing, guiding, and takeoff system uses communication and navigation associated controllers,
- Industrial robots substitute human operation which is ultimate application of control systems,
- Biomedical experimentation and diagnosis employ control systems which work on cellular level, provide multidimensional movement, and aid handicapped with prosthetics,
- Space exploration without rover would have been impossible which operates controller run by radio frequency,
- ... and many more.

Above merely is a handful of numerous applications.

Computer-internet-cell phone trio is opening new window of thinking in all aspects of our life. Academia is no exception. A supplemental tool definitely helps learners of science and engineering fellows to solve their own problems independently thereby bringing digital democracy even in the education sector. Classroom professors can not lay a hand on all sorts of instructional materials or teaching tools because of time constraint or other reason. The core task of a control engineer is twofold; operate controller based system and carry out engineering design. Former is the practical operation, only operating process is involved. The later needs creativity and analysis, in this direction the text would be a perfect read.

Chapter 1 presents a brief introduction to MATLAB's and SIMULINK's getting-started features. Chapter 2 addresses control system implementation in MATLAB. Chapter 3 merely demonstrates control system modeling in SIMULINK. Both chapters 2 and 3 address implementation on single input single output (SISO) and multi input multi output (MIMO) control systems. Time domain control system response is considered in chapter 4 while chapter 5 explicates the SIMULINK counterpart. Frequency response is widely exercised in control system design which is solely taken care of in chapter 6. Root locus and stability of a control system are part and parcel of analysis and design on whose account chapter 7 is devised. Training on control system design is vital in the course plus hand-on and effective learning need specific code writing or modeling which we focus in chapter 8. Moreover appendices A through F explain control system related coding, function, or embedded graphing tool to the context of MATLAB/SIMULINK.

My words of acknowledgement are due to the King Fahd University of Petroleum and Minerals (KFUPM). I am especially appreciative of library facilities, control system reading materials, and MATLAB software that I received from the university.

Mohammad Nuruzzaman

Table of Contents

Chapter 1
Introduction to MATLAB and SIMULINK
- 1.1 What is MATLAB? 1
  - 1.1.1 MATLAB's opening window features 2
  - 1.1.2 How to get started in MATLAB? 5
  - 1.1.3 Some queries about MATLAB environment 9
- 1.2 What is SIMULINK? 11
  - 1.2.1 How can I get into SIMULINK? 12
  - 1.2.2 Where can I build a SIMULINK model? 13
  - 1.2.3 Block manipulation in SIMULINK 14
  - 1.2.4 Basic block categories of SIMULINK 19
  - 1.2.5 Display and Scope blocks of SIMULINK 19
  - 1.2.6 How to get started in SIMULINK? 20
  - 1.2.7 Control system library in MATLAB and SIMULINK 24
- 1.3 How to get help? 25

Chapter 2
Implementing Control System in MATLAB
- 2.1 Defining a SISO control system in MATLAB 27
- 2.2 Defining a MIMO control system in MATLAB 33
- 2.3 Defining series and parallel control systems 37
- 2.4 Feedback control systems 39
- 2.5 Control system from conversion 40
- 2.6 Interconnected control systems 44
- Exercises 55
- Answers 60

Chapter 3
Modeling Control System in SIMULINK
- 3.1 Modeling a SISO control system in SIMULINK 63
- 3.2 Modeling a MIMO control system in SIMULINK 66
- 3.3 Modeling series and parallel systems in SIMULINK 70
- 3.4 Modeling a feedback control system 72
- 3.5 Modeling interconnected control systems 73
- 3.6 SIMULINK model to transfer function 76
- Exercises 79
- Answers 82

Chapter 4
Time Domain Control System in MATLAB
- 4.1 Control system input in MATLAB 87
- 4.2 Periodic control input in MATLAB 90
- 4.3 Control system output in MATLAB 94
- 4.4 MIMO control system output in MATLAB 96

4.5 Step and impulse responses of a control system   100
4.6 First order control system performance   103
4.7 Second order control system performance   105
4.8 Error performance indices   109
Exercises   113
Answers   117

Chapter 5
Time Domain Control System in SIMULINK

5.1 Control input modeling in SIMULINK   123
5.2 Modeling unit step and its derivative inputs   124
5.3 Modeling ramp and its derivative inputs   127
5.4 Modeling sine wave and its derivative inputs   129
5.5 Modeling rectangular wave and its derivative inputs   133
5.6 Modeling triangular wave and its derivative signals   135
5.7 Modeling triggered and user-defined nonperiodic signals   138
5.8 SISO and MIMO control system outputs in SIMULINK   141
5.9 Control system performance in SIMULINK   144
5.10 Error performance indices in SIMULINK   147
Exercises   149
Answers   152

Chapter 6
Control System in Frequency Domain

6.1 Why frequency domain analysis?   155
6.2 Frequency response of a control system   156
6.3 Graphing frequency spectrum of a control system   159
6.4 Pole-zero map of a control system   162
6.5 Natural frequency and damping of a control system   164
6.6 Gain and phase margins of a control system   166
6.7 Nyquist plot of a control system   168
6.8 Nichol's chart of a control system   169
6.9 Control system DC gain and bandwidth   171
Exercises   175
Answers   177

Chapter 7
Control System Root Locus and Stability

7.1 Characteristic equation of a control system   183
7.2 Implementing a Routh table   186
7.3 Implementing a user-defined root locus   188
7.4 Root locus by embedded function   191
7.5 Stability by time domain output   193
7.6 Stability by a pole-zero map   194
7.7 Stability by Routh table   196
7.8 Stability by Nyquist plot   197
7.9 Stability with delay elements   199
Exercises   201
Answers   203

Chapter 8
Control System Projects

    Project 1: Effect of external gain on a second order system 207
    Project 2: Effect of internal gain on performance parameters 209
    Project 3: Identification of damping from time domain control output 210
    Project 4: Effect of a third order pole on performance parameters 212
    Project 5: Characteristics of a phase lag/lead network 215
    Project 6: Control system model order reduction 216
    Project 7: Liquid level control system design 218
    Project 8: DC motor transfer function and its response 220
    Project 9: Root sensitivity to a control system parameter 226
    Project 10: Space telescope pointing control system design 228
    Project 11: Elements of a root locus 230
    Project 12: Designing a phase lag network for compensation 234
    Project 13: Linking gain and phase margins of a control system 236
    Project 14: Linking damping ratio and phase margin 238
    Project 15: Controller of an inverted pendulum 240
    Project 16: Speed control by using a PI controller 243

Appendices

    Appendix A
        Coding in MATLAB 249
    Appendix B
        MATLAB functions exercised in the text 252
    Appendix C
        SIMULINK block links for modeling control systems 256
    Appendix D
        MATLAB functions/statements for control system study 260
    Appendix E
        Some graphing functions of MATLAB 281
    Appendix F
        Creating a function file 288

References 291

Subject Index 293

Chapter 1

Introduction to MATLAB and SIMULINK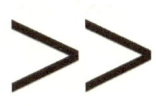

MATLAB is a computing software, which provides the quickest and easiest way to compute scientific and technical problems and visualize the solutions. As worldly standard for simulation and analysis, engineers, scientists, and researchers are becoming more and more affiliated with MATLAB and SIMULINK. The general questionnaires about MATLAB or SIMULINK before one gets started with are contents of this chapter. SIMULINK is designed to function over MATLAB. Much of MATLAB computing approach presupposes that the element to be handled is a vector or matrix. Whereas SIMULINK maps a technical problem into computer model through elementary blocks. Our highlight covers the following:

- MATLAB and SIMULINK features at the command window
- Getting started in MATLAB/SIMULINK from scratch
- Frequently encountered questions on MATLAB/SIMULINK
- Relevant introductory topics and forms of assistance about MATLAB/SIMULINK

1.1 What is MATLAB?

MATLAB is mainly a scientific and technical computing software whose elaboration is matrix laboratory. Command prompt of MATLAB (>>) provides an interactive system. In the workspace of MATLAB, most data element is dealt as a matrix without dimensioning. The package is incredibly advantageous for matrix-oriented computing. MATLAB's easy-to-use platform enables us to compute and manipulate matrices, perform numerical analysis, and visualize different variety of one/two/three dimensional graphics in a matter of second or seconds without conventional programming as conducted in FORTRAN, PASCAL, or C.

1.1.1 MATLAB's opening window features

If you do not have MATLAB installed in your personal computer, contact MathWorks (owner and developer, www.mathworks.com) for the installation CD. If you know how to get in MATLAB and its basics, you can skip the chapter. Assuming the package is installed in your system, run MATLAB from the Start of Microsoft Windows. Let us get familiarized with MATLAB's opening window features. Figure 1.1(a) shows a typical firstly opened MATLAB window. Depending on desktop setting or MATLAB version, your MATLAB window may not look like figure 1.1(a) but descriptions of the features by and large are appropriate.

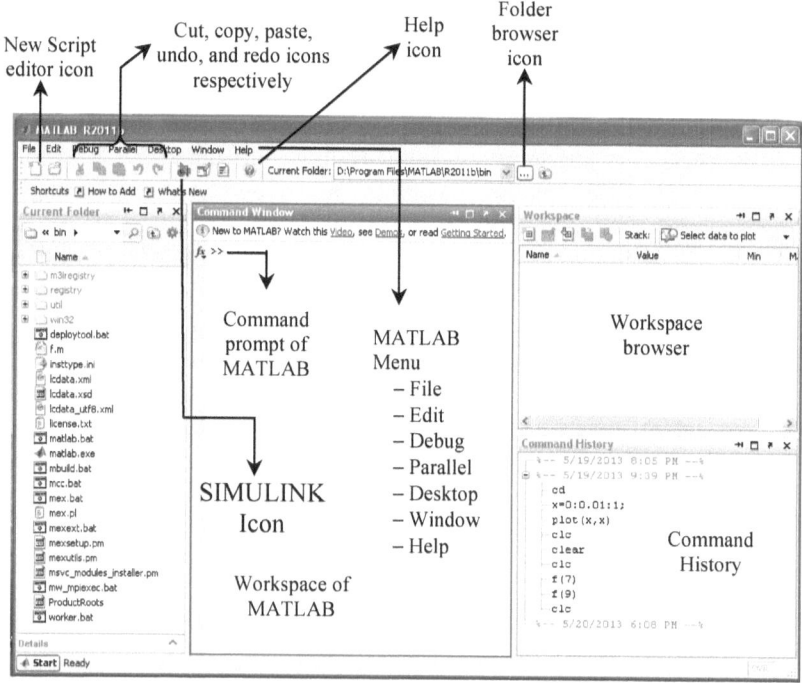

Figure 1.1(a)   Typical features of MATLAB's firstly opened window

♦ Command prompt of MATLAB

Command prompt means that you tell MATLAB to do something from here. As an interactive system, MATLAB responds to user through this prompt. MATLAB cursor will be blinking after >> prompt once you open MATLAB i.e. MATLAB is ready to take your commands. To enter any command, type executable MATLAB statements from keyboard and to execute that, press Enter key (symbol ↵ for 'Hit the Enter Key' operation).

♦ MATLAB Menu

MATLAB is accompanied with seven submenus namely File, Edit, Debug, Parallel, Desktop, Window, and Help. Each submenu has its own features. Use the mouse to click different submenus and their brief descriptions are as follows:

Submenu File: It (figure 1.1(b)) opens a new script or M-file, figure, model, or Graphical User Interface (GUI) layout maker, opens a file which was saved before, loads a saved workspace, imports data from a file,

Figure 1.1(b) Menu File

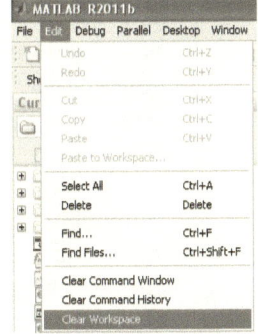

Figure 1.1(c) Menu Edit

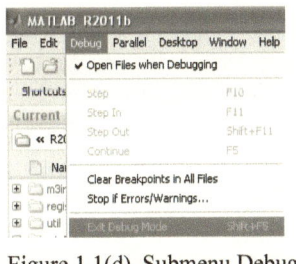

Figure 1.1(d) Submenu Debug

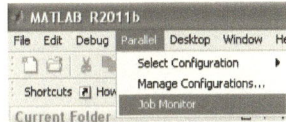

Figure 1.1(e) Submenu Parallel

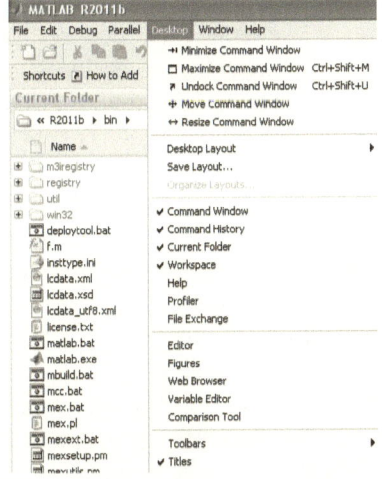

Figure 1.1(f) Submenu Desktop

saves the workspace variables, sets the required path to execute a file, prints the workspace, and keeps provision for changing the command window property.

Submenu Edit: The second submenu Edit (figure 1.1(c)) includes cutting, copying, pasting, undoing, and clearing operations. These operations are useful when you frequently work at command prompt.

Submenu Debug: The submenu Debug (figure 1.1(d)) is mainly related with the text mode or M-file programming.

Submenu Parallel: This submenu (figure 1.1(e)) provides necessary links or tools for parallel computing.

Submenu Desktop: The fifth submenu Desktop (figure 1.1(f)) is accompanied with MATLAB command window viewing functions such as displaying the workspace variable information, current directory information, command history, etc.

Submenu Window: You may open some graphics window from MATLAB command prompt or running some M-files. From the sixth submenu Window (figure 1.1(g)), one can see how many graphics window under MATLAB are open and can switch from one window to other by clicking the mouse to the required window.

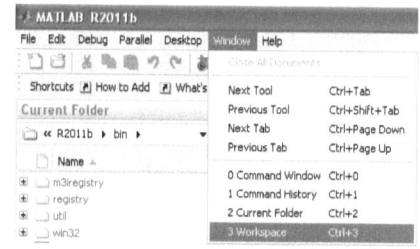

Figure 1.1(g) Submenu Window

Submenu Help: MATLAB holds abundant help facilities. The last submenu shows Help (figure 1.1(h)) in different ways. Latter in this chapter, we mention how one gets specific help. The submenu also provides the easiness to get connected with the MathWorks Website provided that your system is connected to Internet.

◆ **Icons**

Available icons are shown in the icon bar (down the menu bar) of figure 1.1(a). Frequently used operations such as opening a new file, opening an existing file, getting help, etc are found in the icon bar so that the user does not have to go through the menu bar over and over.

◆ **MATLAB workspace**

Workspace (figure 1.1(a)) is the platform of

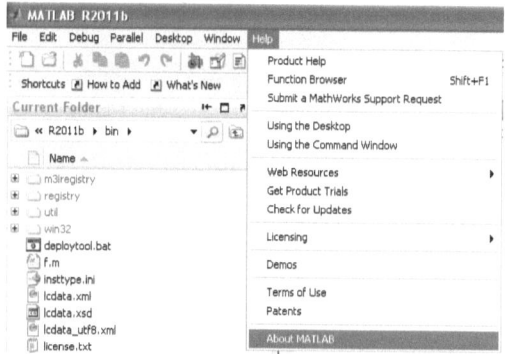

Figure 1.1(h) Submenu Help

MATLAB where one executes MATLAB commands. During execution of commands, one may have to deal with some input and output variables. These variables can be one-dimensional array, multi-dimensional array, characters, symbolic objects, etc. Again to deal with graphics window, we have texts, graphics, or object handles. Workspace holds all those variables or handles for you. As a subwindow of figure 1.1(a), its browser exhibits the types or properties of those variables or handles. If the browser is not seen in the opening window of MATLAB, click the Desktop down Workspace in the menu bar to bring the subwindow (figure 1.1(f)).

♦ MATLAB command history

There is a subwindow in the figure 1.1(a) called Command History which holds all previously used commands at command prompt. Depending on desktop setting, it may or may not appear during the opening of MATLAB. If it does not, click the Command History from figure 1.1(f) under the Desktop.

1.1.2 How to get started in MATLAB?

New MATLAB users face a common question how to get started in MATLAB? This tutorial is for beginners in MATLAB. Here we address the terms under the following bold headings.

♦ How one can enter a vector/matrix

The first step is the user has to be in the command window of MATLAB. Look for the command prompt >> in the command window. Row or column matrices are termed as vectors. We intend to enter the row matrix R=[2   3   4   -2   0] into the workspace of MATLAB. Type the following from the keyboard at the command prompt:

>>R=[2 3 4 -2 0]   ← Arial font set for executable commands i.e. R⇔R

There is one space gap between two elements of the matrix R but no space gap at the edge elements. All elements are placed under the [ ]. Press Enter key after the third brace] from the keyboard and we see

R =

    2   3   4   -2   0

>>   ← command prompt is ready again

It means we assigned the row matrix to the workspace variable R. Whenever we call R, MATLAB understands the whole row matrix. Matrix R is having five elements. Even if R had 100 elements, it would understand the whole matrix that is one of many appreciative features of MATLAB. Next we wish to enter the column matrix $C=\begin{bmatrix} 7 \\ 8 \\ 10 \\ -11 \end{bmatrix}$. Again type the following from the keyboard at the blinking cursor:

>>C=[7;8;10;-11] ↵   you will see (↵ means 'Press the Enter Key'),

C =

    7
    8
   10
  -11

>>   ← command prompt is ready again

This time we also assigned the column matrix to the workspace variable C. For the column matrix, there is one semicolon ; between two consecutive elements of the matrix C but no space gap is necessary. As another option, the matrix C could have been entered by writing C=[7 8 10 -11]'. The operator ' of keyboard is matrix transposition operator in MATLAB. As if you entered a row matrix but at the end just the transpose operator ' is attached. After that the rectangular matrix $A=\begin{bmatrix} 20 & 6 & 7 \\ 5 & 12 & -3 \\ 1 & -1 & 0 \\ 19 & 3 & 2 \end{bmatrix}$ is to be entered:

>>A=[20 6 7;5 12 -3;1 -1 0;19 3 2] ↵   you will see,

```
A =
     20    6    7
      5   12   -3
      1   -1    0
     19    3    2
```

Two consecutive rows of A are separated by semicolon ; and consecutive elements in a row are separated by one space gap. Instead of typing all elements in a row, one can type the first row, press Enter key, the cursor blinks in the next line, type the second row, and so on.

◆ How one can use the colon and semicolon operators

The operators semicolon ; and colon : have special significance in MATLAB. Most MATLAB statements and M-file programming use these two operators almost in every line. Generation of vectors can easily be performed by the colon operator no matter how many elements we need. Let us carry out the following at the command prompt to see the importance of the colon operator:

>>A=1:4 ↵   you will see,

```
A =
     1   2   3   4
```
 ← We created a vector A or row matrix where A=[1 2 3 4]

Let us interact with MATLAB by the following commands:

>>R=1:3:10 ↵   you will see,

```
R =
     1   4   7   10
```
 ← We created a vector or row matrix R whose elements form an arithmetic progression with first element 1, last element 10, and common difference or increment 3

Vector with decrement can also be generated:

>>C=[0:-2:-10]' ↵   you will see,

```
C =
      0
     -2
     -4
     -6
     -8
    -10
```
 ← We created a vector or column matrix C whose consecutive elements have the decrement 2 with the first element 0 and the last element −10

MATLAB is also capable of producing vectors whose elements are decimal numbers. Let us form a row matrix R whose first element is 3, last element is 6, and increment is 0.5 which we accomplish as follows:

>>R=3:0.5:6 ↵   you will see,

R =
          3.0000   3.5000   4.0000   4.5000   5.0000   5.5000   6.0000

Then, what is the use of semicolon operator? Append a semicolon at the end in the last command and execute that:
    >>R=3:0.5:6; ↵   you will see,
    >>   ← Assignment is not shown

Type R at the command prompt and press Enter:
    >>R ↵

R =
          3.0000   3.5000   4.0000   4.5000   5.0000   5.5000   6.0000

It indicates that the semicolon operator prevents MATLAB from displaying the contents of the workspace variable R.

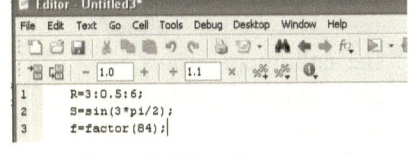

◆ **How one can call a built-in MATLAB function**

Figure 1.2(a) Last three executed statements are typed in the M-file editor of MATLAB

In MATLAB, thousands of M-files or built-in function files are embedded. Knowing descriptions of the function, numbers of input and output arguments, and nature of the arguments is mandatory in order to execute a built-in function. Let us start with a simplest example. We intend to find $\sin x$ for $x = \frac{3\pi}{2}$ which should be -1. The MATLAB counterpart (appendix A) of $\sin x$ is sin(x) where x can be any real or complex number in radians and can be a matrix too. The angle $\frac{3\pi}{2}$ is written as 3*pi/2 (π is coded by pi) and let us perform it as follows:
    >>sin(3*pi/2) ↵

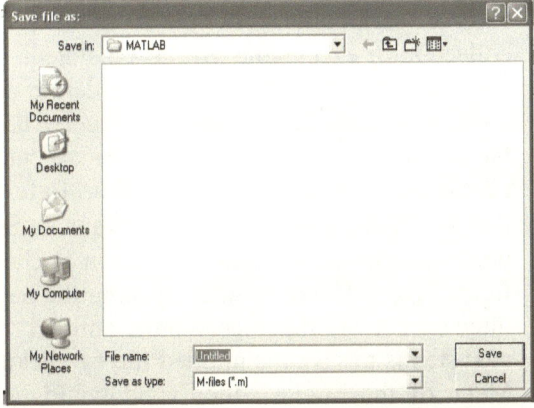

Figure 1.2(b) Save dialog window for naming the M-file

    ans =
          -1

By default the return from any function is assigned to workspace **ans**. If you wanted to assign the return to S, you would write **S=sin(3*pi/2);**. As another example, let us factorize the integer 84 (84=2×2×3×7). The MATLAB built-in function **factor** finds the factors of an integer and the implementation is as follows:
    >>f=factor(84) ↵

f =

    2   2   3   7

Output of the **factor** is a row matrix which we assigned to workspace f in fact the f can be any user-given name. Thus you can call any other built-in function from the command prompt provided that you have the knowledge about the calling of inputs to and outputs from the function.

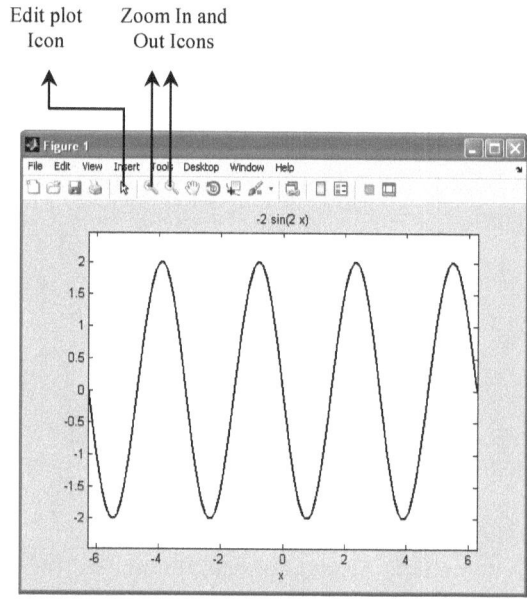

Figure 1.2(c) Graph of $-2\sin 2x$ versus x

✦ **How one can open and execute an M-file**

This is the most important start up for the beginners. An M-file can be regarded as a text or script file. A collection of executable MATLAB statements are the contents of an M-file. Ongoing discussion made you familiarize with entering a matrix, computing a sine value, and factorizing an integer. These three executions took place at the command prompt. They can be executed from an M-file as well. This necessitates opening the M-file editor. Referring to figure 1.1(b), you find the link for M-file editor as File → New → Script and click it to see the new untitled script editor. Another option is click the New M-file editor icon of figure 1.1(a). However after opening the new script editor, we typed the last three executable statements in the untitled file as shown in figure 1.2(a). The next step is to save the untitled file by clicking the Save icon or from the File Menu of the M-file editor window. Figure 1.2(b) presents the File Save dialog window. We typed the file name as **test** (can be any name of your choice) in the slot of File name in the window. The script file has the extension .m but we do not type .m only the file name is enough. After saving the file, let us move on to MATLAB command prompt and conduct the following:

>>test ↵
>>   ← command prompt is ready again

It indicates that MATLAB executed the M-file by name **test** and is ready for next command. We can check calling the assignees whether the previously performed executions occurred exactly as follows:

>>R ↵

```
R =
    3.0000  3.5000  4.0000  4.5000  5.0000  5.5000  6.0000
>>S ↵                                 >>f ↵

S =                                   f =
    -1                                    2    2    3    7
```

This is what we found before. Thus one can run any executable statements in script file. The reader might ask in which folder or path the **test** was saved. Figure 1.1(a) shows one slot for the **Current Folder** in upper middle portion of the window. That is the location of your file. If you want to save the script file in other folder or directory, change path by clicking the path browser icon before saving the file. When you call the **test** or any other file from the command prompt, the prompt must be in the same directory where the file is in or its path must be defined to MATLAB.

♦ Input and output arguments of a function file

MATLAB is a collection of thousands of script files. Some files are executed without any return and some return results which are called function files (appendix F). You have seen the use of function **sin(x)** before, which has one input argument **x**. The statement **test(x,y)** means that the **test** is a function file which has two input arguments – **x** and **y**. Again the **test(x,y,z)** means the **test** is a function file which needs three input arguments – **x**, **y**, and **z**. Similar style also follows for the return but under the third brace. The **[a,b]=test(x,y)** means there are two output arguments from the **test** which are **a** and **b** and the **[a,b,c]=test(x,y)** means three returns from the **test** which are **a**, **b**, and **c**.

♦ How one can plot a graph

MATLAB is very convenient for plotting different sorts of graphs. The graphs are plotted either from mathematical expression or from data. Let us plot the function $y = -2\sin 2x$. MATLAB function **ezplot** plots y versus x type graph taking the expression as its input argument. MATLAB code (appendix A) for the $-2\sin 2x$ is **-2*sin(2*x)**. The functional code is input argumented by using single inverted comma hence we conduct the following at command prompt:

>>ezplot('-2*sin(2*x)') ↵

Figure 1.2(c) presents the outcome from above execution. The window in which the graph is plotted is called MATLAB figure window. Any graphics is plotted in the figure window, which has its own menu (such as File, Edit, etc) as shown in figure 1.2(c).

1.1.3 Some queries about MATLAB environment

Users need to know the answers to some questions when they start working in MATLAB. MATLAB environment related some queries are presented in the sequel:

How to change the numeric format?

When you perform any computation at the command prompt, the output is returned up to four decimal display due to short numeric format which is the default one. There are other numeric formats too. To reach the numeric format dialog box, the clicking operation sequence is MATLAB command window $\Rightarrow$ File $\Rightarrow$ Preferences $\Rightarrow$ Command Window $\Rightarrow$ Text Display $\Rightarrow$ Numeric Format (select from the popup menu e.g. long).

How to change the font or background color settings?

One might be interested to change the background color or font color while working in the command window. The clicking sequence is MATLAB command window $\Rightarrow$ File $\Rightarrow$ Preferences $\Rightarrow$ Colors.

How to delete some/all variables from the workspace?

In order to delete all variables present at the workspace, the clicking sequence is MATLAB command window $\Rightarrow$ Edit $\Rightarrow$ Clear Workspace (figure 1.1(c)). If you want to delete a particular workspace variable, select the concern variable by using the mouse pointer in the workspace browser (assuming that it is open like the figure 1.2(d)) and then rightclick $\Rightarrow$ delete.

How to clear workspace but not the variables?

Once you conduct some sessions at the command prompt, monitor screen keeps all interactive sessions. You can clear the screen contents without removing the variables by command clc or performing the clicking operation MATLAB command window $\Rightarrow$ Edit $\Rightarrow$ Clear Command Window (figure 1.1(c)).

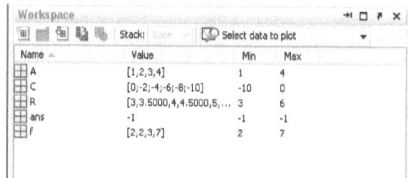

Figure 1.2(d) Workspace browser displays variable information

How to know the current path?

In the upper portion of figure 1.1(a), the Current Folder bar is located that indicates in which path the command prompt is or execute cd (abbreviation for the current directory) at the command prompt.

How to see different variables at the workspace?

There are two ways of viewing this – either use the command who or look at the workspace browser (like figure 1.2(d)) which exhibits information about workspace variables for example R is the name of a variable which holds some values. One can view, change, or edit the contents of a variable by doubleclicking the concern variable situated in the workspace browser.

How to enter a long command line?

MATLAB statements can be too long to fit in one line. Giving a break in the middle of a statement is accomplished by the ellipsis (three dots are called ellipsis). We show that considering the entering of vector x=[1:3:10] as follows:
```
>>x=[1:3: . . . ↵
```

```
             10] ↵
   x =
             1   4   7   10
```
Typing takes place in two lines and there is one space gap before the ellipsis.

⌗ Editing at the command prompt

This is advantageous specially for those who work frequently in the command window without opening a script file. Keyboard has different arrow keys marked by ← ↑ → ↓. One may type a misspelled command at the command prompt causing error message to appear. Instead of retyping the entire line, press uparrow (for previous line) or downarrow (for next line) to edit the MATLAB statement. Or you can reexecute any past statement this way. For example we generated a row vector 1 through 10 with increment 2 and assigned the vector to x. The necessary command is x=1:2:10. Mistakenly you typed x+1:2:10. The response is as follows:
```
    >>x+1:2:10 ↵
    ??? Undefined function or variable 'x'.
```
You discovered the mistake and want to correct that. Press ↑ key to see,
```
    >>x+1:2:10
```
Edit the command going to the + sign by using the left arrow key or mouse pointer. At the prompt, if you type x and press ↑ again and again, you see used commands that start with x.

⌗ Saving and loading data

User can save workspace variables or data in a binary file having the extension .mat. Suppose you have the matrix $A = \begin{bmatrix} 3 & 4 & 8 \\ 0 & 2 & 1 \end{bmatrix}$ and wish to save A in a file by the name data.mat. Let us carry out the following:
```
    >>A=[3 4 8;0 2 1]; ↵        ← Assigning the A to A
```
Now move on to the workspace browser (figure 1.2(d)) and you see the variable A including its information located in the subwindow. Bring mouse pointer on A, rightclick the mouse, and click the **Save As**. The Save dialog window appears and type only **data** (not data.mat) in the slot of **File name**. If it is necessary, you can save all workspace variables by using the same action but clicking **File ⇒ Save Workspace As** (figure 1.1(b)). One retrieves the data file by clicking the menu **File ⇒ Import Data** (figure 1.1(b)). Another option is use the command **load data** at the command prompt.

⌗ How to delete a file from the command prompt?

Let us delete just mentioned data.mat by executing the command **delete data.mat** at the command prompt.

⌗ How to see the data held in a variable?

Figure 1.2(d) presents some variable information in which you find R. Doubleclick the R or your variable in the workspace browser and find the matrix contents of R in a data sheet.

1.2 What is SIMULINK?

SIMULINK is an additional part of MATLAB which provides an easeful way to model, simulate, and analyze many dynamic systems which are characterized by some inputs and outputs. Without opening MATLAB we can not turn SIMULINK operational.

Elaboration of SIMULINK is simulation and link. A particular input-output relationship can be assigned to some block. One can interpret that SIMULINK is a vast collection of this kind of blocks. Although the blocks stand for simple mathematical relationship but being concatenated they build a much complicated system. Initially SIMULINK was intended particularly to handle the linear time invariant continuous systems. With the progress of time the discrete time systems as well as the hybrid ones surfaced in SIMULINK to adapt it to more pragmatic modeling of the real world's dynamic systems. In a simplistic way if MATLAB is a world of matrices, then SIMULINK is a world of blocks. In most scientific and engineering systems three types of constitutive elements are seen – source, system, and sink. For example in electrical engineering, applied DC voltage to a circuit, $R-C$ filter, and voltmeter correspond to the source-system-sink terminology. SIMULINK is the best tool for technical analysis if we can characterize a scientific problem in terms of source, system, and sink.

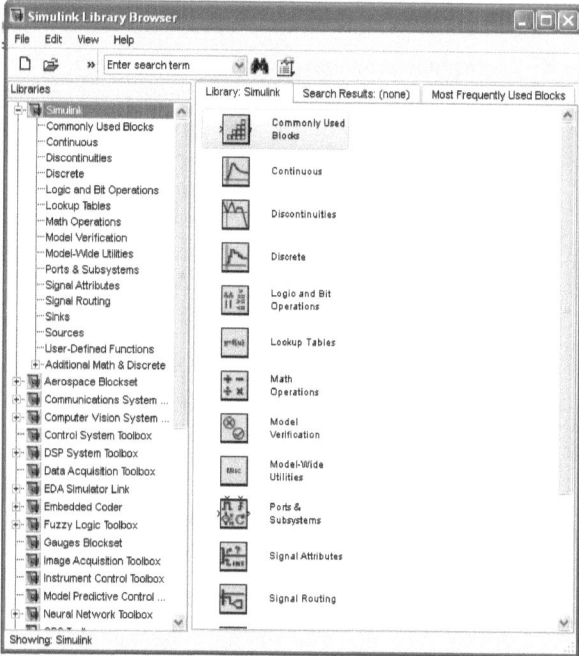

Figure 1.3(a) SIMULINK library contents

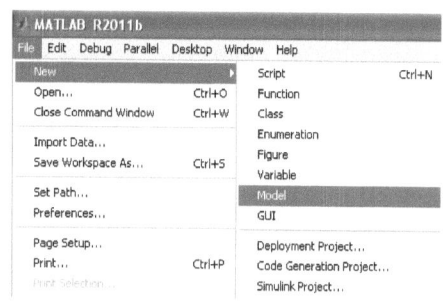

Figure 1.3(b) New file options in MATLAB

1.2.1 How can I get into SIMULINK?

Since SIMULINK is an extension of MATLAB, first we have to get into MATLAB. Both the MATLAB command prompt and its menu bar provide means of getting into SIMULINK. Figure 1.1(a) shows the indication of command prompt. Either you type **simulink** at the command prompt and press enter or click the SIMULINK library browser icon shown in figure 1.1(a). SIMULINK is an aggregation of functional blocks arranged in a tree structure which you see like figure 1.3(a) by just mentioned either action.

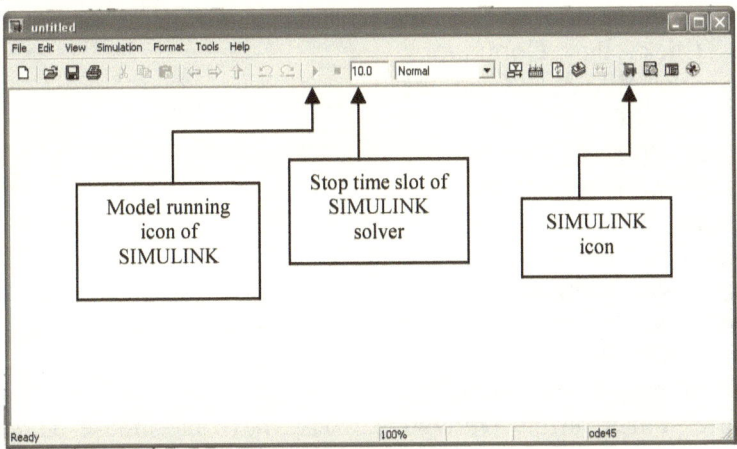

Figure 1.3(c) A newly opened SIMULINK model file

In the figure 1.3(a) you find different families of block sets. For example **Aerospace Blockset** keeps source/system/sink analysis blocks for the aerospace engineering. Again the **Communications System** keeps source/system/sink analysis blocks for communications engineering and so on. If you scroll down the library by using mouse, you find one family by the name **Control System Toolbox** which is solely dedicated for robust control system problems unfortunately that family is for professional or industrial use not for class-discussed problems.

1.2.2 Where can I build a SIMULINK model?

Like every software there should be a file where we can build our SIMULINK model. The model we intend to build is entirely problem dependent. Referring to MATLAB Command Window of figure 1.1(a) if you click the **File** menu down **New** of MATLAB, you see the pulldown menu as shown in figure 1.3(b). Click the **Model** in the pulldown menu and you see the newly untitled SIMULINK model file opened like figure 1.3(c). This is the platform where we build any SIMULINK model.

The SIMULINK library browser icon is also found in the menu bar of the untitled model file (seen in figure 1.3(c)). The reader should have the knowledge about any particular block's function and its input-output descriptions before bringing it to the untitled model file from SIMULINK library. A SIMULINK model is defined as interconnected library blocks to work out a scientific/engineering problem.

Let us bring a block from the SIMULINK library in the untitled model file. To perform such action, click the SIMULINK library browser icon located in the menu bar of the untitled model file thus the SIMULINK library of figure 1.3(a) appears and the cursor is residing in SIMULINK. The right half part of the window in figure 1.3(a) displays the subfamily blocksets for example **Commonly Used Blocks, Continuous,** and so on. Click the **Continuous** down the **SIMULINK** on the left part of the window in figure 1.3(a). We see various blocks available under the subclass **Continuous** like the figure 1.3(d) for example **Derivative, Integrator, State-Space,** etc. We wish to get the **Derivative** block into our untitled model file. Bring the

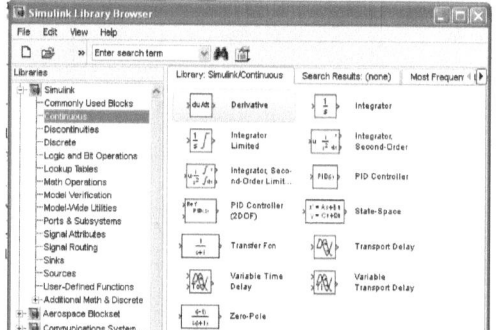

Figure 1.3(d) Different blocks availability under the subfamily **Continuous**

mouse pointer on the **Derivative** block, move the mouse pointer keeping your finger pressed in the left button of the mouse to any convenient area of the untitled model, and release the left button of the mouse. Now you see the **Derivative** block in the untitled model file as shown in figure 1.3(e). Another way of bringing the block is click the **Derivative**, rightclick the mouse, see the **Add to untitled** in the popup, click the **Add to untitled,** and find the block in your model file. Therefore we say the link for **Derivative** block is **SIMULINK → Continuous → Derivative.** We maintain this style of locating a block in appendix C. However we have been successful in bringing the **Derivative** block in the untitled model file. Now we can save the model by any convenient name in the working directory. By the way MATLAB source code file has the extension .m but the SIMULINK model file does .mdl. Keep in mind that *one must know the link of a block to bring it from the SIMULINK library to a model file.*

1.2.3 Block manipulation in SIMULINK

During SIMULINK model building process, some manipulations of the blocks are essential to construct a seemly, well-placed, and well-devised model, most frequently encountered ones of which are in the following.

⛶ How to select a block?

Let us say you brought the **Derivative** block in an untitled model file following the link **Simulink → Continuous → Derivative.** Bring the mouse pointer on the **Derivative** block and click the left button of the mouse to see the selection as shown in figure 1.3(f).

🗗 **How to detect the input and output ports of a block?**

Referring to the **Derivative** block of figure 1.3(e), we see that the left and right hand sides of the block contain the symbol >. One can identify the input and output ports of the block as presented in figure

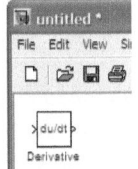

Figure 1.3(e) **Derivative** block in the untitled model file

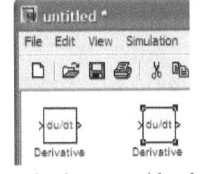

without selection   with selection

Figure 1.3(f) **Derivative** block with and without selection

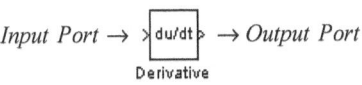

Figure 1.3(g) **Derivative** block mentioning input and output ports

Figure 1.3(h) Blocks with multiple input and output ports

1.3(g). The number of input and output ports is not always 1, the blocks **Real-Imag to Complex** and **Complex to Magnitude-Angle** of figure 1.3(h) have two input and two output ports respectively.

Before flipping   *After flipping*

Figure 1.4(a) Flipping the **Derivative** block

🗗 **How to delete a block?**

You just brought the **Derivative** block in an untitled model file but want to delete that. There are several options for this. First select the block, then press the **Delete** button from the keyboard, click the **Cut** icon in the model menu bar, or click the **Cut** followed by the rightclick of the mouse on the block.

🗗 **How to flip a block?**

We have the ongoing **Derivative** block in our model file. We wish to flip the block like figure 1.4(a). First we bring the mouse pointer on the block, click the right button of the mouse, and then click the **Flip Block** via **Format**. The action is shown in figure 1.4(b).

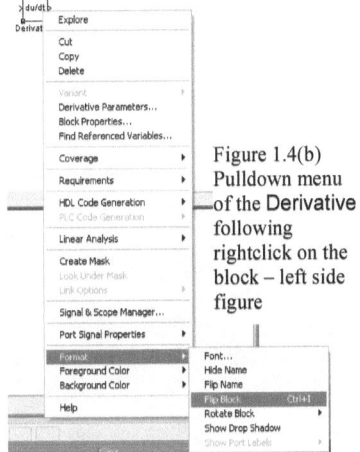

Figure 1.4(b) Pulldown menu of the **Derivative** following rightclick on the block – left side figure

🗗 **How to rotate a block?**

Suppose we wish to rotate just mentioned **Derivative** block. Bring mouse pointer on the block, click the right button of the mouse, and click the

Rotate Block via **Format** (figure 1.4(b) shows the action too). You see the change as shown in figure 1.4(c). Figure 1.4(c) seen rotation indicates the operation by 90^0 clockwise. If you intend to rotate the block 270^0 clockwise (indicated in figure 1.4(d)), you need the operation three times.

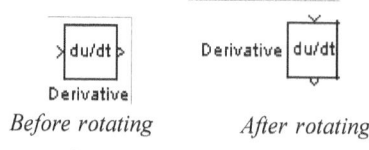

Before rotating   After rotating

Figure 1.4(c) Rotating the **Derivative** block by 90^0 clockwise

▣ How to remove the block name?

The name of **Derivative** block can be removed as presented in figure 1.4(e) by clicking the **Hide Name** of figure 1.4(b). A clumsy model might give a better look on removing block names if the reader is well acquainted with the blocks' operation.

Figure 1.4(d) 270^0 clockwise rotation of the **Derivative** block

Figure 1.4(e) The **Derivative** block without the block name

▣ How to enlarge or contract a block?

Let us bring a **Gain** block as shown in figure 1.5(a) following the link Simulink → Math Operations → Gain in an untitled SIMULINK model file, **Gain** has the default gain 1. If we have five or six digits gain, the default size will not allow to display that. Doubleclick the block to see its parameter window like figure 1.5(e), let us enter the **Gain** slot value of the figure 1.5(e) from default 1 to 2700 (as a four digit example) by using the keyboard, and click OK. The block displays the inside gain as shown in figure 1.5(b). Select the block, bring your mouse pointer on the upper right square target to see the figure 1.5(c), move the mouse pointer to the right keeping the left button of the mouse pressed, and release the left button of the mouse. You should see the enlarged **Gain** block of figure 1.5(d). In a similar way for an oversize block, we can reduce the block size by moving towards inside of the block after selection.

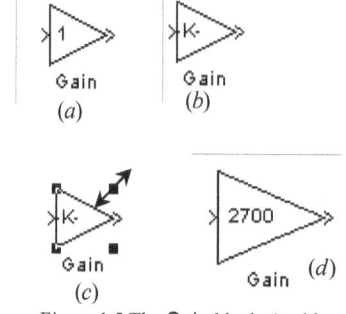

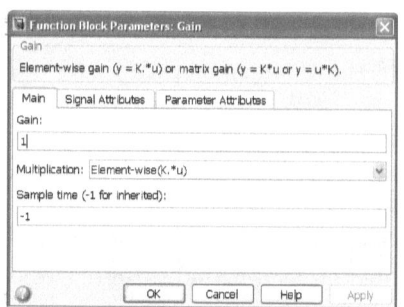

Figure 1.5 The **Gain** block a) with default gain, b) with gain 2700, c) with selection, and d) after enlargement

Figure 1.5(e) Block parameter window of **Gain** block

Control System Analysis & Design in MATLAB and SIMULINK

🗗 How to rename a block?

During SIMULINK model building process, it may be necessary to use the same kind of block twice or more. Then it requires renaming the block for identification. Let us say we brought a Derivative block in an untitled model file as shown in figure 1.3(e). We wish to write just **D** as the block name instead of **Derivative** (like figure 1.5(f)). Bring the mouse pointer on the word **Derivative**, click left button of the mouse (word is selected), and delete the other letters except **D** by using the **Delete** button of keyboard, bring the mouse pointer outside the block, and leftclick it. After completely deleting the name **Derivative**, you can even enter any word of your choice from the keyboard.

Figure 1.5(f) Derivative block by the name D

Derivative block in a model

Figure 1.5(g) Derivative block with the annotation

🗗 How to include the annotation to a block?

Let us say we have the Derivative block in the untitled model file as shown in figure 1.3(e) and wish to write the line **Derivative block in a model** down the block. Bring mouse pointer at the desired position in the model file, doubleclick the mouse to see

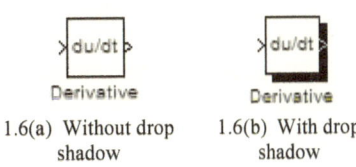

1.6(a) Without drop shadow   1.6(b) With drop shadow

Figures 1.6(a)-(b) Derivative block without and with drop shadow

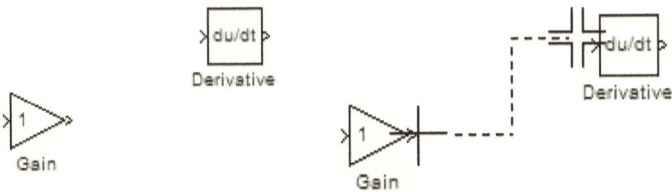

Figure 1.6(c) The Derivative and Gain blocks are residing in a SIMULINK model file

Figure 1.6(d) Connection phase of the two blocks

the blinking cursor, type the **Derivative block in a model** from the keyboard, bring mouse pointer out of the block, and click the left button of the mouse. Figure 1.5(g) shows the action we performed. Once typed, we can even drag the whole text to move anywhere in the model file.

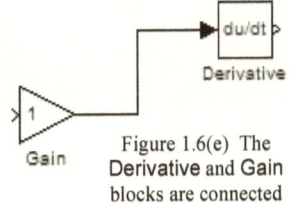

Figure 1.6(e) The Derivative and Gain blocks are connected

🗗 How to add drop shadow to a block?

Some reader might be interested to see the drop shadow form of SIMULINK block rather than plain shape. Let us say we have the Derivative block in an untitled model file (figure 1.3(e)). Rightclick on the block to see the pulldown menu of figure 1.4(b) and click the **Show Drop Shadow** via **Format**. The necessary change is depicted in figures 1.6(a)-(b). To remove the shadow, again rightclick on the block and click the **Hide Drop Shadow** via **Format**.

🗗 **How to change SIMULINK model file background color?**

Rightclick anywhere in the model file and see the **Screen Color** in the prompt menu. From the popup of **Screen Color**, you can choose any background color for the model.

🗗 **How to copy a block within a SIMULINK model?**

Select the block, click **Copy** icon in the model menu bar (this action is called copy in the clipboard), and paste it as many times as you want.

🗗 **How to connect two blocks?**

This manipulation is very important in the sense that we frequently need to connect blocks in a model while working in SIMULINK in subsequent chapters. The reader is familiar with **Gain** and **Derivative** blocks from previous discussions. Let us say the two blocks are residing in a SIMULINK model file as shown in figure 1.6(c). We intend to connect the two blocks. Connection of the blocks must be correct syntactically. The output port of any block can only be connected to the input port of any other block but not to the output port of others. The same syntax is also true for the input port. However bring the mouse pointer on the output port of **Gain** block, see the single cross target as shown in figure 1.6(d), press the left button of mouse, move the single cross target anywhere in the model keeping your finger pressed, bring the single cross target close to the input port of **Derivative** block, see the double cross target as shown in the figure 1.6(d), and release the left button of the mouse. You should find the two blocks connected as shown in figure 1.6(e).

🗗 **What is a parameter or block parameter window?**

In the following chapters we are going to mention the term parameter window frequently. After bringing any block in a SIMULINK model file and

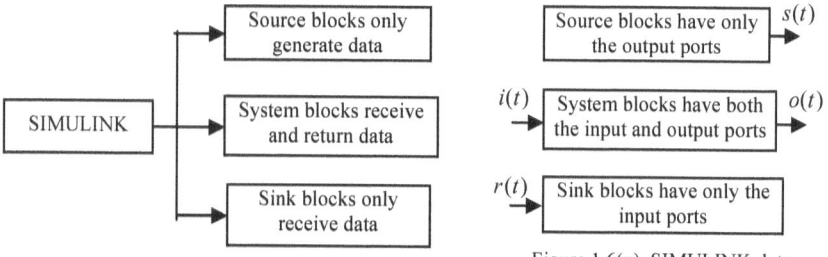

Figure 1.6(f) Basic block types in SIMULINK

Figure 1.6(g) SIMULINK data flow in various blocks

then doubleclicking it, always do we see a prompt or dialog window in which we find one or more slots for value or parameter taking depending on purpose of the block. That dialog window is termed as the parameter window for example the window for **Gain** block in figure 1.5(e).

1.2.4 Basic block categories of SIMULINK

By now we know that SIMULINK is a vast collection of blocks. These blocks follow three basic characteristics. Some blocks only generate data which are called source blocks, some blocks receive data, perform mathematical operation depending on the problem, and then return data which are called system blocks, and the third types only receive data which are called sink blocks. Figure 1.6(f) presents the basic block types found in SIMULINK. Figure 1.6(g) shows the data flow as a function of t where $s(t)$ generated, $i(t)$ input and $o(t)$ output, and $r(t)$ received correspond to source, system, and sink blocks respectively.

It is extremely important to mention that all block functional data whether source, system, or sink present in SIMULINK model shares the common t variation. This is called the state or t dependency of SIMULINK block data. Figure 1.6(h) illustrates this sort of dependency assuming all linear data.

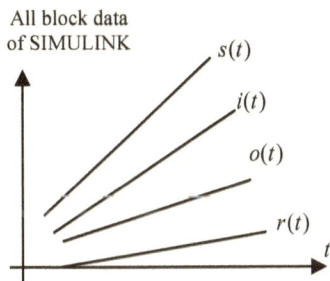

Figure 1.6(h) State or t dependency of SIMULINK data

1.2.5 Display and Scope blocks of SIMULINK

A practical model contains dozens of blocks which are interconnected by functional lines in SIMULINK. When a model is being run, the blocks **Display** and **Scope** (appendix C for link and outlook) show how the functions are changing. The functional data flow or computation may or may not be seen during the run time because it happens so rapidly – in a fraction of second. It also depends on the problem whether it is time consuming.

The **Display** block is convenient only for showing a single scalar or some matrix data output at the end of simulation. The block is designed to show instantaneous value flowing through the functional line which it is connected to. Once SIMULINK has finished a simulation, the block shows the last value. Default size of the block is for a single scalar. For matrix data output one needs to enlarge the block to view all in it.

If the turnout of a SIMULINK block is in the form of a long row or column matrix which may hold hundreds of data elements, it is not feasible to see the results through the **Display** block. The graphical plot is a better way to observe the output. The **Scope** in all sense mimics an oscilloscope that essentially displays the signal variation with time. The **Scope** has two axes – horizontal and vertical. The horizontal and vertical axes simulate the independent and dependent variables respectively. The horizontal axis does not have to be time even though it is originated in that name, any physical quantity such as displacement, frequency, speed or other can be assigned to horizontal axis.

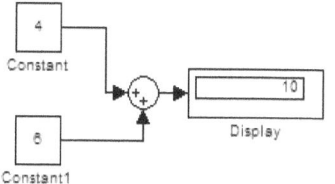

1.2.6 How to get started in SIMULINK?

Figure 1.7(a) Adding two constants and displaying the result

In previous sections the reader has gone through bringing and connecting blocks in a SIMULINK model file. Now we present simple modeling lessons for beginners in SIMULINK aiming to illustrate simulation style in this platform. Whatever operations such as manipulations, computations, assignments, or comparisons are carried out in conventional software can be conducted in SIMULINK through various blocks and functional lines. Since most algorithms are hidden in functional lines and blocks of SIMULINK, initially one might feel it complicated. Most model building in SIMULINK happens through mouse operation rather than writing source codes.

Let us go through the following three tutorials as a quick start in SIMULINK.

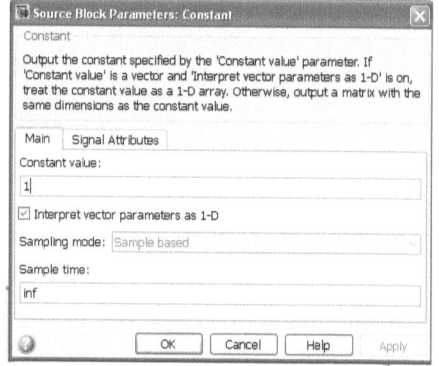

✦ **Tutorial one**

Two numbers are to be added – 4 and 6. The output should be 10. This is the problem statement.

The first question is where we should keep the numbers. In SIMULINK every programming aspect happens through blocks. You find a block called **Constant**

Figure 1.7(b) Block parameter window for **Constant**

through the link SIMULINK → Sources → Constant (figure 1.3(a)). Open a new SIMULINK model file (subsection 1.2.2) and bring the **Constant** block in the untitled model file as we did before. Default value in the block is 1. Doubleclick the block to see its parameter window like figure 1.7(b) and enter 4 in the slot of **Constant value** from keyboard after deleting the default 1 but leaving the other parameters unchanged in the window. It means the first number 4 is going to be generated by the **Constant** block. Similarly bring another **Constant** block from the same link and enter 6 as the **Constant value** after doubleclicking it. You see the latter block by the name **Constant1**. If you bring one more **Constant** block in the model file,

SIMULINK names that as **Constant2**. This style of naming is followed for all other blocks.

However we need a **Sum** block to add the two numbers that can be reached via SIMULINK → Math Operations → Sum. Bring the **Sum** block in the model file. To see the computation, we need a **Display** block whose link is SIMULINK → Sinks → Display and bring the block in the model file. Place the four blocks relatively and connect them (subsection 1.2.3) according to figure 1.7(a). Referring to figure 1.3(c), you find an inactivated start simulation icon whose symbol is ▶. The symbol remains inactivated when no blocks are present in the model. Click the start simulation icon ▶ at the icon bar in the model and SIMULINK responds showing the summation 10 in the **Display** block like figure 1.7(a). You can also run the model file from the menu bar of SIMULINK by clicking first the **Simulation** and then the **Start** in the pulldown menu.

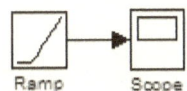

Figure 1.7(c) **Ramp** block connected with **Scope**

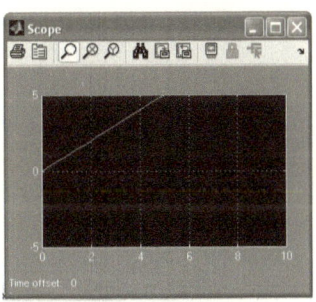

Figure 1.7(d) **Scope** block shows the ramp function

MATLAB command window provides another alternative for running a SIMULINK model from its command prompt. Let us say we have the SIMULINK model saved by the name **test.mdl**. To run it from the command prompt, we carry out the following (**sim** is a built-in command for running a SIMLINK model file and the file name must be under quote):

>>sim('test') ↵

This action would show **10** in the **Display** block too.

✦ **Tutorial two**

We intend to generate the function $y(t) = t$ and view the generation in earlier mentioned **Scope** – this is the problem statement.

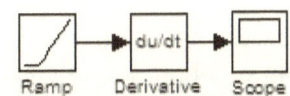

Figure 1.7(e) **Derivative** block differentiates the output generated by the **Ramp** and **Scope** shows the **Derivative** output

The function is a straight line passing through the origin and has a unity slope which is also known as the ramp function. There is a block by the name **Ramp** (appendix C for link and outlook) in SIMULINK which generates t data by default. That means it is a source block as depicted in figure 1.6(f) or 1.6(g). Bring one **Ramp** and one **Scope** blocks in a new SIMULINK model file following the link mentioned in appendix C and connect the two blocks as shown in figure 1.7(c). Run the model by clicking the start simulation icon

▶ at icon bar in the model and doubleclick the **Scope** to view figure 1.7(d) presented curve which displays our wanted straight line plot. The horizontal and vertical axes of the **Scope** correspond to the t and $y(t)$ data respectively so the **Ramp** generated the $y(t)$ and the **Scope** just displayed that. You may inspect that the line of figure 1.7(d) is passing through the points (0,0) and (5,5) confirming the generation with correct slope.

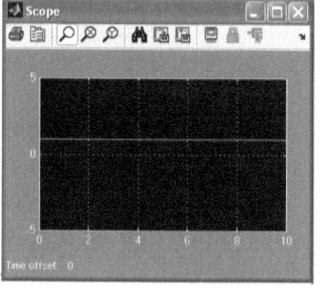

◆ **Tutorial three**

If we differentiate the ramp function $y(t)=t$ with respect to t, we should get $\frac{dy(t)}{dt}=1$. We intend to simulate this mathematical operation.

Figure 1.7(f) **Scope** output from the model in figure 1.7(e)

The reader is familiar with **Derivative** block from subsection 1.2.3. The block differentiates any input signal to its input port and returns the derivative signal at its output port both in continuous sense. Insert the **Derivative** between the **Ramp** and **Scope** blocks of figure 1.7(c) so that we have the model in figure 1.7(e). Select the **Ramp** or **Scope** block of figure 1.7(c), use the left or right arrow key from the keyboard so that the space is enough between the two blocks to accommodate the **Derivative** block. When you bring the **Derivative** block, drop the block keeping its input and output ports in line with the connection line between the **Ramp** and **Scope**. SIMULINK is so smart that it connects the **Derivative** automatically like figure 1.7(e). On forming the model, click the start simulation icon ▶ at the icon bar and doubleclick **Scope** to view output like figure 1.7(f) which is essentially a straight line parallel to horizontal axis and located at 1 in the vertical axis of **Scope**. In other words vertical and horizontal axes of the **Scope** refer to $\frac{dy(t)}{dt}$ and t data respectively. That is what we expected from SIMULINK.

Note: Models in figures 1.7(a), 1.7(c), or 1.7(e) have connecting lines between various blocks. Data flowing through these lines is functional data not the t data. It is the style of SIMULINK that the t data in all these blocks is common. For example in figure 1.7(e), $y(t)$ data flows from the **Ramp** to **Derivative** and $\frac{dy(t)}{dt}$ data flows from the **Derivative** to **Scope** blocks.

◆ **Interval entering or t information in SIMULINK**

In tutorial 2 the **Scope** (figure 1.7(d)) shows $y(t)=t$ versus t graph and in the tutorial 3 the **Scope** (figure 1.7(f)) shows $\frac{dy(t)}{dt}=1$ versus t graph.

We mentioned that the horizontal axis of both **Scope**s represents t variation. The t variation seen in both **Scope**s is between 0 and 10 because that is the default setting. Mathematically we can say that the $y(t)$ or $\frac{dy(t)}{dt}$ is graphed over the interval $0 \le t \le 10$. In SIMULINK term, lower and upper bounds of the interval $0 \le t \le 10$ are called the **Start time** and **Stop time** respectively.

What if we wish to enter other interval for example $-3 \le t \le 7.5$ instead of the default $0 \le t \le 10$? Just mentioned **Start time** and **Stop time** then become -3 and 7.5 respectively which need to be entered. There is a window called simulation parameter window which keeps provision for entering the interval description, differential equation solver type (like stiff or nonstiff), step size of the computation (like adaptive or fixed), etc. The next question is how to reach to that window? Figure 1.3(c) shows a newly opened SIMULINK model file and you find menu **Simulation** in the menu bar of window. If you click the **Simulation** menu, find one submenu by the name **Configuration Parameters** in the pulldown menu. Click the **Configuration Parameters** and the action lets you see the simulation parameter window of figure 1.7(g). In upper portion of the window there are two slots for the **Start time** and **Stop time** respectively. There we enter -3 and 7.5 from keyboard deleting the default values for lower and upper bounds of the interval $-3 \le t \le 7.5$ respectively.

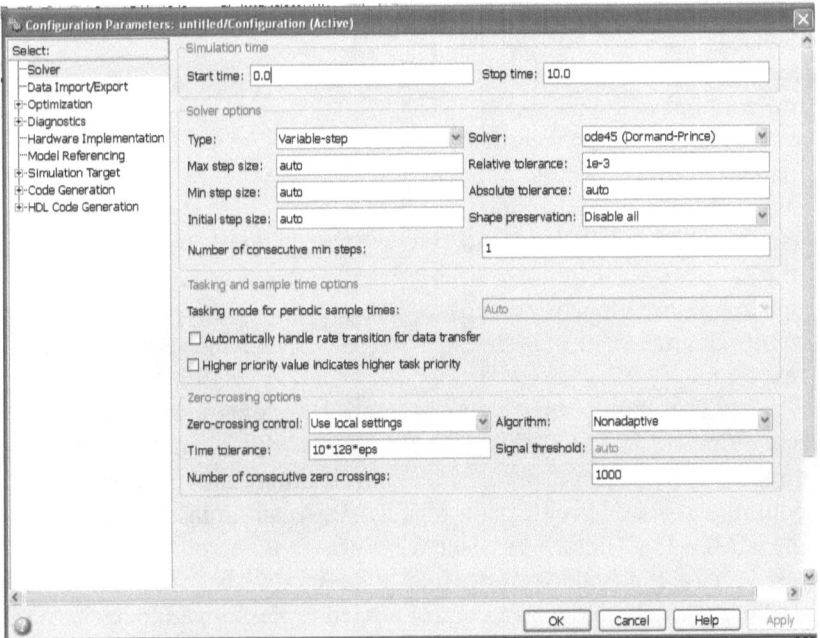

Figure 1.7(g) Simulation parameter window of SIMULINK

If the lower bound is 0 and upper bound is other than 10, you do not even need to open the parameter window. In the untitled model file of figure 1.3(c), there is a slot for **Stop time** as indicated by an arrow. For example if the interval is $0 \le t \le 15$, we just enter 15 in the **Stop time** slot of figure 1.3(c) without opening the parameter window.

Note: There are many parameters in the simulation parameter window of figure 1.7(g) whose discussions are beyond the scope of the text. We suggest you do not change any other parameter unless you know about it. SIMULINK approach of modeling any dynamic problem is completely numerical. Also computer always works on discrete data instead of continuous one even though we carry out the simulation in continuous sense.

1.2.7 Control system library in MATLAB and SIMULINK

Control system is a gigantic branch of electrical or systems engineering and library contents of MATLAB and SIMULINK are so. Control system library is included both in MATLAB and SIMULINK. Built-in functions of MATLAB and block-sets of SIMULINK are devised to handle robust control system problems for professional use. Professional way of explaining theoretical matters may not be appreciative to a beginner. We select the functions and blocks whose executions or operations are closely related to introductory control system text terms.

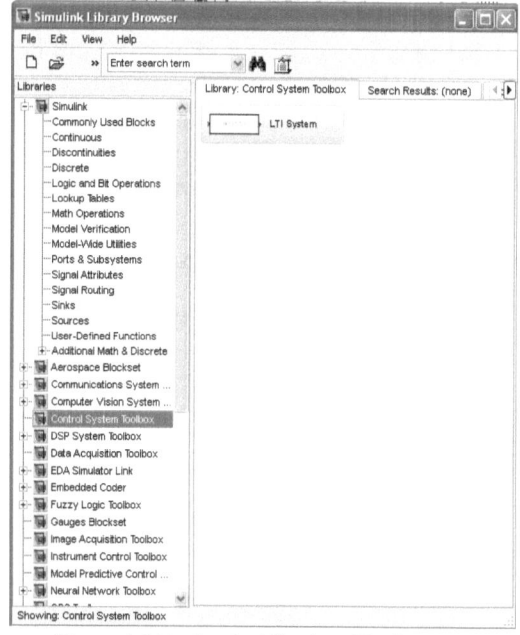

Figure 1.8(a) **Control System Toolbox** of SIMULINK

The library that contains control system related functions has the name **control** in MATLAB. But that library alone is not enough to solve all control system problems. Other libraries may be necessary for solving control associated problems in MATLAB. Again in the figure 1.3(a) you find the SIMULINK library browser window. If you scroll down using mouse in left half of the window, you find a library by the name **Control System Toolbox** which holds most control related professional blocks in SIMULINK (figure 1.8(a)). But for elementary level we will be using figure 1.3(d) shown **Continuous** and other subfamilies for control system problems.

1.3 How to get help?

Help facilities in MATLAB are plentiful. One may access to information about a MATLAB function or SIMULINK block in a variety of ways. Command **help** finds the help of a particular function file. There is an embedded function by name **tf** which defines control system transfer function. You can have command prompt help regarding the **tf** as follows:

>>help tf ⏎   ← Function name without the argument

    tf Construct transfer function or convert to transfer function.

       Construction:
         SYS = tf(NUM,DEN) creates a continuous-time transfer function SYS with
         numerator NUM and denominator DEN. SYS is an object of type tf when
         NUM,DEN are numeric arrays, of type GENSS when NUM,DEN depend on
                                ⋮

One disadvantage of this method is that the user has to know the exact file name of a function. For a novice this facility may not be appreciative.

Casually we know partial name of a function or try to check whether the function exists by that name. Suppose we intend to see whether any function by the name **tf** exists. We execute the following by intermediacy of command **lookfor** (no space gap between **look** and **for**) to see all possible functions bearing the file name **tf** or having the file name **tf** partly:

>>lookfor tf ⏎

    tf   - Construct transfer function or convert to transfer function.
    adaptfilt   - Adaptive Filter Implementation.
    AbstractPortfolio   - Abstract portfolio object for portfolio optimization and analysis.
    Portfolio   - Portfolio object for mean-variance portfolio optimization and
                                ⋮

The return is having all possible matches of functions containing the word **tf**. Now the command **help** can be conducted to go through a particular one for example the second one is **adaptfilt** and we execute **help adaptfilt** to see its description at the command prompt.

Control system library help we view by the following:

>>help control ⏎

    Control System Toolbox
    Version 9.2 (R2011b) 08-Jul-2011

    General.
      ctrlpref   - Set Control System Toolbox preferences.
      InputOutputModel - Overview of input/output model objects.
      DynamicSystem   - Overview of dynamic system objects.
      lti   - Overview of linear time-invariant system objects.

    Graphical User Interfaces.
      ltiview   - LTI Viewer (time and frequency response analysis).
      sisotool   - SISO Design Tool (interactive compensator tuning).
                                ⋮

MATLAB exhibits a long list of functions under the library **control** in the last execution. Again if you execute **help ctrlpref** (for example for the first one) at the command prompt, you see its relevant description.

In order to have window form help, click the Help icon (i.e. ?) of figure 1.1(a) and MATLAB responds with the opening Help window of figure 1.8(b). As the figure shows, help is available content-based or index-based. If you have some search word for MATLAB or SIMULINK, you can search that through the Search of figure 1.8(b). This help form is better

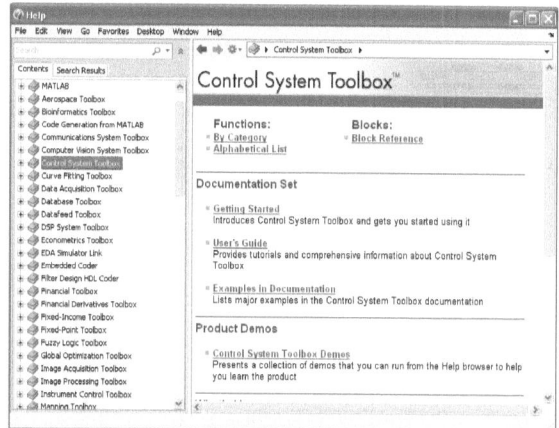

Figure 1.8(b) General Help window for MATLAB and SIMULINK

when one navigates MATLAB/SIMULINK's capability not looking for a particular function/block.

Under **Control System Toolbox** what functions are available and their help you can view also through this window which is explicit in figure 1.8(b).

Hidden algorithm or mathematical expression is often necessary whose assistance we can have through search option from MathWorks Website provided that our PC is connected to internet (figure 1.1(h)).

However we close the introductory discussion on MATLAB and SIMULINK with this.

Chapter 2

Implementing Control System in MATLAB $H(s)$

The subject matter in this chapter is to address the basics of control system entering into MATLAB. Our illustration starts with an elementary transfer function which is the building block of a realistic control system. As happens in MATLAB, most transfer functions are entered through matrix oriented data and embedded function depending on the control system complexity. A distinct link is explained on how to exercise large interconnected control system. However know-how details of the chapter highlight the following:

- ✦ ✦ How to define a single input-single output control system
- ✦ ✦ How to define a multiple input-multiple output control system
- ✦ ✦ Control systems connected in series, parallel, and feedback
- ✦ ✦ MATLAB techniques to implement an interconnected system

2.1 Defining a SISO control system in MATLAB

A single input single output (SISO) continuous control system may be defined in different ways – from transfer function in polynomial form, from transfer function in pole-zero-gain form, etc. Each of the forms has dedicated built-in function to define the SISO in MATLAB, which we will explain in detail in the following. These prototype defining functions are useful in combining a large interconnected system.

A continuous control system or plant is often characterized by a transfer function $H(s)$ in Laplace or s domain. All about entering the $H(s)$ information into MATLAB is in this section.

◆ Defining a transfer function by polynomial form

The control system $H(s)$ is supposed to be expressible in numerator and denominator polynomial forms. MATLAB built-in function **tf** (abbreviation for the transfer function) is applied to define the $H(s)$ with the syntax tf(numerator polynomial coefficients as a row matrix, denominator polynomial coefficients as a row matrix) where the polynomial coefficients must be in descending order and any missing coefficient is set to 0.

Suppose a continuous control system is represented by the transfer function $H(s) = \dfrac{7s^3 - 7s + 42}{s^4 - 118s^2 - 240s}$. The example $H(s)$ has the polynomial coefficient representation as [7 0 −7 42] and [1 0 −118 −240 0] for the numerator and denominator respectively. Having known so, we define the system as follows:

>>H=tf([7 0 -7 42],[1 0 -118 -240 0]) ↵

Transfer function:
  7 s^3 - 7 s + 42
  ----
  s^4 - 118 s^2 - 240 s

In above execution we assigned the return from the **tf** to the workspace **H** (can be any user-supplied variable). If we append a semicolon at the end of the statement i.e. H=tf([7 0 -7 42],[1 0 -118 -240 0]);, the functional popup would not be displayed. However the variable **H** holds the $H(s)$ information as a system.

◆ Defining a transfer function by pole-zero form

Not always is the transfer function in numerator-denominator polynomial form. When the transfer function numerator or denominator is given in factored or pole-zero-gain form, we employ MATLAB built-in function **zpk** with the syntax zpk(zeroes as a row matrix, poles as a row matrix, gain as a single number).

For example the $\begin{Bmatrix} \text{zeroes}: 2, 3, -1 \\ \text{poles}: 5, 0, 8, 6 \\ \text{gain}: 7 \end{Bmatrix}$ forms the control system function $H(s)$ as $\dfrac{7(s-2)(s-3)(s+1)}{(s-5)s(s-8)(s-6)}$ which we enter as follows:

>>z=[2 3 -1]; ↵   ← Assigning the zeroes as a row matrix to **z**, **z** is user-chosen
>>p=[5 0 8 6]; ↵ ← Assigning the poles as a row matrix to **p**, **p** is user-chosen
>>k=7; ↵   ← Assigning the gain as a scalar to **k**, **k** is user-chosen

```
>>H=zpk(z,p,k) ↵  ← Calling the zpk with the mentioned syntax and assigned the
                   return from zpk to H where H is user-chosen variable
Zero/pole/gain:
7 (s-2) (s-3) (s+1)
-----------------------
s (s-5) (s-6) (s-8)
```

We could have executed all commands in one line as **H=zpk([2 3 -1],[5 0 8 6],7)** instead of intermediate assigning. Appending a semicolon at the end of the statement i.e. **H=zpk(z,p,k);** does not show the functional popup. The last variable H holds the $H(s)$ information as a system from the pole-zero-gain description.

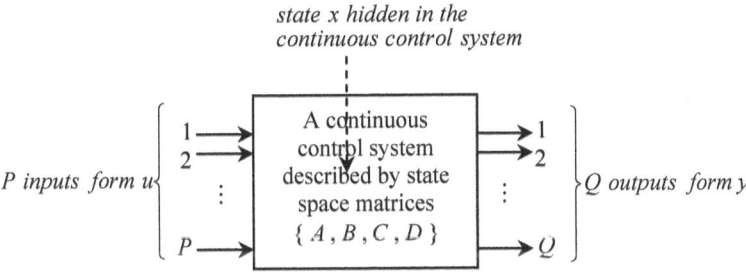

Figure 2.1(a) State space model of a continuous control system

◆ Defining a transfer function by state space form

A continuous control system $H(s)$ is also modeled by the state space form, governing equation of which is $\begin{cases} \dot{x} = Ax + Bu \\ y = Cx + Du \end{cases}$ where $\{A, B, C, D\}$ are called the state space matrices. The state x (in general is a column vector) is hidden in the continuous control system and is associated with energy storage elements inside the system.

The matrix A carries system parameter information. For an electric circuit, the A is purely a function of resistance, inductance, and capacitance. In general orders of the A, B, C, and D are $N \times N$, $N \times P$, $Q \times N$, and $Q \times P$ respectively, all matrices have real constant elements, P and Q are the numbers of inputs and outputs respectively, and P, Q, and N are integers. Figure 2.1(a) shows the schematics of the state space representation.

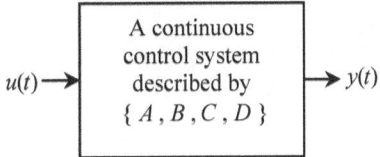

Figure 2.1(b) A single input - single output continuous control system

MATLAB built-in function **ss** (abbreviation for the state space) defines a continuous control system from its state space matrix representation with the syntax **ss**(A, B, C, D). Matrix entering discussion is seen in section 1.1.2.

A single input and single output control system $H(s)$ like the figure 2.1(b) is characterized by $A = \begin{bmatrix} -2 & -1 \\ 2 & 3 \end{bmatrix}$, $B = \begin{bmatrix} -1 \\ -2 \end{bmatrix}$, $C = [-2 \quad 1]$, and $D = [-1]$ which we wish to define.

Let us go through the following in this regard:
```
>>A=[-2 -1;2 3]; B=[-1;-2]; C=[-2 1]; D=-1; ↵
>>H=ss(A,B,C,D) ↵   ← Calling ss with the mentioned syntax and the return is
                      assigned to H where H is user-chosen variable
a =                          ← Meaning A
        x1   x2
   x1   -2   -1
   x2    2    3
b =                          ← Meaning B
        u1
   x1   -1
   x2   -2
c =                          ← Meaning C
        x1   x2
   y1   -2    1
d =                          ← Meaning D
        u1
   y1   -1
Continuous-time model.
```

In last implementation the first command line is basically the entering of given matrices to like name variables for example A to A. Usage of H= ss(A,B,C,D); does not show the matrix popup. The **x1** and **x2** mean two states. The **u1** and **y1** mean one input and one output respectively. The workspace H holds the continuous control system information of figure 2.1(b) which we wanted.

Note that for a SISO system state number can be any (which is here 2) and B and C must be one column and one row matrices respectively. If you had 4 state variables, the size of A would be 4×4 for the SISO.

◆ Defining a system by frequency response data

If a control system is represented by $H(s)$, replacing the Laplace variable s by $j\omega$ provides frequency response of the system. Built-in function frd defines a SISO continuous control system from $H(j\omega)$ data with the syntax frd($H(j\omega)$ data as a row matrix, ω data as a row matrix) over $\omega \geq 0$.

As an example the $H(j\omega)$ is 1, $1+j2$, $1+j3$, and $1+j3$ at $\omega = 0, 10, 20,$ and $30\ rad/\sec$ respectively and we wish to form a control system based on these frequency domain data.

Carry out the following for that:

>>w=[0 10 20 30]; ↵   ← Entering ω values as a row matrix to w where w is user-chosen

>>V=[1 1+j*2 1+j*3 1+j*3]; ↵   ← Entering $H(j\omega)$ values as a row matrix to V where V is user-chosen

>>H=frd(V,w) ↵   ← Forming the system from $H(j\omega)$ data

From input 1 to:

Frequency(rad/s)	output 1
0	1
10	1+2i
20	1+3i
30	1+3i

Continuous-time frequency response data model.

In above execution the H is a user-chosen variable which mimics the $H(s)$ system. Usage of the command H=frd(V,w); does not show the functional popup.

✦ Defining from other forms

Apart from previously explained forms, sometimes we come across mixed form transfer and second order prototype functions in control system study. These two forms are addressed in the following.

Mixed form transfer function:

Mixed form like $H(s) = \dfrac{s+42}{(s^4 + 3s^2 - 240)(s+2)(2s-1)}$ is manipulated until it takes previously mentioned forms i.e. numerator-denominator, pole-zero-gain, or state space. Polynomial multiplication is required in such system function which is conducted by using the built-in **conv** with the syntax **conv**(first polynomial as a row matrix, second polynomial as a row matrix) but maintaining descending power of s and setting missing coefficient to 0 in the representation.

The mixed function had better be in numerator-denominator form. Multiplication of the denominator factor $(s+2)(2s-1)$ then takes place by **conv([1 2],[2 -1])** with that the whole $(s^4 + 3s^2 - 240)(s+2)(2s-1)$ is obtained by **conv([1 0 3 0 -240],conv([1 2],[2 -1]))**. Knowing so, the complete system H we form by **H=tf([1 42],conv([1 0 3 0 -240],conv([1 2],[2 -1])))** where earlier explained **tf** is applied. In order to avoid typing mistakes, intermediate assignee variables could have been used.

Second order system from damping ratio and natural frequency:
A second order prototype system is defined in terms of damping ratio ζ and natural frequency ω_n which takes the form $H(s) = \dfrac{1}{s^2 + 2\zeta\omega_n s + \omega_n^2}$. Built-in function **ord2** defines a second order prototype system with syntax **ord2(** ω_n, ζ **)** but does not construct an object as the **tf**. Taking example of $\omega_n = 2\,rad/\sec$ and $\zeta = 0.5$, we first obtain the numerator and denominator polynomial coefficients through two output arguments (section 1.1.2) by writing **[N,D]=ord2(2,0.5)**; where the **N** and **D** are user-supplied variables for numerator and denominator polynomial coefficients respectively and then form the prototype second order system by writing **H=tf(N,D)**; where **H** holds the $H(s)$.

✦ Defining a random order system

For quick entering of a higher order system, this built-in function can be exercised. It is not feasible that we type again and again a large system for study. The built-in function **rmodel** keeps several options for arbitrary order control system generation with the syntax **rmodel(user-supplied order)**. The output argument number of **rmodel** can be two or four. If it is two, we get random numerator-denominator polynomial coefficients as return. If it is four, we get state space matrices as the return.

Let us generate a random third order system in numerator-denominator polynomial form as follows:

>>[N,D]=rmodel(3); ↵ ← N and D hold random numerator and denominator
                                polynomial coefficients respectively from the rmodel
                                where N and D are user-supplied variables
>>H=tf(N,D) ↵   ← Forming the control system from last obtained N and D by tf,
                               H holds the system, H is user-chosen

Transfer function:
  -1.213 s^2 - 1.118 s - 0.3409
----
s^3 + 1.851 s^2 + 61.66 s + 19.03

Above execution says that you have $H(s) = \dfrac{-1.213s^2 - 1.118s - 0.3409}{s^3 + 1.851s^2 + 61.66s + 19.03}$ available in H. When you run above statements, you find another third order system transfer function due to randomness.

What if we require state space third order random system generation? Just execute the following:

>>[A,B,C,D]=rmodel(3); ↵ ← A, B, C, and D hold random state space matrices
                                     from the rmodel where A, B, C, and D are
                                     user-supplied variables

```
>>H=ss(A,B,C,D) ⏎    ← Forming the control system from last obtained A, B, C,
                       and D by ss, H holds the system, H is user-chosen
a =
            x1         x2         x3
    x1    -0.3684     0.2027    0.1493
    x2    -0.2364    -0.6478    0.515
    x3     0.08665   -0.5292   -0.5992

b =
            u1
    x1    -0.1364
    x2     0.1139
    x3     0

c =
            x1         x2         x3
    y1      0        -0.09565   -0.8323

d =
            u1
    y1    0.2944
Continuous-time model.
```

Above execution says that we formed the third order random system function $H(s)$ where the state space matrices are $A = \begin{bmatrix} -0.3684 & 0.2027 & 0.1493 \\ -0.2364 & -0.6478 & 0.515 \\ 0.08665 & -0.5292 & -0.5992 \end{bmatrix}$, $B = \begin{bmatrix} -0.1364 \\ 0.1139 \\ 0 \end{bmatrix}$, $C = [0 \quad -0.09565 \quad -0.8323]$, and $D = [0.2944]$. When you run above statements, you find other third order state space matrices.

2.2 Defining a MIMO control system in MATLAB

Most built-in functions described so far keep provision for entering multi input multi output (MIMO) control system. For the MIMO control system the transfer function $H(s)$ is in matrix form rather than a single one as in SISO. The concept of cell array is a prerequisite for MIMO definition whose reference is seen in appendix D.15. The numbers of rows and columns of $H(s)$ indicate the output and input numbers of $H(s)$ respectively.

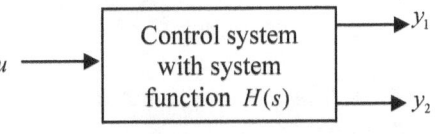

Figure 2.2(a) A single input – two output continuous control system

✦ Defining a single input multi output system

Figure 2.2(a) shows a single input two output MIMO where $H_1(s) = \frac{4}{3s-7}$ and $H_2(s) = \frac{4}{3s^2-5}$ and $H(s) = \begin{bmatrix} H_1(s) \\ H_2(s) \end{bmatrix}$. The $H_1(s)$ and $H_2(s)$ refer to y_1 to u and y_2 to u respectively. We wish to define this MIMO in MATLAB.

In last section we mentioned how to enter the transfer function in polynomial coefficient form, the same style is applicable here for both the numerator and denominator of each system. Let us enter all numerator and denominator polynomial coefficients as follows:

>>N1=4; D1=[3 -7]; N2=4; D2=[3 0 -5]; ↵   ← N1, D1, etc user-chosen variables

The **N1**, **N2**, **D1**, and **D2** correspond to the numerator of $H_1(s)$, the numerator of $H_2(s)$, the denominator of $H_1(s)$, and the denominator of $H_2(s)$ respectively. The **tf** defines this sort of system with the syntax tf({numerators of the component system functions as a column matrix separated by semicolon},{denominators of the component system functions as a column matrix separated by semicolon}) so we exercise the following:

>>H=tf({N1;N2},{D1;D2}) ↵   ← Forming the system $H(s)$, H holds $H(s)$, H is user-chosen

Transfer function from input to output...

```
          4
#1:   --------              ← Meaning H₁(s)
         3 s - 7


          4
#2:   ------------          ← Meaning H₂(s)
        3 s^2 - 5
```

Meaning $H_1(s)$

Meaning $H_2(s)$

If there were another output, the command would have been H=tf({N1;N2; N3},{D1;D2;D3}) where the N3 and D3 stand for the third output numerator and denominator polynomial coefficients respectively. Appending a semicolon in the last command i.e. H=tf({N1;N2},{D1;D2}); does not show the functional popup.

✦ Defining a multi input single output system

Figure 2.2(b) shows a two input one output MIMO control system where $H_1(s) = \frac{4}{3s-7}$ and $H_2(s) = \frac{4}{3s^2-5}$ and $H(s) = [H_1(s) \; H_2(s)]$. The $H_1(s)$ and $H_2(s)$ refer

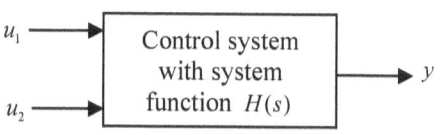

Figure 2.2(b) A single output – two input continuous control system

Control System Analysis & Design in MATLAB and SIMULINK

to y to u_1 and y to u_2 respectively. Define this MIMO in MATLAB.

The syntax we apply is tf({numerator polynomial coefficients of $H_1(s)$, $H_2(s)$, etc as a row matrix separated by comma},{denominator polynomial coefficients of $H_1(s)$, $H_2(s)$, etc as a row matrix separated by comma}) and the input argument polynomial coefficients must be in order therefore we exercise the following:

>>N1=4; D1=[3 -7]; N2=4; D2=[3 0 -5]; ↵ ← N1, D1, etc bear earlier meanings

>>H=tf({N1,N2},{D1,D2}) ↵ ← Forming the system $H(s)$, H holds $H(s)$, H is user-chosen

Transfer function from input 1 to output:
```
     4
  ---------          ← Meaning H₁(s)
   3 s - 7
```
← Meaning $H_1(s)$

Transfer function from input 2 to output:
```
      4
  ------------       ← Meaning H₂(s)
   3 s^2 - 5
```
← Meaning $H_2(s)$

If there were another input, the command would have been H=tf({N1,N2,N3},{D1,D2,D3}) where the N3 and D3 stand for the third input numerator and denominator polynomial coefficients as a row matrix respectively. Appending a semicolon in the tf does not show the functional popup i.e. H= tf({N1,N2,N3},{D1,D2,D3});.

✦ Defining a multi input multi output system

Figure 2.2(c) shows a two input three output MIMO where the system function $H(s)=\begin{bmatrix} H_{11}(s) & H_{12}(s) \\ H_{21}(s) & H_{22}(s) \\ H_{31}(s) & H_{32}(s) \end{bmatrix}$ is consisted of the following: $H_{11}(s)=\dfrac{7}{7s-5}$, $H_{12}(s)=\dfrac{-8}{-s+6}$, $H_{21}(s)=\dfrac{-1}{-s-5}$, $H_{22}(s)=\dfrac{1}{-3s+6}$, $H_{31}(s)=\dfrac{6}{3s-5}$, and $H_{32}(s)=\dfrac{9}{7s+6}$. The $H_{11}(s)$, $H_{12}(s)$, $H_{21}(s)$, $H_{22}(s)$, $H_{31}(s)$, and $H_{32}(s)$ refer to y_1 to u_1, y_1 to u_2, y_2 to u_1, y_2 to u_2, y_3 to u_1, and y_3 to u_2 respectively. Our objective is to define this MIMO in MATLAB.

Figure 2.2(c) A two input – three output continuous control system

The required syntax is tf({numerator polynomial coefficients of $H_{11}(s)$, $H_{12}(s)$, etc as a cell array which follows the order of the numerator

-35-

coefficients of $H(s)$ }, {denominator polynomial coefficients of $H_{11}(s)$, $H_{12}(s)$, etc as a cell array which follows the order of the denominator coefficients of $H(s)$ }) therefore we exercise the following:

>>N11=7; N12=-8; ↵   ← Numerators of $H_{11}(s)$ and $H_{12}(s)$ assigned to N11 and N12 respectively
>>N21=-1; N22=1; ↵   ← Numerators of $H_{21}(s)$ and $H_{22}(s)$ assigned to N21 and N22 respectively
>>N31=6; N32=9; ↵   ← Numerators of $H_{31}(s)$ and $H_{32}(s)$ assigned to N31 and N32 respectively
>>D11=[7 -5]; D12=[-1 6]; ↵ ← Denominators of $H_{11}(s)$ and $H_{12}(s)$ assigned to D11 and D12 respectively
>>D21=[-1 -5]; D22=[-3 6]; ↵ ← Denominators of $H_{21}(s)$ and $H_{22}(s)$ assigned to D21 and D22 respectively
>>D31=[3 -5]; D32=[7 6]; ↵   ← Denominators of $H_{31}(s)$ and $H_{32}(s)$ assigned to D31 and D32 respectively

In the last executions the variables **N11, N12**, etc are user-chosen. All numerator and denominator polynomial coefficients are available in workspace variables. Knowing so, the complete MIMO system is formed by the cell array **H** as follows:

>>H=tf({N11,N12;N21,N22;N31,N32},{D11,D12;D21,D22;D31,D32}) ↵

Transfer function from input 1 to output...

```
        7
#1:  --------                    ← Meaning H₁₁(s)
      7 s - 5
```
← Meaning $H_{11}(s)$

```
        1
#2:  -------                     ← Meaning H₂₁(s)
       s + 5
```
← Meaning $H_{21}(s)$

```
        6
#3:  --------                    ← Meaning H₃₁(s)
      3 s - 5
```
← Meaning $H_{31}(s)$

Transfer function from input 2 to output...

```
        8
#1:  ------                      ← Meaning H₁₂(s)
       s - 6
```
← Meaning $H_{12}(s)$

```
       -1
#2:  ---------                   ← Meaning H₂₂(s)
      3 s - 6
```
← Meaning $H_{22}(s)$

```
        9
#3:  ---------                   ← Meaning H₃₂(s)
      7 s + 6
```
← Meaning $H_{32}(s)$

2.3 Defining series and parallel control systems

Control systems can be connected in series or parallel form. In either case built-in function is embedded in MATLAB to find the equivalent control system. Equivalent means we can replace all constitutive control systems by a single one.

✦ Seriesly connected systems

Figure 2.3(a) shows two control systems $H_1(s) = \dfrac{5s^2 - s + 1}{s^3 - 1}$ and $H_2(s) = \dfrac{5.43(s-3)}{(s-1)(s+4)}$ connected in series. Their equivalent control system is given by $H_{eq}(s) = H_1(s)\,H_2(s) = \dfrac{27.15s^3 - 86.88s^2 + 21.72s - 16.29}{s^5 + 3s^4 - 4s^3 - s^2 - 3s + 4}$ which we wish to implement.

Figure 2.3(a) Two control systems connected in series

MATLAB built-in function **series** helps us compute the series equivalent system with the syntax **series**(system 1 representing variable name, system 2 representing variable name). Section 2.1 explained **tf, zpk, ss**, or other defines any given control system that is how the $H_1(s)$ and $H_2(s)$ are entered as follows:

>>H1=tf([5 -1 1],[1 0 0 -1]); ↵ ← $H_1(s)$ is defined from numerator and denominator coefficients and assigned to workspace H1 where H1 is user-chosen name and H1⇔ $H_1(s)$

>>H2=zpk(3,[1 -4],5.43); ↵ ← $H_2(s)$ is defined from pole-zero-gain and assigned to workspace H2 where H2 is user-chosen name and H2⇔ $H_2(s)$

>>Heq=series(H1,H2) ↵ ← $H_{eq}(s)$ is computed by the **series** and assigned to Heq where Heq is user-chosen name and Heq⇔ $H_{eq}(s)$

Zero/pole/gain:
```
27.15 (s-3) (s^2  - 0.2s + 0.2)
-----------------------------------
   (s-1)^2 (s+4) (s^2  + s + 1)
```

As the execution says the Heq contents are in the pole-zero-gain form. If we exercise the **tf** on Heq, we find the equivalent system as follows:

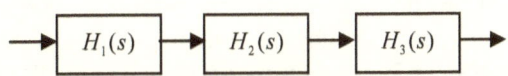

Figure 2.3(b) Three control systems connected in series

>>tf(Heq) ↵
Transfer function:
```
27.15 s^3 - 86.88 s^2 + 21.72 s - 16.29
-------------------------------------------
    s^5 + 3 s^4 - 4 s^3 - s^2 - 3 s + 4
```
← the $H_{eq}(s)$ we expected

Suppose we have three control systems in series like figure 2.3(b) where $H_3(s) = \dfrac{3.2(s-2.2)}{(s+1)(s+1.5)}$ and $H_1(s)$ and $H_2(s)$ are the foregoing ones. With these three control systems, we have the equivalent $H_{eq}(s) = H_1(s)\,H_2(s)\,H_3(s)$ by using the following:

>>H3=zpk(2.2,[-1 -1.5],3.2); ↵ ← $H_3(s)$ is defined from pole-zero-gain and assigned to workspace H3 where H3 is user-chosen name and H3⇔ $H_3(s)$

>>Heq=series(series(H1,H2),H3); ↵

In the last command the inner **series(H1,H2)** finds the series equivalent of $H_1(s)$ and $H_2(s)$ and the outer **series** does the resulting one and $H_3(s)$. The last **Heq** keeps the equivalent of the three control systems. If you intend to view it, execute **tf(Heq)** at the command prompt. Note that **series(H1,H2)** is equivalent to **H1*H2**.

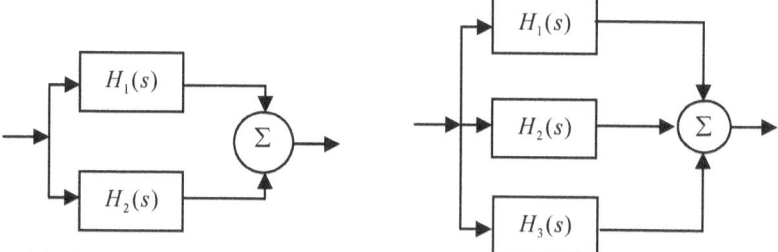

Figure 2.3(c) Two control systems connected in parallel

Figure 2.3(d) Three control systems connected in parallel

✦ Parallelly connected systems

Figure 2.3(c) presents two control systems connected in parallel. Taking the example control systems $H_1(s)$ and $H_2(s)$ of series connection into account, we should have the equivalent control system as $H_{eq}(s) = H_1(s) + H_2(s) = \dfrac{10.43s^4 - 2.29s^3 - 22s^2 + 1.57s + 12.29}{s^5 + 3s^4 - 4s^3 - s^2 - 3s + 4}$. MATLAB built-in function **parallel** determines the equivalent system function of two control systems connected in parallel with the syntax **parallel**(system 1 representing variable name, system 2 representing variable name). We know that the $H_1(s)$ and $H_2(s)$ are stored in the workspace variables H1 and H2 respectively therefore execute the following:

>>Heq=parallel(H1,H2); ↵ ← $H_{eq}(s)$ is computed by the **parallel** and assigned to Heq where Heq is user-chosen name and Heq⇔ $H_{eq}(s)$

Just to view the computation in coefficient form, let us carry out the following:

>>tf(Heq) ↵

Transfer function:
```
10.43 s^4 - 2.29 s^3 - 22 s^2 + 1.57 s + 12.29
-----------------------------------------------
       s^5 + 3 s^4 - 4 s^3 - s^2 - 3 s + 4
```

Figure 2.3(d) depicts three control systems connected in parallel whose equivalent is easily found by exercising Heq=parallel(parallel(H1,H2), H3); at the command prompt based on the foregoing $H_1(s)$, $H_2(s)$, and $H_3(s)$ where the symbols have previously mentioned meanings and mathematically Heq returns $H_1(s) + H_2(s) + H_3(s)$. Note that parallel(H1,H2)⇔H1+H2.

2.4 Feedback control systems

A feedback control system is composed of forward system $G(s)$ and feedback system $H(s)$. Figures 2.4(a) and 2.4(b) show the feedback control system diagrams for negative and positive feedbacks respectively. In each configuration the $R(s)$ and $Y(s)$ represent the system functions of the input and output respectively.

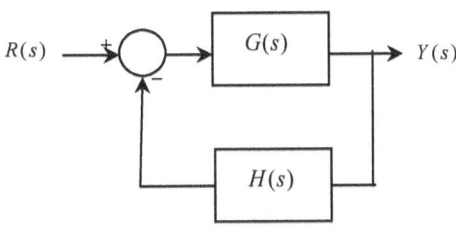

Figure 2.4(a) A negative feedback system

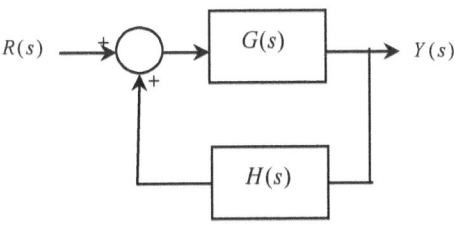

Figure 2.4(b) A positive feedback system

MATLAB built-in function feedback implements both feedback control systems. The syntax we apply is feedback($G(s)$, $H(s)$, type of feedback). The $G(s)$ or $H(s)$ is entered by employing section 2.1 mentioned built-in functions regardless of the numerator-denominator, pole-zero-gain, or state-space form. The type of feedback is +1 for positive feedback and -1 for negative feedback. Let us see the following examples in this regard.

◆ **Negative feedback**

In figure 2.4(a) when $G(s) = \dfrac{3s}{2s^2 + 3s - 4}$ and $H(s) = \dfrac{2}{s+1}$, it is given that the overall control system function is $\dfrac{Y(s)}{R(s)} = \dfrac{G(s)}{1 + G(s)H(s)} = \dfrac{3s^2 + 3s}{2s^3 + 5s^2 + 5s - 4}$

which we wish to obtain.

Applying section 2.1 cited **tf** we enter the $G(s)$ and $H(s)$ as follows:

```
>>G=tf([3 0],[2 3 -4]); ↵     ← G holds $G(s)$, G is user-chosen
>>H=tf(2,[1 1]); ↵            ← H holds $H(s)$, H is user-chosen
```
After that we call the **feedback** with just cited syntax and assign the return from the function to E (some user-chosen variable) as follows:

```
>>E=feedback(G,H,-1) ↵   ← E holds the $\frac{Y(s)}{R(s)}$
```

Transfer function:
```
    3 s^2 + 3 s
-----------------------
2 s^3 + 5 s^2 + 5 s – 4
```

Usage of command E=feedback(G,H,-1); does not display functional popup.

✦ Positive feedback

It is given that the overall control system is $\frac{Y(s)}{R(s)} = \frac{G(s)}{1-G(s)H(s)} = \frac{3s^2+3s}{2s^3+5s^2-7s-4}$ for figure 2.4(b) with negative feedback mentioned system functions which we wish to obtain. All we need is change the third input argument of the function as follows:

```
>>E=feedback(G,H,1) ↵   ← E holds the $\frac{Y(s)}{R(s)}$
```

Transfer function:
```
    3 s^2 + 3 s
-----------------------
2 s^3 + 5 s^2 - 7 s - 4
```

2.5 Control system form conversion

Suppose you entered a control system in one form for instance in pole-zero-gain form and wish to view the system in other form for example in state space. This section demonstrates several examples implementing the conversion.

✦ Example 1: Polynomial to pole-zero-gain form

The $H(s) = \frac{5s^2-s+1}{s^3-1}$ when expressed in pole-zero-gain form we get $H(s) = \frac{5(s^2-0.2s+0.2)}{(s-1)(s^2+s+1)}$. We first enter the $H(s)$ using the **tf** of section 2.1:

```
>>H=tf([5 -1 1],[1 0 0 -1]); ↵
```

Then call **zpk** of the same section to see the pole-zero-gain form:
```
>>zpk(H) ↵
```

Zero/pole/gain:
```
 5 (s^2 - 0.2s + 0.2)
----------------------
   (s-1) (s^2 + s + 1)
```

The problem in last execution is **zpk** expresses a transfer function into lowest possible linear and quadratic factors. Sometimes we may look for all linear factor in that case poles, zeroes, and gain are extracted by the **tf2zp** (abbreviation for transfer function to zero-pole) but the input arguments have to be the polynomial coefficients of $H(s)$ as follows:

>>[z,p,k]=tf2zp([5 -1 1],[1 0 0 -1]) ↵

z =
       0.1000 + 0.4359i
       0.1000 - 0.4359i
p =
       -0.5000 + 0.8660i
       -0.5000 - 0.8660i
       1.0000
k =
       5

The user-supplied output arguments i.e. [z,p,k] stand for the zeroes, poles, and gain which hold the return respectively. The user himself has to write the expression in light of above which is $\dfrac{5(s-0.1-j0.4359)(s-0.1+j0.4359)}{(s-1)(s+0.5-j0.866)(s+0.5+j0.866)}$.

✦ **Example 2: Pole-zero-gain to polynomial form**

Say we have $\dfrac{5(s-1-j2)(s-1+j2)}{(s-1)(s+0.5-j3)(s+0.5+j3)}$ which in polynomial form turns out to be $\dfrac{5s^2 - 10s + 25}{s^3 + 8.25s - 9.25}$. How do we achieve this? The **zp2tf** (abbreviation for zero-pole to transfer function) helps us determine the form. Zeroes are entered as a column matrix, so are the poles and do so by:

>>z=[1+i*2;1-i*2]; p=[1;-0.5+i*3;-0.5-i*3]; k=5; ↵

Exercise the following to extract the numerator and denominator polynomial coefficients (**N** for numerator and **D** for denominator):

>>[N,D]=zp2tf(z,p,k) ↵

N =
     0   5   -10   25
D =
     1.0000   0   8.2500   -9.2500

Then you may exercise the **tf** on above **N** and **D** in order to form the system **H**:

>>H=tf(N,D) ↵

Transfer function:
5 s^2 - 10 s + 25
----
s^3 + 8.25 s - 9.25

◆ **Example 3: State-space to polynomial form**

Suppose we have $A = \begin{bmatrix} -2 & -1 \\ 2 & 3 \end{bmatrix}$, $B = \begin{bmatrix} -1 \\ -2 \end{bmatrix}$, $C = [-2 \quad 1]$, and $D = [-1]$ and wish to obtain its polynomial form, which should be $G(s) = C(sI - A)^{-1}B + D = \dfrac{-s^2 + s - 12}{s^2 - s - 4}$.

The function **ss2tf** (abbreviation for state space to transfer function) makes this conversion possible. It has four input arguments: the state space matrices respectively. Entering of state space matrices takes place by:
>>A=[-2 -1;2 3]; B=[-1;-2]; C=[-2 1]; D=-1; ↵

The ss2tf has two output arguments; the numerator and denominator polynomial coefficients respectively and does not construct an object so after finding numerator (to **N**) and denominator (**D1**) coefficients, the **tf** is exercised:
>>[N,D1]=ss2tf(A,B,C,D); ↵
Since the state space form involves D, we need another variable for the denominator coefficients which is **D1**:
>>tf(N,D1) ↵
Transfer function:
-s^2 + s - 12
----
s^2 - s - 4

◆ **Example 4: Polynomial to state-space form**

A transfer function $\dfrac{5s^2 - 10s + 25}{s^3 + 8.25s - 9.25}$ has the state space representation $A = \begin{bmatrix} 0 & -8.25 & 9.25 \\ 1 & 0 & 0 \\ 0 & 1 & 0 \end{bmatrix}$, $B = \begin{bmatrix} 1 \\ 0 \\ 0 \end{bmatrix}$, $C = [5 \quad -10 \quad 25]$, and $D = [0]$ which we wish to obtain.

Assign the given transfer function polynomial coefficients to some variables (**N** for numerator and **D1** for denominator):
>>N=[5 -10 25]; D1=[1 0 8.25 -9.25]; ↵

MATLAB function **tf2ss** (abbreviation for transfer function to state space) determines the state space matrices. It has two input arguments: the numerator and denominator polynomial coefficients and four output arguments: the state space matrices respectively therefore call the converter as:
>>[A,B,C,D]=tf2ss(N,D1) ↵

A =
       0   -8.2500   9.2500
  1.0000   0   0
       0   1.0000   0

```
B =
     1
     0
     0
C =
     5  -10  25
D =
     0
```

✦ Example 5: Pole-zero-gain to state space form

The transfer function $\dfrac{5(s-1-j2)(s-1+j2)}{(s-1)(s+0.5-j3)(s+0.5+j3)}$ is in pole-zero-gain form which has the state space representation $A = \begin{bmatrix} 1 & 0 & 0 \\ 1 & -1 & -3.0414 \\ 0 & 3.0414 & 0 \end{bmatrix}$, $B = \begin{bmatrix} 1 \\ 0 \\ 0 \end{bmatrix}$, $C = [5\ -15\ -6.9870]$, and $D = [0]$ and we intend to exercise this.

MATLAB function **zp2ss** (abbreviation for zero-pole to state space) helps us exercise the conversion which has the input arguments zeroes, poles, and gain respectively, and entering of which occur like example 2.

>>z=[1+i*2;1-i*2]; p=[1;-0.5+i*3;-0.5-i*3]; k=5; ↵

Output arguments of the **zp2ss** are the four state space matrices so exercise the following at the command prompt:

>>[A,B,C,D]=zp2ss(z,p,k) ↵

```
A =
     1.0000        0         0
     1.0000   -1.0000   -3.0414
          0    3.0414         0
B =
     1
     0
     0
C =
     5.0000  -15.0000  -6.9870
D =
     0
```

✦ Example 6: State space to pole-zero-gain form

With state space matrices $A = \begin{bmatrix} -2 & -1 \\ 2 & 3 \end{bmatrix}$, $B = \begin{bmatrix} -1 \\ -2 \end{bmatrix}$, $C = [-2\ \ 1]$, and $D = [-1]$, the transfer function of a control system has the pole-zero-gain representation $\dfrac{-(s-0.5-j3.4278)(s-0.5+j3.4278)}{(s+1.5616)(s-2.5616)}$. Our objective is to attain this.

MATLAB function **ss2zp** (abbreviation for state space to zero-pole) makes this conversion attainable. Entering of state space matrices occurs by:

```
>>A=[-2 -1;2 3]; B=[-1;-2]; C=[-2 1]; D=-1; ↵
```
The converter has four input arguments: the state space matrices and three output arguments: the zeroes, poles, and gain respectively hence calling with ongoing symbology is conducted by:
```
>>[z,p,k]=ss2zp(A,B,C,D) ↵
```

z =
   0.5000 + 3.4278i
   0.5000 - 3.4278i

p =
   -1.5616
   2.5616

k =
   -1

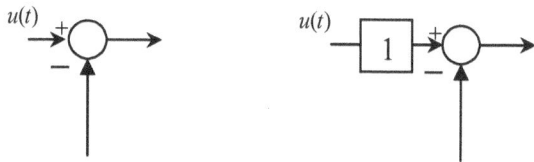

Figure 2.5(a) Organizing the input to summing point

2.6 Interconnected control systems

Series, parallel or feedback system is not the only connection found in control systems. Interconnected control system formation needs different approach and MATLAB functions. This approach is effective when we have large interconnected control system with many transfer functions. There is an algorithm while constructing the interconnected control system, which we explain now.

(a) In an interconnected system find all system functions, we label them by S_1, S_2, S_3, etc.

(b) From input to output there must a system function for every branch in the control system. If some branch does not have a system function, we may assume that as 1 without the loss of generality. Specially a branch feeding to summing point sometimes needs this, illustration of which is in figure 2.5(a).

(c) Enter the system functions one at a time by earlier **tf**, **zpk**, or **ss**. Let us say the transfer functions of the interconnected system are assigned to the variables S1, S2, S3, etc respectively. Variables are user-chosen.

(d) Form an unconnectedly stacked system by placing S_1, S_2, S_3, etc on top of the other where the placement order is user-defined. Thus you get a system like the figure 2.5(b). Every transfer function may have single/multiple input and output. In figure 2.5(b) the second system S_2 has two inputs and one output. Again the third system in the figure has one input and two outputs.

(e) Build a stacked system on figure 2.5(b) employing the built-in command **append** which applies the syntax **append(S_1, S_2, S_3, etc)** and assign the stacked system to some user-chosen variable **S**.

(f) Number the inputs and outputs of the stacked system sequentially. For example in the figure 2.5(b) the inputs are numbered by 1 through 4, so are the outputs by the same.

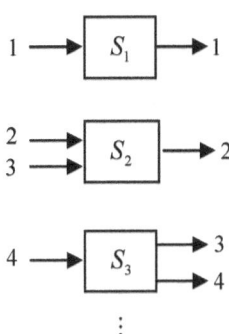

Figure 2.5(b) Unconnectedly stacked system from all transfer functions

(g) Look for the external input(s) of the given interconnected system from the number of unconnectedly stacked system and put them as a row matrix (call the matrix as **I**). The **I** is a user-chosen variable.

(h) Look for the external output(s) of the given interconnected system from the number of unconnectedly stacked system and put them as a row matrix (call the matrix as **O**). The **O** is a user-chosen variable.

(i) Form the interconnection matrix **C** where **C** is a user-chosen variable. The **C** is mainly for summing and pick off points. Every row of **C** refers to a component of stacked system in figure 2.5(b). The first element in **C** is the input number and subsequent elements refer to output from the other systems. Suppose S_1 gets its input from S_2 and S_3 where S_2 is in direct feeding and S_3 is in negative feedback and assume that they follow the numbering of figure 2.5(b). Then the row for S_1 is [1 2 −3 −4]. For any pick off point we take the output as 0. Similarly we have to finish the row style description for other transfer functions in the interconnection. Enter **C** as a rectangular matrix. The elements in each row in **C** are made equal in number

-45-

by padding zero in case any row element number is unequal with that of the other.
(j) The complete system is constructed by another embedded function called **connect** which has the syntax **connect**(stacked system, interconnection matrix, inputs, outputs). Obliged by the earlier symbology, one may call the transfer function constructor as **connect(S,C,I,O)** and assign the return to some user-chosen variable **OS** which represents the overall system.

Note that this sort of system formation is possible if the transfer function numerator degree is at most equal to that of the denominator. Let us see the following examples implementing the algorithm.

✦ Example 1: A feedback system

Consider the negative feedback system of section 2.4. We wish to obtain the same system function by using this approach that will solidify our concept.

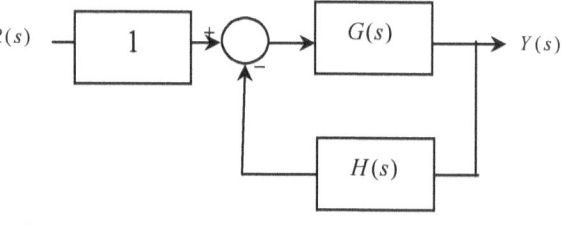

Figure 2.5(c) Feedback system of figure 2.4(a)

With a unity gain the feedback system of figure 2.4(a) is rearranged for feeding to the computing which we see in figure 2.5(c). With that the unconnectedly stacked system along with the input and output numbering is shown in figure 2.5(d).

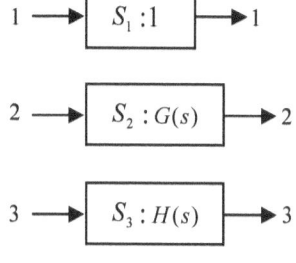

Figure 2.5(d) Unconnectedly stacked system of figure 2.5(c)

Let us define (get $G(s)$ and $H(s)$ from section 2.4) the three systems as follows:

>>S1=1; ↵   ← S1 ⇔ S_1
>>S2=tf([3 0],[2 3 -4]); ↵   ← S2 ⇔ S_2
>>S3=tf(2,[1 1]); ↵   ← S3 ⇔ S_3

Form the stacked system using above and assign that to **S** by:
>>S=append(S1,S2,S3); ↵

Now we have to look into the interconnection. Compare figure 2.5(d) to 2.5(c), we see one summing and one pick off points. Input of S_2 is coming

from output of S_1 and S_3 with S_3 negative feedback so the connection row matrix for S_2 is [2 1 −3]. Again S_3 is from the pick off point so its row matrix is [3 2 0]. Knowing so the connection matrix **C** is formed by:
>>C=[2 1 -3;3 2 0]; ↵

Overall input $R(s)$ is going into S_1 so from figure 2.5(d), it is the input number 1 hence form the **I** by:
>>I=1; ↵

Overall output $Y(s)$ is coming out of S_2 so from figure 2.5(d), it is the output number 2 hence form the **O** by:
>>O=2; ↵

Finally construct the overall system **OS** by:
>>OS=connect(S,C,I,O) ↵

```
Transfer function:
   1.5 s^2 + 1.5 s
---------------------------------
s^3 + 2.5 s^2 + 2.5 s - 2
```

Above function we found in section 2.4 for the negative feedback case. You need to divide section 2.4 numerator and denominator by 2 in order to get identical function. Well rationalization does not change a control system.

♦ **Example 2: A series system**
In figure 2.3(b) three control systems are connected in series with $H_1 = \frac{s+1}{s^2+1}$, $H_2 = \frac{3s}{2s^2+3s-4}$, and $H_3 = \frac{2}{s+1}$. It is given that the equivalent system function is $H_{eq} = \frac{6s^2+6s}{2s^5+5s^4+s^3+s^2-s-4}$ which we wish to verify by using this interconnection approach.

Figure 2.5(e) Unconnectedly stacked system for series connection in figure 2.3(b)

Figure 2.5(e) depicts the unconnectedly stacked system for the series connection of figure 2.3(b). Let us enter the three systems by:
>>S1=tf([1 1],[1 0 1]); ↵   ← S1⇔ S_1
>>S2=tf([3 0],[2 3 -4]); ↵   ← S2⇔ S_2
>>S3=tf(2,[1 1]); ↵   ← S3⇔ S_3

Form the stacked system by:
>>S=append(S1,S2,S3); ↵

Input of given system is the input number 1 of figure 2.5(e) so enter the input by:

```
>>I=1; ↵
```
Output of given system is the output number 3 of figure 2.5(e) so enter the output by:
```
>>O=3; ↵
```
Compare figure 2.5(e) to 2.3(b), input 2 of S_2 is from output 1 of S_1 and input 3 of S_3 is from output 2 of S_3 so interconnection matrix is entered by:
```
>>C=[2 1;3 2]; ↵
```
The overall system is finally formed by:
```
>>OS=connect(S,C,I,O) ↵
```

Transfer function:
```
      3 s^2 + 3 s - 7.494e-016
    -----------------------------------------
    s^5 + 2.5 s^4 + 0.5 s^3 + 0.5 s^2 - 0.5 s - 2
```

As you see, the return is rationalized by 2 besides there is an additional number -7.494×10^{-16} in the numerator. Given the numerical nature of the computing, the additional number is extremely small compared to other polynomial coefficients so ignore that.

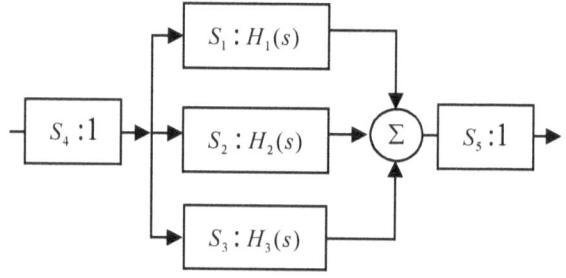

✦ Example 3: A parallel system

In figure 2.3(d) three control systems are connected in parallel. With the transfer functions of example 2 the equivalent system is $H_{eq} =$

$$\frac{4.5s^4 + 8s^3 + 1.5s^2 + 2s - 6}{s^5 + 2.5s^4 + 0.5s^3 + 0.5s^2 - 0.5s - 2}$$

which we intend to obtain by ongoing approach.

With unity transfer functions one at input and one at output the rearranged system is depicted in figure 2.5(f) along with the system labeling. Also is the unconnectedly stacked system in conjunction with input-output number in figure 2.5(g).

Enter the five systems using earlier symbology as follows:

Figure 2.5(f) Three parallel control systems of figure 2.3(d) with unity transfer functions

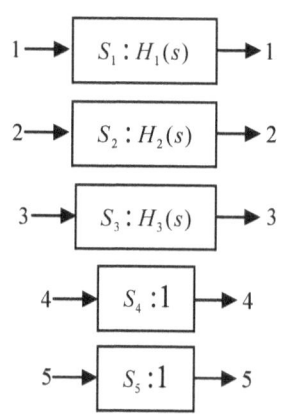

Figure 2.5(g) Stacked version of parallel system in figure 2.5(f)

-48-

```
>>S1=tf([1 1],[1 0 1]); ↵
>>S2=tf([3 0],[2 3 -4]); ↵
>>S3=tf(2,[1 1]); ↵
>>S4=1; S5=1; ↵
```
Form the stacked system by:
```
>>S=append(S1,S2,S3,S4,S5); ↵
```
Overall input of given system is the input number 4 of figure 2.5(g) so enter the input by:
```
>>I=4; ↵
```
Overall output of given system is the output number 5 of figure 2.5(g) so enter the output by:
```
>>O=5; ↵
```
Compare figure 2.5(g) to 2.5(f), input 1 of S_1 is from output 4 of S_4 which is a pick off point therefore the row matrix for S_1 input is [1 4 0]. Similarly for S_2 and S_3 we get [2 4 0] and [3 4 0] respectively. Again the S_5 input is composed of outputs from S_1, S_2, and S_3 (all three additive) so the row matrix for the S_5 is [5 1 2 3]. Since the maximum number of row elements is 4, we pad the last three row matrices by 0 hence the complete interconnection matrix is $\begin{bmatrix} 1 & 4 & 0 & 0 \\ 2 & 4 & 0 & 0 \\ 3 & 4 & 0 & 0 \\ 5 & 1 & 2 & 3 \end{bmatrix}$ which needs the following entering:
```
>>C=[1 4 0 0;2 4 0 0;3 4 0 0;5 1 2 3]; ↵
```
At last the overall system is formed by:
```
>>OS=connect(S,C,I,O) ↵
Transfer function:
4.5 s^4 + 8 s^3 + 1.5 s^2 + 2 s - 6
-----------------------------------------------------------
s^5 + 2.5 s^4 + 0.5 s^3 + 0.5 s^2 - 0.5 s - 2
```

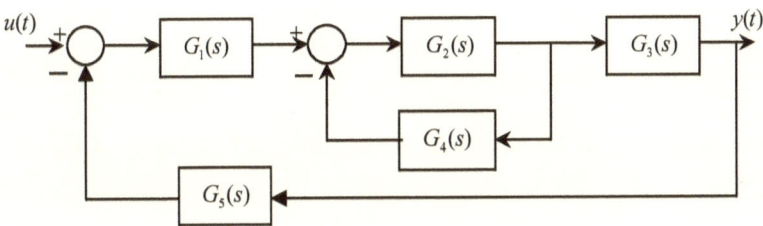

Figure 2.6(a) An interconnected control system

◆ **Example 4: Single input-output interconnected system**

Figure 2.6(a) shows a single input-output interconnected control system with $G_1(s) = 20$, $G_2(s) = \dfrac{0.3}{s^2 + 5s + 12}$, $G_3(s) = \dfrac{s + 0.01}{(s + 0.1)(s + 0.02)}$, $G_4(s) =$

$\frac{s}{s+1}$, and $G_5(s) = \{A, B, C, D\}$ where $A = 1$, $B = 1.1$, $C = 10^{-5}$, and $D = 0.5$. We intend to define this control system.

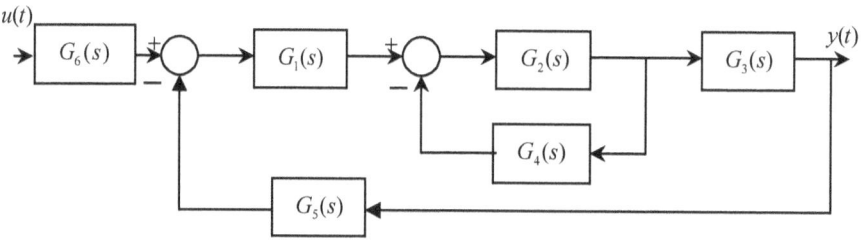

Figure 2.6(b) Control system of figure 2.6(a) with unity transfer function

Injection of unity transfer function is required (here at the input: $G_6(s) = 1$) as exercised in previous examples. Figure 2.6(b) indicates the injection. We find the relevant stacked system along with the numbered input-output in figure 2.6(c). Let us enter the control system elements following ongoing symbology:

>>S1=20; ↵
>>S2=tf(0.3,[1 5 12]); ↵
>>S3=zpk(-0.01,[-0.1 -0.02],1); ↵
>>S4=tf([1 0],[1 1]); ↵
>>A=1; B=1.1; C=1e-5; D=0.5; ↵
>>S5=ss(A,B,C,D); ↵
>>S6=1; ↵

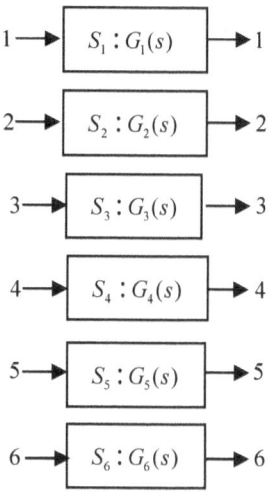

Figure 2.6(c) Stacked form of system in figure 2.6(b)

One forms the stacked system by:
>>S=append(S1,S2,S3,S4,S5,S6); ↵

Overall input of given system is the input number 6 of figure 2.6(c) so enter the input by:
>>I=6; ↵

Overall output of given system is the output number 3 of figure 2.6(c) so enter the output by:
>>O=3; ↵

For the connection matrix, we discover the following:
input of $G_1(s)$ from outputs $G_6(s)$ and $-G_5(s)$ i.e. [1 6 −5],
input of $G_2(s)$ from outputs $G_1(s)$ and $-G_4(s)$ i.e. [2 1 −4],
input of $G_3(s)$ from output $G_2(s)$ on pick off i.e. [3 2 0],
input of $G_4(s)$ from output $G_2(s)$ on pick off i.e. [4 2 0], and

input of $G_5(s)$ from output $G_3(s)$ on pick off i.e. [5 3 0].
Hence the complete interconnection matrix is entered by:
>>C1=[1 6 -5;2 1 -4;3 2 0;4 2 0;5 3 0]; ↵

We used the variable **C1** because **C** is used for the $G_5(s)$. Eventually we get the overall system by:
>>OS=connect(S,C1,I,O); ↵

At this point if any system is in state space form, the overall return will be in state space form too. In other examples we have seen the OS as transfer functions. The last OS is in state space form because of $G_5(s)$. If we wish to view the overall transfer function, exercise the tf on the OS:
>>tf(OS) ↵
Transfer function:

```
        6 s^3 + 0.06 s^2 - 6 s - 0.06
-----------------------------------------------------------
s^6 + 5.12 s^5 + 11.9 s^4 - 0.934 s^3 - 12.58 s^2 - 4.451 s - 0.054
```

i.e. $\dfrac{Y(s)}{U(s)}$ of figure 2.6(a) is $\dfrac{6s^3 + 0.06s^2 - 6s - 0.06}{s^6 + 5.12s^5 + 11.9s^4 - 0.934s^3 - 12.58s^2 - 4.451s - 0.054}$.

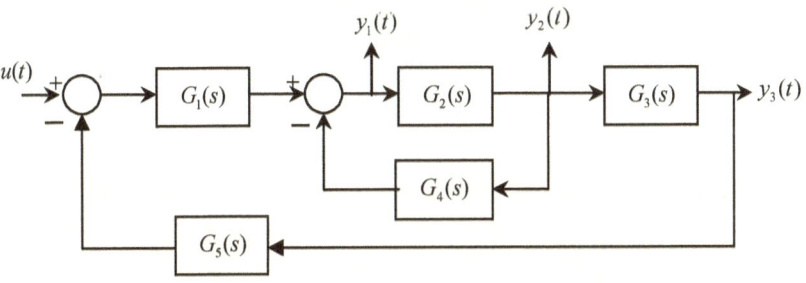

Figure 2.6(d) A single input-three output control system

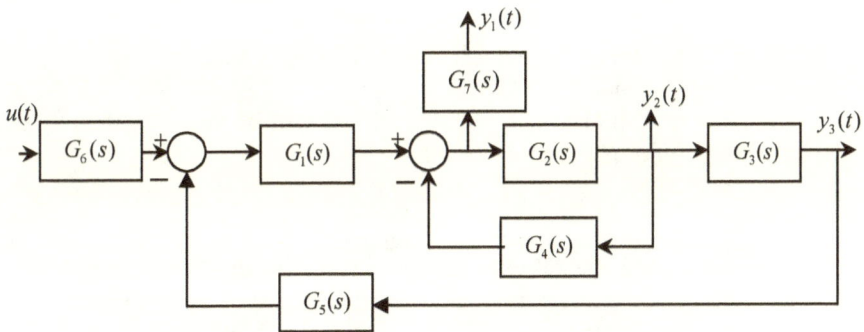

Figure 2.6(e) Injection of unity transfer functions to single input-three output control system of figure 2.6(d)

-51-

◆ **Example 5: Single input-multiple output interconnected system**

From example 4 interconnected system we take three outputs as indicated in figure 2.6(d). Our objective is to implement this system.

In the figure outputs $y_2(t)$ and $y_3(t)$ are also the outputs of $G_2(s)$ and $G_3(s)$ respectively hence only the $u(t)$ and $y_1(t)$ need unity transfer function injection, inclusion of which is in figure 2.6(e) where $G_7(s)=1$.

The $G_7(s)$ needs to be appended in figure 2.6(c) which you find in figure 2.6(f). From example 4 we need to enter S1 through S6. After that we have to enter the unity transfer function for $G_7(s)$:

>>S7=1; ↵

Making all components available, we form the stacked system by:

>>S=append(S1,S2,S3,S4,S5,S6,S7); ↵

Overall input of figure 2.6(d) is the same as in figure 2.6(a) so:

>>I=6; ↵

Now overall output is not single instead triple. The outputs $y_1(t)$, $y_2(t)$, and $y_3(t)$ are the output numbers 7, 2, and 3 from $G_7(s)$, $G_2(s)$, and $G_3(s)$ in figure 2.6(f) respectively. The O is not a scalar anymore instead a row matrix indicating the output numbers hence:

>>O=[7 2 3]; ↵

Figure 2.6(f) Stacked form of system in figure 2.6(e)

The interconnection matrix of example 4 needs addition of a row for the $G_7(s)$ which we illustrate by:

input of $G_7(s)$ from outputs $G_1(s)$ and $-G_4(s)$ i.e. [7 1 −4].

Taking the addition into account, the interconnection matrix is entered as:

>>C1=[1 6 -5;2 1 -4;3 2 0;4 2 0;5 3 0;7 1 -4]; ↵

Eventual formation of the overall system is conducted by:

>>OS=connect(S,C1,I,O); ↵

As mentioned in the last example the output will be in state space form from which the polynomial form is seen by:

>>tf(OS) ↵

```
Transfer function from input to output...
       20 s^6 + 102.4 s^5 + 232 s^4 - 73.4 s^3 - 251.6 s^2 - 29 s - 0.48
#1:  ----------------------------------------------------------------------
       s^6 + 5.12 s^5 + 11.9 s^4 - 0.934 s^3 - 12.58 s^2 - 4.451 s - 0.054

             6 s^4 + 0.72 s^3 - 5.988 s^2 - 0.72 s - 0.012
#2:  ----------------------------------------------------------------------
       s^6 + 5.12 s^5 + 11.9 s^4 - 0.934 s^3 - 12.58 s^2 - 4.451 s - 0.054
```

#3:
$$\frac{6s^3 + 0.06 s^2 - 6 s - 0.06}{s^6 + 5.12 s^5 + 11.9 s^4 - 0.934 s^3 - 12.58 s^2 - 4.451 s - 0.054}$$

As the return shows we have three outputs for $y_1(t)$, $y_2(t)$, and $y_3(t)$ respectively. For instance the second output of figure 2.6(d) has the transfer function $\dfrac{Y_2(s)}{U(s)} = \dfrac{6s^4 + 0.72s^3 - 5.988s^2 - 0.72s - 0.012}{s^6 + 5.12s^5 + 11.9s^4 - 0.934s^3 - 12.58s^2 - 4.451s - 0.054}$.

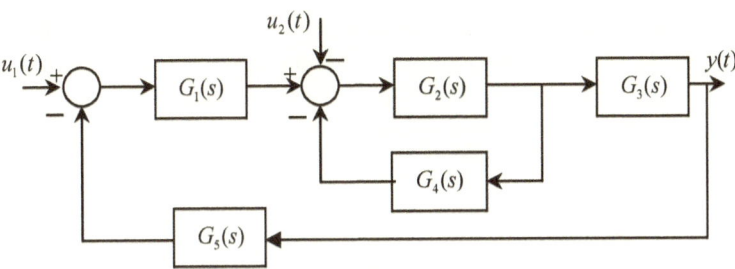

Figure 2.7(a) The control system of figure 2.6(a) with two inputs

❖ Example 6: Multiple input-single output interconnected system

Example 4 interconnected control system is now modified for two inputs with the same component transfer functions. Figure 2.7(a) depicts the modification. We wish to implement this system.

Like $u(t)$ of example 4 we need to inject two unity transfer functions one for $u_1(t)$ and the other for $u_2(t)$ (i.e. $G_6(s)=1$ and $G_7(s)=1$). Doing so yields the figure 2.7(b). Stacked version of the system is identical to that of figure 2.6(f). Enter systems **S1** through **S6** from example 4 and then **S7** by:
>>S7=1; ↵

Like before the stacked system is composed of:
>>S=append(S1,S2,S3,S4,S5,S6,S7); ↵

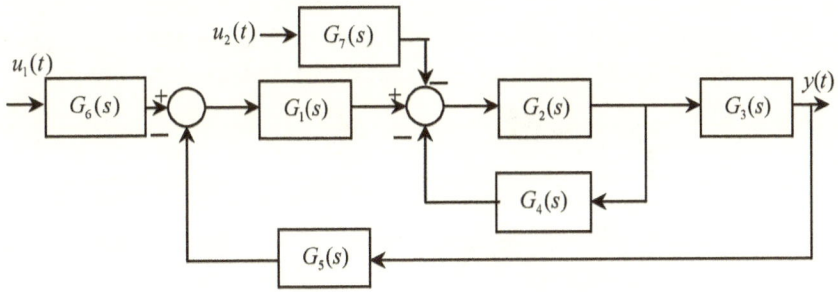

Figure 2.7(b) Figure 2.7(a) inputs with unity transfer functions

Overall inputs are not single for this problem instead double; from inputs of $G_6(s)$ and $G_7(s)$ respectively consequently that information is put into by:
>>I=[6 7]; ↵
Overall output is like example 4:
>>O=3; ↵
Compared to example 4, only $G_2(s)$ is different; other interconnections are the same as those in example 4. The $G_2(s)$ is given in figure 2.7(a) as
input of $G_2(s)$ from $G_1(s)$, $-G_4(s)$, and $-G_7(s)$ i.e. [2 1 −4 −7].

Above is the highest number element row matrix so after padding other rows by zeroes, we get the interconnection matrix by:
>>C1=[1 6 -5 0;2 1 -4 -7;3 2 0 0;4 2 0 0;5 3 0 0]; ↵
The rest commands we require are:
>>OS=connect(S,C1,I,O); ↵
>>tf(OS) ↵

Transfer function from input 1 to output:
   6 s^3 + 0.06 s^2 - 6 s - 0.06
----
s^6 + 5.12 s^5 + 11.9 s^4 - 0.934 s^3 - 12.58 s^2 - 4.451 s - 0.054

Transfer function from input 2 to output:
   -0.3 s^3 - 0.003 s^2 + 0.3 s + 0.003
----
s^6 + 5.12 s^5 + 11.9 s^4 - 0.934 s^3 - 12.58 s^2 - 4.451 s - 0.054

Undoubtedly above return is for two inputs $u_1(t)$ and $u_2(t)$ respectively. For instance the second input of figure 2.7(a) has the transfer function $\frac{Y(s)}{U_2(s)} =$
$\frac{-0.3s^3 - 0.003s^2 + 0.3s + 0.003}{s^6 + 5.12s^5 + 11.9s^4 - 0.934s^3 - 12.58s^2 - 4.451s - 0.054}$.

You can combine the techniques of examples 5 and 6 for multi-input multi-output interconnected control system however this example brings an end to this chapter.

Exercises

1. Use built-in MATLAB function to define each of the following SISO continuous control systems and verify the definition by inspecting the functional popup:

   (a) $H(s) = \dfrac{12}{32s^2 + 5}$ (b) $H(s) = \dfrac{2}{s}$ (c) $H(s) = \dfrac{4}{s(s+1)}$

   (d) $H(s) = \dfrac{-3s^2 + 12}{8s^5 + 32s^2 - 240s - 98}$ (e) $H(s) = \dfrac{45}{7s(s+5)(s+4)}$

   (f) $H(s) = \dfrac{4(s+1)}{s(s+5+j6)(s+5-j6)}$ (g) $H(s) = \begin{Bmatrix} zeroes: -1, -2, -3 \\ poles: 0,\ j2, -j2,\ 2 \\ gain: 3.2 \end{Bmatrix}$

   (h) $H(s) = \{A, B, C, D\}$ where $A = \begin{bmatrix} -2 & -1 & 0 \\ 2 & 3 & 2 \\ 1 & 2 & 5 \end{bmatrix}$, $B = \begin{bmatrix} 1 \\ -2 \\ 3 \end{bmatrix}$, $C = [1 \ -2 \ 1]$, and $D = [2]$ (i) $H(j\omega) = 2, 3+j2, 3+j3,$ and $4+j4$ at $\omega = 0, 5, 10,$ and $15\ rad/sec$ respectively.

2. Define each of the following SISO control systems in MATLAB: (a) $H(s) = \dfrac{s^2 + 42}{(2s^3 + 3s + 40)(s+1)(3s+1)}$ (b) $H(s) = \dfrac{s^2 + 4}{(2s+1)^4}$ (c) $H(s) =$ a second order system with $\omega_n = 2.5\ rad/sec$ and $\zeta = 0.77$ (d) $H(s) =$ a random fourth order system   (e) $H(s)$ in terms of state space matrices for a random fourth order system.

3. Form a MIMO system employing MATLAB built-in function for the following: one input and three outputs where the system functions are output 1 to input; $H_1(s) = \dfrac{4}{4s^3 + 3}$, output 2 to input; $H_2(s) = \dfrac{s^2 + 4}{(2s+1)^4}$, and output 3 to input; $H_3(s) = \dfrac{2}{s}$.

4. Form a MIMO system employing MATLAB built-in function for the following: three inputs and one output where the system functions are output to input 1; $H_1(s) = \dfrac{4}{4s^3 + 3}$, output to input 2; $H_2(s) = \dfrac{s^2 + 4}{(2s+1)^4}$, and output to input 3; $H_3(s) = \dfrac{2}{s}$.

5. Form a MIMO system employing MATLAB built-in function for the following: three inputs and two outputs where the system functions are output 1 to input 1; $H_{11}(s) = \dfrac{4}{4s^3 + 3}$, output 1 to input 2; $H_{12}(s) = \dfrac{9}{s(s+1)}$, output 1 to input 3; $H_{13}(s) = \dfrac{2}{s}$, output 2 to input 1; $H_{21}(s) = \dfrac{2s}{s+5}$, output 2

to input 2; $H_{22}(s) = \dfrac{1}{s^2+7}$, and output 2 to input 3; $H_{23}(s) = \dfrac{6.7}{(s+3)(s+5)}$ and overall control system is $\begin{bmatrix} H_{11}(s) & H_{12}(s) & H_{13}(s) \\ H_{21}(s) & H_{22}(s) & H_{23}(s) \end{bmatrix}$.

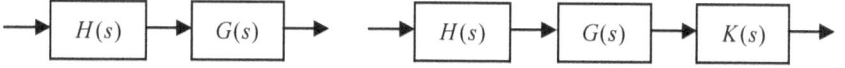

Figure E.2(1) Two control systems connected in series

Figure E.2(3) Three control systems connected in series

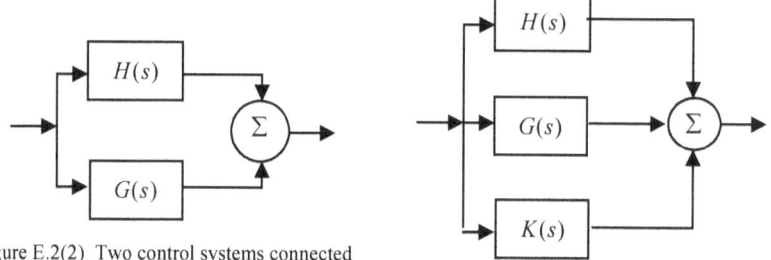

Figure E.2(2) Two control systems connected in parallel

Figure E.2(4) Three control systems connected in parallel

6. Figure E.2(1) shows two control systems connected in series where $H(s) = \dfrac{3}{4.6s+2}$ and $G(s) = \dfrac{1.4}{3s^2+2.5}$. Verify that their equivalent system function is $H_{eq}(s) = \dfrac{4.2}{13.8s^3+6s^2+11.5s+5}$.

7. Figure E.2(2) shows two control systems connected in parallel where $H(s) = \dfrac{14s-8}{(s+2)(s-3)}$ and $G(s) = \dfrac{1.4}{3s^2+2.5}$. Verify that their equivalent system function is $H_{eq}(s) = \dfrac{42s^3-22.6s^2+33.6s-28.4}{3s^4-3s^3-15.5s^2-2.5s-15}$.

8. Figure E.2(3) shows three control systems connected in series where $H(s) = \dfrac{14s-8}{(s+2)(s-3)}$, $G(s) = \dfrac{1.4}{3s^2+2.5}$, and $K(s) = \dfrac{8}{3s(s+2.5)}$. Verify that their equivalent system function is $H_{eq}(s) = \dfrac{156.8s-89.6}{9s^6+13.5s^5-69s^4-123.8s^3-63.75s^2-112.5s}$.

9. Figure E.2(4) shows three control systems connected in parallel where $H(s) = \dfrac{14(s-5)}{s+2}$, $G(s) = \dfrac{s-5}{s^2+2.5}$, and $K(s) = \dfrac{1}{3s}$. Verify that their equivalent system function is $H_{eq}(s) = \dfrac{14s^4-68.67s^3+32.67s^2-184.2s+1.667}{s^4+2s^3+2.5s^2+5s}$.

10. Determine the output to input equivalent system on each of the following control systems:

(a) negative feedback system of figure 2.4(a) with $G(s) = \dfrac{-7s}{s^2 - 2s - 3}$ and $H(s) = \dfrac{3}{s}$,

(b) positive feedback system of figure 2.4(b) with $G(s) = \dfrac{-7(s+5)}{s(s+3)(s+2)}$ and

$H(s) = \{A, B, C, D\}$ where $A = \begin{bmatrix} -2 & -1 \\ 2 & 3 \end{bmatrix}$, $B = \begin{bmatrix} -1 \\ -2 \end{bmatrix}$, $C = \begin{bmatrix} -2 & 1 \end{bmatrix}$, and $D = [-1]$,

(c) a negative feedback system in series with another control system as in

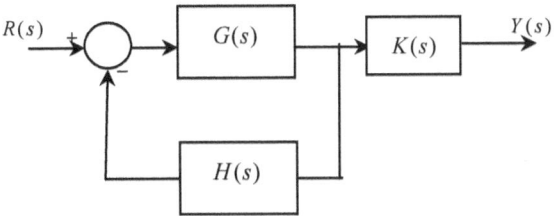

Figure E.2(5) A negative feedback system in series with another system

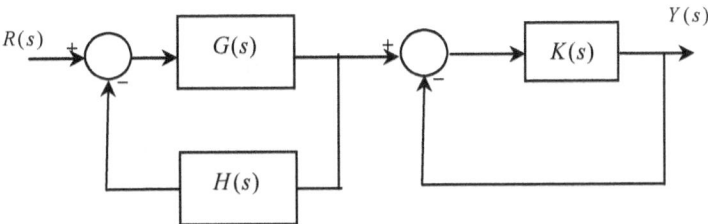

Figure E.2(6) A control system

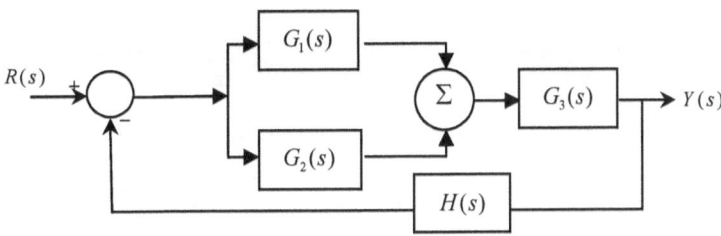

Figure E.2(7) A control system

figure E.2(5) where $G(s) = \dfrac{-7s}{s+4}$, $H(s) = -3$, and $K(s) = s$,

(d) the control system of figure E.2(6) with system functions of part (c), and

(e) the control system of figure E.2(7) with $G_1(s) = \dfrac{-7s}{s+4}$, $G_2(s) = \{A, B, C, D\}$ where $A = \begin{bmatrix} -2 & -1 \\ 2 & 3 \end{bmatrix}$, $B = \begin{bmatrix} -1 \\ -2 \end{bmatrix}$, $C = [-2 \quad 1]$, and $D = [-1]$, $H(s) = -3$, and $G_3(s) = s$.

11. Perform the following control system conversions: (a) $\dfrac{7s^3 - 7s + 1}{s^4 - 1}$ to pole-zero-gain both mixed and all linear (b) $\dfrac{8(s-1)(s-4)}{(s+4)(s+5-j2)(s+5+j2)}$ to polynomial (c) $A = \begin{bmatrix} -2 & -1 & 0 \\ 2 & 3 & 2 \\ 1 & 2 & 5 \end{bmatrix}$, $B = \begin{bmatrix} -1 \\ -2 \\ -3 \end{bmatrix}$, $C = [-2 \quad 1 \quad 7]$, and $D = [-3]$ to polynomial (d) $\dfrac{s^3 - 11s + 5}{s^4 + 8s - 9}$ to state space form (e) to state space form in part (b) (f) to pole-zero-gain form in part (c).

12. Figure E.2(8) presents a single input single output interconnected control system where the component transfer functions are $G_1(s) = \dfrac{2s}{s+1}$,

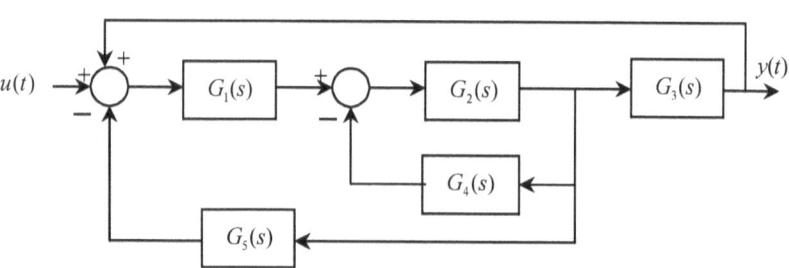

Figure E.2(8) An interconnected control system

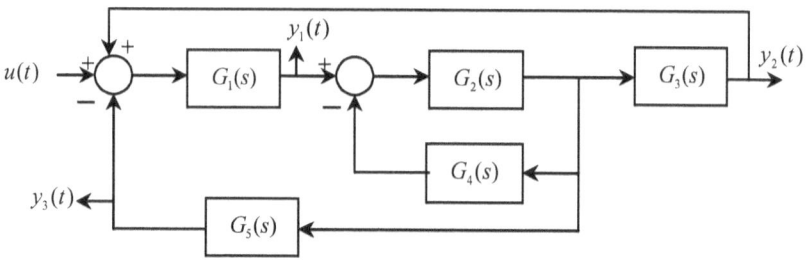

Figure E.2(9) Control system of figure E.2(8) with three outputs

$G_2(s) = \dfrac{5}{s^2 + 6s + 10}$, $G_3(s) = \dfrac{s+0.1}{s+0.2}$, $G_4(s) = \{A, B, C, D\}$ where $A = 0.5$, $B = 2$, $C = 1$, and $D = 0.7$, and $G_5(s) = 5$. Define this control system into MATLAB and determine the overall transfer function.

13. The control system of figure E.2(8) is now modified for three outputs, which is shown in figure E.2(9). Consider the component transfer functions of last problem. Define this control system in MATLAB and determine the overall transfer functions.

14. The interconnected control system of figure E.2(10) has three inputs. Define this control system in MATLAB and determine the overall transfer functions considering the functional elements of problem (12) and $G_6(s) = 6$.

15. The control system of figure E.2(10) is now modified for three inputs and two outputs, which is shown in figure E.2(11). Consider the component transfer functions of problem (12) and $G_6(s) = 6$. Define this control system in MATLAB and determine the overall transfer functions.

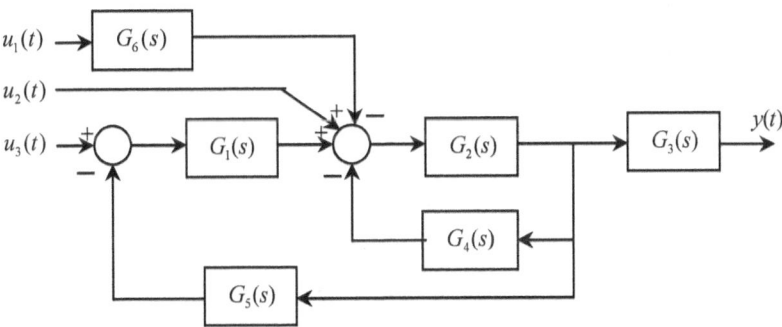

Figure E.2(10) A three input one output control system

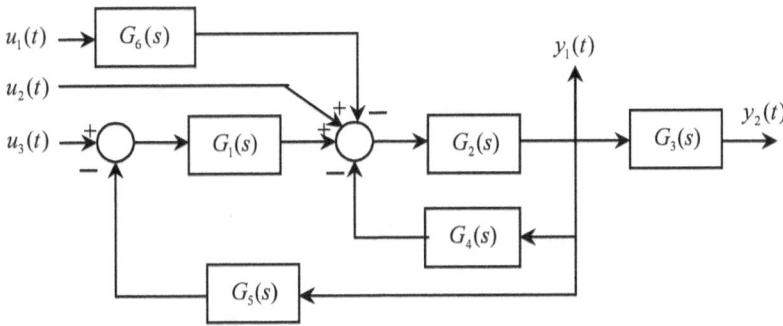

Figure E.2(11) A three input two output control system

Answers:

(1) (a) H=tf(12,[32 0 5]) (b) H=tf(2,[1 0]) (c) H=zpk([ ],[0 -1],4) (d) H=tf([-3 0 12],[8 0 0 32 -240 -98]) (e) H=zpk([ ],[0 -5 -4],45/7) (f) H=zpk(-1,[0 -5+6i -5-6i],4) (g) H=zpk([-1 -2 -3],[0 2i -2i 2],3.2) (h) A=[-2 -1 0;2 3 2;1 2 5]; B=[1;-2;3]; C=[1 -2 1]; D=2; H=ss(A,B,C,D) (i) w=0:5:15; V=[2 3+2i 3+3i 4+4i]; H=frd(V,w).
hint: section 2.1

(2) (a) N=[1 0 42]; D=conv(conv([2 0 3 40],[1 1]),[3 1]); H=tf(N,D) (b) H=zpk([2i -2i],[-1/2 -1/2 -1/2 -1/2],1/16) (c) [N,D]=ord2(2.5,0.77); H=tf(N,D) (d) [N,D]=rmodel(4); H=tf(N,D) (e) [A,B,C,D]=rmodel(4); H=ss(A,B,C,D).
hint: section 2.1

(3) N1=4; D1=[4 0 0 3]; N2=[1 0 4]; D2=conv(conv(conv([2 1],[2 1]),[2 1]),[2 1]); N3=2; D3=[1 0]; H=tf({N1;N2;N3},{D1;D2;D3})
hint: section 2.2

(4) N1=4; D1=[4 0 0 3]; N2=[1 0 4]; D2=conv(conv(conv([2 1],[2 1]),[2 1]),[2 1]); N3=2; D3=[1 0]; H=tf({N1,N2,N3},{D1,D2,D3})
hint: section 2.2

(5) N11=4; N12=9; N13=2; N21=[2 0]; N22=1; N23=6.7; D11=[4 0 0 3]; D12=[1 1 0]; D13=[1 0]; D21=[1 5]; D22=[1 0 7]; D23=conv([1 3],[1 5]);
H=tf({N11,N12,N13;N21,N22,N23},{D11,D12,D13;D21,D22,D23})
hint: section 2.2

(6) H=tf(3,[4.6 2]); G=tf(1.4,[3 0 2.5]); E=series(H,G)
hint: section 2.3

(7) H=tf([14 -8],conv([1 2],[1 -3])); G=tf(1.4,[3 0 2.5]); E=parallel(H,G); tf(E)   hint: section 2.3

(8) H=tf([14 -8],conv([1 2],[1 -3])); G=tf(1.4,[3 0 2.5]); K=tf(8,[3 7.5 0]); E=series(series(H,G),K); tf(E)   hint: section 2.3

(9) H=zpk(5,-2,14); G=tf([1 -5],[1 0 2.5]); K=tf(1,[3 0]); E=parallel(parallel(H,G),K); tf(E)   hint: section 2.3

(10) (a) $\dfrac{-7s^2}{s^3 - 2s^2 - 24s}$ Command: G=tf([-7 0],[1 -2 -3]); H=tf(3,[1 0]); E=feedback(G,H,-1)

(b) $\dfrac{-7s^3 - 28s^2 + 63s + 140}{s^5 + 4s^4 - 10s^3 - 54s^2 - 73s - 420}$ Command: G=zpk(-5,[0 -2 -3],-7); A=[-2 -1;2 3]; B=[-1;-2]; C=[-2 1]; D=-1; [N1,D1]=ss2tf(A,B,C,D); H=tf(N1,D1); E=feedback(G,H,1); tf(E)

(c) $\dfrac{-0.31818s^2}{s + 0.1818}$ Command: G=tf([-7 0],[1 4]); H=zpk([ ],[ ],-3); K=zpk(0,[ ],1); E=series(feedback(G,H,-1),K)

(d) $\dfrac{-0.31818s^2}{(s+1)(s+0.1818)}$ Command: G=tf([-7 0],[1 4]); H=zpk([ ],[ ],-3); K=zpk(0,[ ],1); E=series(feedback(G,H,-1),feedback(K,1,-1))

(e) $\dfrac{-0.3333s^4 + 0.1667s^3 + 0.8333s^2 - 2s}{s^4 - 0.4583s^3 - 2.375s^2 + 5.667s - 0.6667}$

Command: G1=tf([-7 0],[1 4]); A=[-2 -1;2 3]; B=[-1;-2]; C=[-2 1];
D=-1; [N1,D1]=ss2tf(A,B,C,D); G2=tf(N1,D1); H=zpk([ ],[ ],-3);
G3=zpk(0,[ ],1); E=feedback(series(parallel(G1,G2),G3),H,-1);
tf(E)
hint: sections 2.1-2.4

(11) (a) G=tf([7 0 -7 1],[1 0 0 0 -1]);zpk(G) $\Rightarrow \dfrac{7(s+1.065)(s-0.919)(s-0.146)}{(s+1)(s-1)(s^2+1)}$

[z,p,k]=tf2zp([7 0 -7 1],[1 0 0 0 -1]) $\Rightarrow \dfrac{7(s+1.065)(s-0.919)(s-0.146)}{(s+1)(s-1)(s+j)(s-j)}$

(b) z=[1;4];p=[-4;-5+2i;-5-2i];k=8;[N,D]=zp2tf(z,p,k);tf(N,D) $\Rightarrow$
$\dfrac{8s^2-40s+32}{s^3+14s^2+69s+116}$

(c) A=[-2 -1 0;2 3 2;1 2 5]; B=[-1;-2;-3]; C=[-2 1 7]; D=-3;
[N,D1]=ss2tf(A,B,C,D); tf(N,D1) $\Rightarrow \dfrac{-3s^3-3s^2-27s+39}{s^3-6s^2-3s+14}$

(d) [A,B,C,D]=tf2ss([1 0 -11 5],[1 0 0 8 -9]) $\Rightarrow$

$A=\begin{bmatrix} 0 & 0 & -8 & 9 \\ 1 & 0 & 0 & 0 \\ 0 & 1 & 0 & 0 \\ 0 & 0 & 1 & 0 \end{bmatrix}$, $B=\begin{bmatrix} 1 \\ 0 \\ 0 \\ 0 \end{bmatrix}$, $C=[1\ 0\ -11\ 5]$, and $D=[0]$

(e) z=[1;4];p=[-4;-5+2i;-5-2i];k=8;[A,B,C,D]=zp2ss(z,p,k) $\Rightarrow$

$A=\begin{bmatrix} -4 & 0 & 0 \\ 1 & -10 & -5.3852 \\ 0 & 5.3852 & 0 \end{bmatrix}$, $B=\begin{bmatrix} 1 \\ 0 \\ 0 \end{bmatrix}$, $C=[8\ -120\ -37.1391]$, and

$D=[0]$

(f) A=[-2 -1 0;2 3 2;1 2 5]; B=[-1;-2;-3]; C=[-2 1 7]; D=-3;
[z,p,k]=ss2zp(A,B,C,D) $\Rightarrow$
$\dfrac{-3(s+1.0686-j3.2076)(s+1.0686+j3.2076)(s-1.1373)}{(s+1.5722)(s-6.1163)(s-1.4559)}$

hint: section 2.5

(12) $\dfrac{Y(s)}{U(s)} = \dfrac{10s^3-4s^2-0.5s}{s^5+6.7s^4+57.3s^3+5.95s^2+1.5s+0.65}$ hint: section 2.6

(13) $\dfrac{Y_1(s)}{U(s)} = \dfrac{2s^5+11.4s^4+23.2s^3+10.7s^2+1.3s}{s^5+6.7s^4+57.3s^3+5.95s^2+1.5s+0.65}$

$\dfrac{Y_2(s)}{U(s)} = \dfrac{10s^3-4s^2-0.5s}{s^5+6.7s^4+57.3s^3+5.95s^2+1.5s+0.65}$

$\dfrac{Y_3(s)}{U(s)} = \dfrac{50s^3-15s^2-5s}{s^5+6.7s^4+57.3s^3+5.95s^2+1.5s+0.65}$ hint: section 2.6

(14) $\dfrac{Y(s)}{U_1(s)} = \dfrac{-30s^3-18s^2+13.5s+1.5}{s^5+6.7s^4+67.3s^3+1.95s^2+s+0.65}$

$\dfrac{Y(s)}{U_2(s)} = \dfrac{5s^3+3s^2-2.25s-0.25}{s^5+6.7s^4+67.3s^3+1.95s^2+s+0.65}$

$$\frac{Y(s)}{U_3(s)} = \frac{10s^3 - 4s^2 - 0.5s}{s^5 + 6.7s^4 + 67.3s^3 + 1.95s^2 + s + 0.65}$$ hint: section 2.6

(15) Overall system function is $H(s) = \begin{bmatrix} H_{11}(s) & H_{12}(s) & H_{13}(s) \\ H_{21}(s) & H_{22}(s) & H_{23}(s) \end{bmatrix}$ where

$$H_{11}(s) = \frac{Y_1(s)}{U_1(s)} = \frac{-30s^2 - 15s + 15}{s^4 + 6.5s^3 + 66s^2 - 11.25s + 3.25}$$

$$H_{21}(s) = \frac{Y_2(s)}{U_1(s)} = \frac{-30s^3 - 18s^2 + 13.5s + 1.5}{s^5 + 6.7s^4 + 67.3s^3 + 1.95s^2 + s + 0.65}$$

$$H_{12}(s) = \frac{Y_1(s)}{U_2(s)} = \frac{5s^2 + 2.5s - 2.5}{s^4 + 6.5s^3 + 66s^2 - 11.25s + 3.25}$$

$$H_{22}(s) = \frac{Y_2(s)}{U_2(s)} = \frac{5s^3 + 3s^2 - 2.25s - 0.25}{s^5 + 6.7s^4 + 67.3s^3 + 1.95s^2 + s + 0.65}$$

$$H_{13}(s) = \frac{Y_1(s)}{U_3(s)} = \frac{10s^2 - 5s}{s^4 + 6.5s^3 + 66s^2 - 11.25s + 3.25}$$

$$H_{23}(s) = \frac{Y_2(s)}{U_3(s)} = \frac{10s^3 - 4s^2 - 0.5s}{s^5 + 6.7s^4 + 67.3s^3 + 1.95s^2 + s + 0.65}$$

hint: section 2.6

Chapter 3

Modeling Control System in SIMULINK

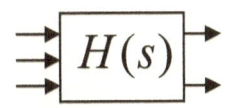

Chapter 2 demonstrated control system implementation has one drawback, the user has to manipulate component or overall transfer function in its entirety. SIMULINK modeling differs from that. Predesigned blocks of SIMULINK eliminate computing on transfer functions. Since model connection line delicately implements the computing, barely does the reader have to go through computing thereby providing comfort. SIMULINK modeling approach is much appreciated when we build up large interconnected control systems. Our effort to elucidate the SIMULINK modeling covers on:

- ✦ ✦ Single input single output (SISO) control system modeling
- ✦ ✦ Multi input multi output (MIMO) control system modeling
- ✦ ✦ Modeling on control system with specific connection
- ✦ ✦ Modeling on interconnected SISO/MIMO control system

3.1 Modeling a SISO control system in SIMULINK

A SISO continuous control system is modeled based on given transfer function which is usually in polynomial, pole-zero-gain, state space, or mixed form. Each of the forms has predesigned block whose link the reader finds in appendix C. The reader also needs the basics of SIMULINK which is explained in section 1.2. These basic blocks are effective in modeling a large interconnected system.

✦ Transfer function from numerator-denominator form

If a transfer function is in numerator-denominator form, the block **Transfer Fcn** models the transfer function. Open a new SIMULINK model file and get the **Transfer Fcn** block in the model file following the link mentioned in appendix C. The block's icon appearance is seen in figure 3.1(a). Doubleclick the block to see its parameter window like figure 3.1(d) and here in this window we enter the transfer function data.

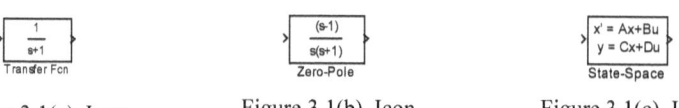

Figure 3.1(a) Icon appearance of **Transfer Fcn** block

Figure 3.1(b) Icon appearance of **Zero-Pole** block

Figure 3.1(c) Icon appearance of **State-Space** block

Figure 3.1(d) Block parameter window of the **Transfer Fcn**

Figure 3.1(e) Block parameter window of the **Zero-Pole**

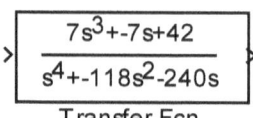

Figure 3.1(f) **Transfer Fcn** models $H(s)$

Figure 3.1(g) Block parameter window of the **State-Space**

For example the transfer function $H(s) = \dfrac{7s^3 - 7s + 42}{s^4 - 118s^2 - 240s}$ is to be modeled.

We enter the numerator and denominator polynomial coefficients as a row matrix in the slots of

Numerator and Denominator of figure 3.1(d) respectively where polynomial coefficients must be in descending order and any missing coefficient is set to 0. The example $H(s)$ has the polynomial coefficient representation [7 0 -7 42] and [1 0 -118 -240 0] for numerator and denominator respectively. Having known so, we enter these two row matrices in the respective slot of the figure 3.1(d) deleting the default values.

One important point is to be kept in mind that the degree of the numerator polynomial must not be more than that of denominator. However upon entering the row matrices you find the block appearance changed. You see **den(s)** in the denominator because the expression overfits the default block size. Enlarge the block (subsection 1.2.3) to see its contents like the figure 3.1(f) i.e. **Transfer Fcn** of figure 3.1(f) models the $H(s)$.

✦ Transfer function from pole-zero-gain form

If a transfer function is in pole-zero-gain form, the **Zero-Pole** block models the system. Open a new SIMULINK model file and get the block in the model file following the link mentioned in appendix C. The block's icon appearance is seen in figure 3.1(b). Doubleclick the block to see its parameter window like the figure 3.1(e) and here in this window we enter the transfer function data.

Figure 3.1(h) **Zero-Pole** models $H(s)$

For example we intend to model the system $H(s) = \dfrac{7(s-2)(s-3)(s+1)}{(s-5)s(s-8)(s-6)}$.

This transfer function has $\begin{cases} \text{zeroes}: 2, 3, -1 \\ \text{poles}: 5, 0, 8, 6 \\ \text{gain}: 7 \end{cases}$. In figure 3.1(e) you find the slots for **Zeros, Poles,** and **Gain** and in these slots we enter given pole-zero-gain data as a row matrix therefore enter [2 3 -1], [5 0 8 6], and 7 in the parameter window deleting the default values respectively. In doing so you see the block contents as **zeros(s)** or **poles(s)** due to overfitting of entered data to the default block size. Enlarge the block by the mouse to see its appearance as shown in figure 3.1(h) which models the given transfer function $H(s)$.

✦ Transfer function in state-space form

When a transfer function is given in state-space form, we employ the block **State-Space** to model the system. Figure 3.1(c) shows the icon appearance of the block, get the block in a new SIMULINK model file. Doubleclick the block to see its parameter window like figure 3.1(g). Here in this window we enter given matrix (subsection 1.1.2) information of the transfer function. The reader is referred to section 2.1 for more about the state-space form.

Suppose a transfer function $H(s)$ is characterized by the state-space matrices $\{A,B,C,D\}$ where $A = \begin{bmatrix} -2 & -1 \\ 2 & 3 \end{bmatrix}$, $B = \begin{bmatrix} -1 \\ -2 \end{bmatrix}$, $C = \begin{bmatrix} -2 & 1 \end{bmatrix}$, and $D = [-1]$, which we intend to model.

Deleting default values from the parameter window of figure 3.1(g) by using keyboard, we enter given matrix information as [-2 -1;2 3], [-1;-2], [-2 1], and -1 in the slots of **A, B, C,** and **D** for A, B, C, and D respectively which still keeps the block appearance like figure 3.1(c) thereby modeling the $H(s)$.

◆ Transfer function in mixed form

When given transfer functions are in mixed form i.e. combination of pole-zero-gain and numerator-denominator, section 2.1 illustrated **conv** is used for polynomial multiplication in order to define as the numerator-denominator form clearly to be fit in the **Transfer Fcn** block.

Mixed form system like $H(s) = \dfrac{s+42}{(s^4 + 3s^2 - 240)(s+2)(2s-1)}$ has the numerator and denominator polynomial coefficients [1 42] and conv([1 0 3 0 -240],conv([1 2],[2 -1])) which are entered in the slots of the **Numerator**

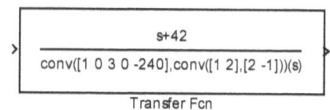

Figure 3.1(i) Modeling mixed form system by **Transfer Fcn**

and **Denominator** of the parameter window in figure 3.1(d) respectively. Having entered the coefficients and enlarged the block, we see the system $H(s)$ modeled like figure 3.1(i).

Since SIMULINK interacts with MATLAB, we can put the numerator and denominator coefficients at the workspace first if they are long and call the coefficients afterwards in the respective slot in the parameter window. For this example let us carry out the following:

>>n=[1 42]; ↵   ← Assigning numerator $s+42$ to n
>>d1=[1 0 3 0 -240]; ↵   ← Assigning coefficients of $s^4 + 3s^2 - 240$ to d1
>>d2=conv([1 2],[2 -1]); ↵ ← Assigning multiplication of $(s+2)(2s-1)$ to d2
>>d=conv(d1,d2); ↵   ← Forming the complete denominator and assigned to d

In above executions the **n, d1, d2,** and **d** are user-chosen variables. After that you can enter just **n** and **d** in the slots of **Numerator** and **Denominator** of parameter window in figure 3.1(d) respectively.

3.2 Modeling a MIMO control system in SIMULINK

Since SIMULINK provides object oriented modeling on control systems, MIMO implementation the way described in section 2.2 is not effective here. SIMULINK blocks are already equipped with inputs and outputs. Block input-output can directly follow the MIMO required input-output. Of coarse MIMO modeling is problem dependent. Nevertheless some combining technique might be necessary to fulfill the MIMO requirement. Let us see the following examples on MIMO modeling.

✦ Modeling a single input multi output MIMO from transfer function coefficients

From last section we know that the **Transfer Fcn** models a transfer function, the same can be used for single input multi output but to a limited extent. The denominator polynomial has to be common for all transfer functions. For example $H(s) = \begin{bmatrix} H_1(s) \\ H_2(s) \end{bmatrix}$ with $H_1(s) = \dfrac{4}{3s-7}$ and $H_2(s) = \dfrac{4}{3s^2 - 5}$ can not be modeled.

Suppose we wish to model $H(s) = \begin{bmatrix} H_1(s) \\ H_2(s) \end{bmatrix}$ for $H_1(s) = \dfrac{4}{3s^3 - 7}$ and $H_2(s) = \dfrac{4s+1}{3s^3 - 7}$ by using **Transfer Fcn**.

The numerator polynomial coefficients for all component transfer functions are entered as a rectangular matrix. Every row of the rectangular matrix refers sequentially to the component transfer function of $H(s)$. If any coefficient is missing, that is preceded by 0. For the above example the common denominator polynomial coefficient is [3 0 0 -7] i.e. $3s^3 - 7$. The numerator polynomial coefficients are [4] and [4 1] for 4 and $4s+1$ respectively. In order to feed into **Transfer Fcn** we will write the numerator coefficients as [0 4;4 1].

Open a new SIMULINK model file (section 1.2) and get the **Transfer Fcn** block in the model file following the link mentioned in appendix C. Doubleclick the block and enter [0 4;4 1] and [3 0 0 -7] in the slots of **Numerator** and **Denominator** in parameter window (figure 3.1(d)) respectively.

Figure 3.2(a) Modeling single input two output $H(s)$

The next question is how to separate the two outputs? The block **Demux** helps us achieve that. The default output number of **Demux** is two, which will separate the $H(s)$ components. Figure 3.2(a) presents the modeling of $H(s)$. The upper and lower output ports of **Demux** in figure 3.2(a) refer to $H_1(s)$ and $H_2(s)$ respectively.

What if we have a single input three output transfer function? For instance $H(s) =$ $\begin{bmatrix} H_1(s) \\ H_2(s) \\ H_3(s) \end{bmatrix}$ with ongoing $H_1(s)$ and $H_2(s)$ and

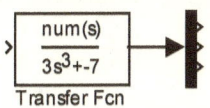

Figure 3.2(b) Modeling single input three output $H(s)$

additional $H_3(s) = \dfrac{s^2}{3s^3 - 7}$.

For this modeling we have to consider the numerator coefficients as [0 0 4], [0 4 1], and [1 0 0] for $H_1(s)$, $H_2(s)$, and $H_3(s)$ respectively because the highest degree is occurring in $H_3(s)$. As a rectangular matrix the numerators should be entered as [0 0 4;0 4 1;1 0 0] in the parameter window of figure 3.1(d). Not to mention the denominator coefficients are identical. Since the default output of the **Demux** is two, you need to doubleclick the block and change its output number from default 2 to 3. With that modification your model should look like figure 3.2(b). In the figure upper, middle, and lower ports of **Demux** correspond to $H_1(s)$, $H_2(s)$, and $H_3(s)$ respectively.

✦ Modeling a MIMO control system by state space form

This is the most important implementation for a MIMO control system. The **Transfer Fcn** or **Zero-Pole** block has some limitation, each of which is suitable for split type transfer function. For the interconnected MIMO they are not suitable. State space approach takes care of interconnected system conveniently.

Bottomline is the four state space matrices must be available for a MIMO control system. Concerning sections 2.1 and 2.2, the state space matrix size determines the input or output number. Following examples illustrate certain MIMO implementation through **State-Space** block.

Multi input single output MIMO:
Consider the state space matrices of a control system as follows: $A = \begin{bmatrix} -1 & 1 \\ 2 & 1 \end{bmatrix}$, $B = \begin{bmatrix} -1 & 4 \\ -2 & -4 \end{bmatrix}$, $C = [-1 \quad 1]$, and $D = [-1 \quad 3]$.

The dimension of D is 1×2 hence the values of P and Q are 2 and 1 respectively. Referring to figure 2.1(a), one infers that there should be two inputs and one output, so should be the modeling.

Open a new SIMULINK model file and get the **State-Space** block in the model file following the link mentioned in appendix C. Doubleclick the block and enter [-1 1;2 1], [-1 4;-2 -4], [-1 1], and [-1 3] in the slots of A, B,

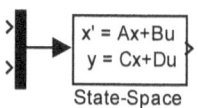

Figure 3.2(c) Model of two input one output MIMO

C, and D in parameter window of figure 3.1(g) respectively. For combining the two inputs we need a **Mux** block. Get a **Mux** block in the model and connect the **Mux** along with the **State-Space** like figure 3.2(c) which models the MIMO.

What if we have three inputs and one output?, an example of which is $A = \begin{bmatrix} -1 & 1 \\ 2 & 1 \end{bmatrix}$, $B = \begin{bmatrix} -1 & 4 & 1 \\ -2 & -4 & 0 \end{bmatrix}$, $C = [-1 \quad 1]$, and $D = [-1 \quad 3 \quad 5]$. In parameter window of figure 3.1(g) we need to enter these matrices. Since the default input number of **Mux** is two, doubleclick the block and change its input number from 2 to 3.

Single input multi output MIMO:
State space matrices of a single input three output control system are: $A = \begin{bmatrix} -2 & -1 \\ 2 & 3 \end{bmatrix}$, $B = \begin{bmatrix} -1 \\ -2 \end{bmatrix}$, $C = \begin{bmatrix} -1 & 0 \\ 2 & 3 \\ 0 & 1 \end{bmatrix}$, and $D = \begin{bmatrix} -2 \\ 3 \\ 1 \end{bmatrix}$.

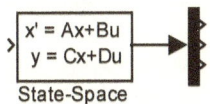

Figure 3.2(d) Model of one input three output MIMO

The dimension of D is 3×1 so there are three outputs. Enter the state space matrices like before. In order to split the output we need a **Demux** block. Also change the output number of the **Demux** from default 2 to 3. Figure 3.2(d) shows the model.

Multi input multi output MIMO:
A two input three output MIMO is given by $A = \begin{bmatrix} -2 & -1 \\ 2 & 3 \end{bmatrix}$, $B = \begin{bmatrix} -1 & 2 \\ -2 & 0 \end{bmatrix}$, $C = \begin{bmatrix} -1 & 0 \\ 2 & 2 \\ 0 & 1 \end{bmatrix}$, and $D = \begin{bmatrix} -2 & 0 \\ -4 & 1 \\ 1 & 0 \end{bmatrix}$.

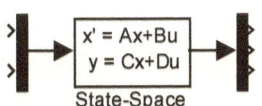

Figure 3.2(e) Model of two input three output MIMO

Both the **Mux** and **Demux** are necessary for the input and output respectively. Enter the matrix descriptions as done before. Figure 3.2(e) presents the MIMO modeling.

◆ Way forward for a MIMO control system

The best solution for a MIMO modeling is obtain the state space matrices for the control system and feed them to **State-Space** block. You can convert the MIMO from polynomial or pole-zero-gain form to state space by applying **tf2ss** or **zp2ss** in MATLAB and just enter the state space matrix names in the parameter window of figure 3.1(g). Nevertheless if any output is required from the internal point of the block diagram, there is no

3.3 Modeling series and parallel systems in SIMULINK

Modeling a series or parallel control system is straightforward. Inspect the component transfer function and look for which form it is given and choose **Transfer Fcn, State Space,** or **Zero-Pole** block (appendix C) accordingly. Here we are not going to see the equivalent transfer function instead construct just the model. Let us see the following modeling.

◆ Seriesly connected control systems

Figure 2.3(a) shows two control systems $H_1(s) = \dfrac{5s^2 - s + 1}{s^3 - 1}$ and $H_2(s) = \dfrac{5.43(s-3)}{(s-1)(s+4)}$ connected in series which we wish to implement.

The $H_1(s)$ and $H_2(s)$ fit to **Transfer Fcn** and **Zero-Pole** respectively (section 3.1). In order to construct the model, we carry out the following:

⇒ Open a new simulink model file, bring a **Transfer Fcn** block for $H_1(s)$ in the model file, doubleclick the **Transfer Fcn**, and enter its **Numerator** and **Denominator** coefficients as [5 -1 1] and [1 0 0 -1] in the parameter window of figure 3.1(d) respectively.

⇒ Get a **Zero-Pole** block for $H_2(s)$ in the model file, doubleclick the block (figure 3.1(e)) to enter its **Zeros, Poles,** and **Gain** as 3, [1 -4], and 5.43 in the parameter window respectively.

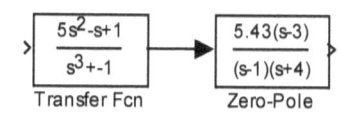

Figure 3.3(a) SIMULINK model of series connection in figure 2.3(a)

⇒ Connect the two blocks like figure 3.3(a) which represents the SIMULINK model of seriesly connected system in figure 2.3(a).

As a three system series connection, consider the figure 2.3(b) depicted control system where $H_3(s) = \{A, B, C, D\}$ with $A = \begin{bmatrix} -2 & -1 \\ 2 & 3 \end{bmatrix}$, $B = \begin{bmatrix} -1 \\ -2 \end{bmatrix}$, $C = [-2\ 1]$, and $D = [-1]$ and $H_1(s)$ and $H_2(s)$ are the above one.

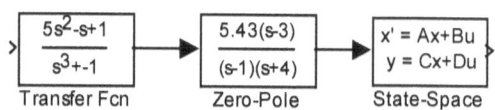

Figure 3.3(b) SIMULINK model of series connection in figure 2.3(b)

All we need is a **State-Space** block for $H_3(s)$ in figure 3.3(a). Bring the block in the model file and enter the matrix descriptions as [-2 -1;2 3], [-1;-2], [-2 1], and [-1] for A, B, C, and D in the parameter window of figure 3.1(g) respectively. Connect the block like figure 3.3(b) which represents the SIMULINK model of series three control systems in figure 2.3(b).

✦ Parallelly connected control systems

Figure 2.3(c) depicts two control systems connected in parallel where $H_1(s)$ and $H_2(s)$ are the ongoing one. We wish to model the parallel system.

Earlier we discussed the modeling of $H_1(s)$ and $H_2(s)$ by **Transfer Fcn** and **Zero-Pole** blocks respectively which you see in figure 3.3(a). Summation of the two systems occurs by the **Sum** block. Bring the **Sum** block in the model of figure 3.3(a) and deleting the connection line between the blocks, reconnect the blocks like figure 3.3(c) which models the parallel system of figure 2.3(c).

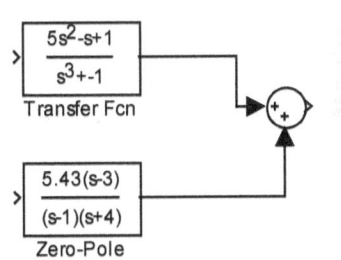

Figure 3.3(c) SIMULINK model of parallel connection in figure 2.3(c)

Again figure 2.3(d) depicts three control systems connected in parallel where $H_1(s)$, $H_2(s)$, and $H_3(s)$ are taken from the series counterpart. We wish to model this parallel system too.

In figure 3.3(c) we just need to include **State-Space** block for $H_3(s)$. The **Sum** block has two inputs by default. Doubleclick the **Sum** to change its **List of signs** from default ++ to +++ for three input taking. Reconnect the blocks of figure 3.3(c) along with the **State-Space** like figure 3.3(d) which shows the SIMULINK model of three systems connected in parallel.

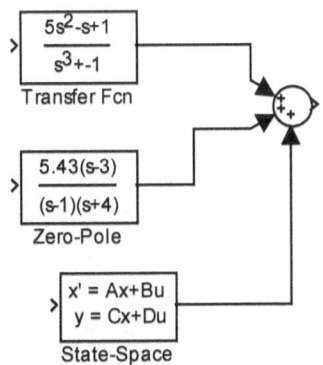

Figure 3.3(d) SIMULINK model of parallel connection in figure 2.3(d)

The input port of above modeling whether series or parallel will be connected to overall input of the given system, so will be the output port. While entering the matrix description of **State-Space**, make sure that the description is for single input single output.

-71-

3.4 Modeling a feedback control system

Feedback control system is addressed in section 2.4. Figures 2.4(a) and 2.4(b) show the basic negative and positive feedback systems respectively. Each feedback system requires mainly three blocks; for summing point, $G(s)$, and $H(s)$. The $G(s)$ or $H(s)$ is modeled by section 3.1 quoted **Transfer Fcn, Zero-Pole,** or **State-Space** depending on given feedback system. The **Sum** block implements the summing point.

◆ Negative feedback

We wish to model the figure 2.4(a) shown negative feedback control system when $G(s) = \dfrac{7s^3 - 7s + 42}{s^4 - 118s^2 - 240s}$ and $H(s) = \dfrac{4(s+2)}{s-1}$.

The $G(s)$ and $H(s)$ are modeled by **Transfer Fcn** and **Zero-Pole** because of given expressions respectively. Anyhow go through following for the modeling.

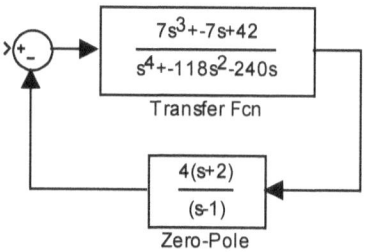

Figure 3.4(a) Modeling the negative feedback system of figure 2.4(a)

Open a new SIMULINK model file, get one **Transfer Fcn**, one **Zero-Pole**, and one **Sum** blocks in the model file following the links of appendix C. Doubleclick the **Transfer Fcn** and enter its **Numerator** and **Denominator** as [7 0 -7 42] and [1 0 -118 -240 0] for modeling $G(s)$ in the parameter window respectively. Then doubleclick the **Zero-Pole** and enter its **Zeros, Poles,** and **Gain** as -2, 1, and 4 for modeling $H(s)$ in the parameter window respectively.

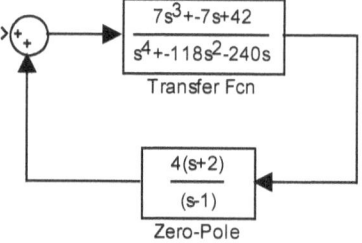

Figure 3.4(b) Modeling the positive feedback system of figure 2.4(b)

Enlarge each of the **Transfer Fcn** and **Zero-Pole** to see its contents.

The **Sum** by default has two inputs, which are related to **List of signs** in the parameter window (indicated by **++**). On doubleclicking the **Sum** if we modify that to **+-**, then the negative port receives the feedback signal. Nevertheless the feedback signal flow in figure 2.4(a) is from right to left so rightclick on the **Zero-Pole** and click the **Flip Block** under the **Format** popup to have the proper orientation of the block in the model. Place so designed blocks relatively and connect them like the figure 3.4(a) which essentially models the negative feedback system.

✦ Positive feedback

Figure 2.4(b) shown positive feedback system we wish to exercise too with ongoing $G(s)$ and $H(s)$.

All we need is keep the Sum block input port sign as it is. Figure 3.4(b) shows the SIMULINK model of the positive feedback system.

3.5 Modeling interconnected control systems

In previous sections we implemented mainly known forms of interconnections like series, parallel, or feedback. Many control systems do not follow the particular three connections. Here the idea is construct the model as they are given in the block diagram. Any sort of MIMO can be modeled too by this way. Following two examples demonstrate how to realize interconnected control systems.

System functions: $G_1(s) = 20$, $G_2(s) = \dfrac{0.3}{s^2 + 5s + 12}$, $G_3(s) = \dfrac{s + 0.01}{(s + 0.1)(s + 0.02)}$, $G_4(s) = s$, and $G_5(s) = \{A, B, C, D\}$ where $A = 1$, $B = 1.1$, $C = 10^{-5}$, and $D = 0.5$

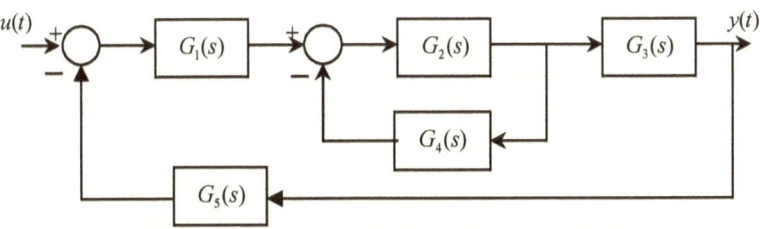

Figure 3.5(a) An interconnected control system

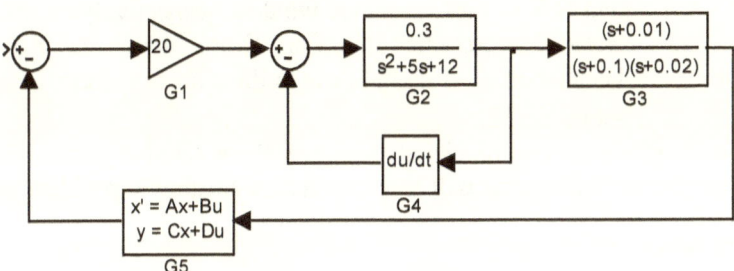

Figure 3.5(b) SIMULINK model for the control system of figure 3.5(a)

✦ Example 1

Figure 3.5(a) shows an interconnected control system which is composed of five modular transfer functions as placed on top of the figure. We intend to construct SIMULINK model for this system.

The type of given transfer function dictates the type of block selection for this reason we model $G_1(s)$, $G_2(s)$, $G_3(s)$, $G_4(s)$, and $G_5(s)$ by **Gain, Transfer Fcn, Zero-Pole, Derivative,** and **State-Space** blocks respectively. Laplace domain s is equivalent to derivative in time domain which is used for the $G_4(s)$. Figure 3.5(b) shows the SIMULINK model of control system in figure 3.5(a). In order to construct the model, we carry out the following:

⇒ Open a new simulink model file, bring a **Gain** block (appendix C) in the model file, doubleclick the **Gain**, enter its **Gain** as 20 in the parameter window, and rename (section 1.2.3) the block as G1 just to make sense with given name.

⇒ Get one **Sum** block in the model file, doubleclick the block to change its **List of signs** from default ++ to + −, and copy and paste the block to get another one in the model file (because there are two summing points in figure 3.5(a)).

⇒ Get one **Transfer Fcn** block in the model file, doubleclick the block to set its **Numerator** and **Denominator** as 0.3 and [1 5 12] for $G_2(s)$ respectively, and rename the block as G2.

⇒ Get one **Zero-Pole** block in the model file for $G_3(s)$, doubleclick the block to enter its **Zeros, Poles,** and **Gain** as -0.01, [-0.1 - 0.02], and 1 respectively, and rename the block as G3.

⇒ Get one **Derivative** block in the file for $G_4(s)$, rightclick on the **Derivative**, click Flip Block in Format popup because of $G_4(s)$ signal flow in the control system, and rename the block as G4.

⇒ Get one **State-Space** block in the model file for $G_5(s)$, doubleclick the block to enter its **A, B, C,** and **D** as 1, 1.1, 1e-5, and 0.5 in the parameter window respectively, rightclick on the **State-Space**, click the **Flip Block** under the **Format** popup because of $G_5(s)$ signal flow in the control system, and rename the block as G5.

Finally place various blocks and connect them relatively according to figure 3.5(b). Input and output will be connected according to problem description.

✦ Example 2

Figure 3.5(c) depicts a MIMO with three inputs and three outputs. We wish to implement the MIMO in SIMULINK. Regard the component transfer functions as follows:

$$G_1(s) = \frac{0.7(s+0.01)}{(s+0.1)(s+0.02)}, \quad G_2(s) = \frac{0.3}{s^2+5s+12}, \quad G_3(s) = \frac{2.4}{s+0.7}, \quad G_4(s) = 5,$$

$G_5(s) = 0.01$, and $G_6(s) = \dfrac{0.1}{s}$.

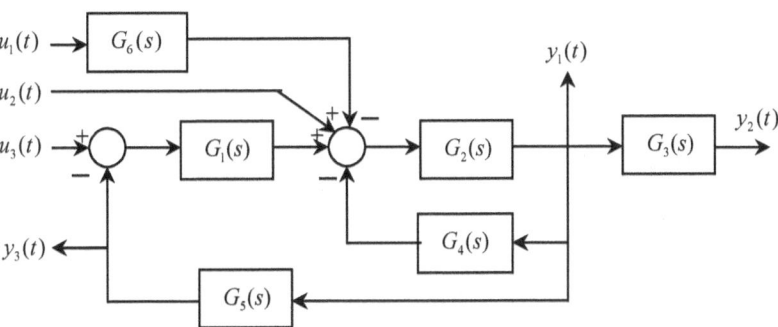

Figure 3.5(c) A three input three output control system

By inspection the block selection should be **Zero-Pole, Transfer Fcn, Transfer Fcn, Gain, Gain,** and **Transfer Fcn** for $G_1(s)$, $G_2(s)$, $G_3(s)$, $G_4(s)$, $G_5(s)$ and $G_6(s)$ respectively. As a procedure we describe the modeling as follows:

⇒ Open a new simulink model file, bring a **Zero-Pole** block in the model file to model $G_1(s)$, doubleclick the block, enter its **Zeros, Poles,** and **Gain** as -0.01, [-0.1 -0.02], and 0.7 in the parameter window respectively, resize the block to see its contents, and rename the block as **G1**.

⇒ Get one **Transfer Fcn** block in the model file, doubleclick the block to set its **Numerator** and **Denominator** as 0.3 and [1 5 12] for $G_2(s)$ respectively, and rename the block as **G2**.

⇒ Copy and paste **G2** in the model file which appears as **G3**. Doubleclick **G3** to set its **Numerator** and **Denominator** as 2.4 and [1 0.7] for $G_3(s)$ respectively.

⇒ Get one **Gain** block in the model file, doubleclick the block to change its **Gain** to 5 for $G_4(s)$, rename the block as **G4**, copy and paste the **G4** to get another one for $G_5(s)$ which appears as **G5**, doubleclick **G5** to set its **Gain** as 0.01, and resize **G5** to see its contents.

⇒ Rightclick on each of **G4** and **G5** and click the **Flip Block** in **Format** popup because each signal flow is from the right to left in given control system.

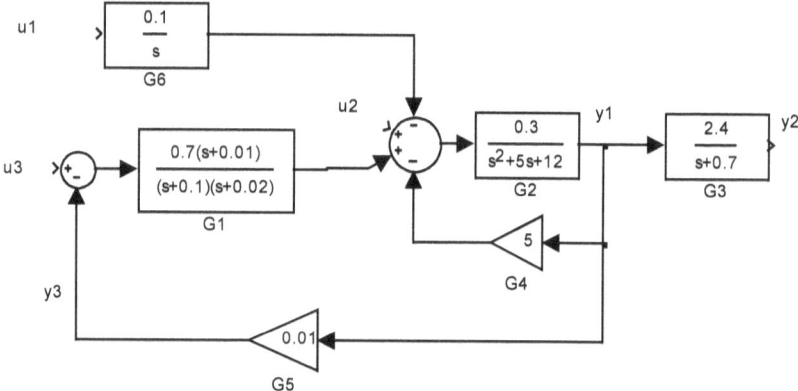

Figure 3.5(d) SIMULINK model for control system of figure 3.5(c)

⇒ Get one **Sum** block for the left summing point in the model file, doubleclick the block to change its **List of signs** from default ++ to + −, copy and paste the block to get another one for the middle summing point, and doubleclick the block to change its **List of signs** to -++- because there are four inputs towards the summing point. The last **Sum** block has to be enlarged because the block accommodates four signals.

⇒ Copy and paste G3 for $G_6(s)$ in the model file which appears as G6. Doubleclick the **G6** to set its **Numerator** and **Denominator** as 0.1 and [1 0] respectively.

Place various blocks and connect them relatively according to figure 3.5(d) which is the model of the interconnected system. Finally annotate (section 1.2.3) the input and output ports or functional points by y1, y2, etc for example y1 for $y_1(t)$.

3.6 SIMULINK model to transfer function

Given the SIMULINK model of an interconnected control system, one might be interested to compute the transfer function between various ports of the control system. There are embedded tools to determine the transfer functions in MATLAB. In order to determine the transfer function, we observe the following steps:

  (1) Construct the SIMULINK model like section 3.5.
  (2) Connect input port block (**In1**) to the provided input. For multiple input ports you can copy and paste the block.
  (3) Connect output port block (**Out1**) to the required output. For multiple output ports you can copy and paste the block.
  (4) Save the model file by some name say **test**.

(5) Call the embedded function linmod with the file name as input argument i.e. linmod('test') but the return from the function is in state space form so four user-supplied variables or output arguments are required to get the return i.e. one needs to exercise [A,B,C,D]=linmod('test') where A, B, C, and D have the state space meanings.
(6) Exercise tf(ss(A,B,C,D)) to view the transfer function, chapter 2 for ss or tf details.

In the sequel we demonstrate two examples on SIMULINK model to transfer function obtaining.

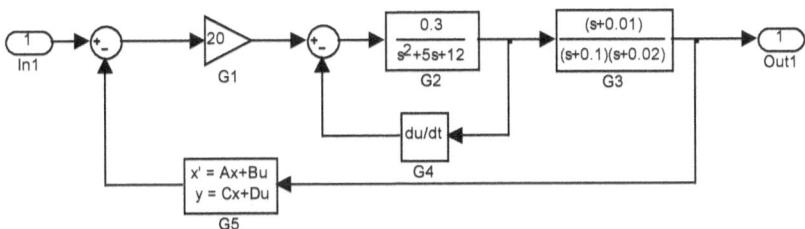

Figure 3.6(a) SIMULINK model of figure 3.5(b) with input and output ports

✦ Example 1

In last section example 1 renders modeling on single input single output whose model you see in figure 3.5(b). We wish to determine $y(t)$ to $u(t)$ transfer function i.e. $\dfrac{Y(s)}{U(s)}$ from the model.

Get one In1 and one Out1 blocks in the model file following the link mentioned in appendix C and connect them in the ports of input and output respectively, the result of which is in figure 3.6(a). Save the file by the name test and keep the test opened. Then execute the following at the MATLAB command prompt:

>>[A,B,C,D]=linmod('test'); ↵

Ignore the warning. State space matrices are available hence execute the following:

>>tf(ss(A,B,C,D)) ↵
Transfer function:

6 s^2 - 5.94 s - 0.06
----
s^5 + 4.12 s^4 + 7.482 s^3 - 5.152 s^2 - 1.366 s - 0.024

One easily reads off the transfer function as $\dfrac{Y(s)}{U(s)} =$

$$\dfrac{6s^2 - 5.94s - 0.06}{s^5 + 4.12s^4 + 7.482s^3 - 5.152s^2 - 1.366s - 0.024}.$$

-77-

✦ Example 2

Example 2 of last section handles three inputs and three outputs and the model you see in figure 3.5(d). We wish to determine the transfer functions from the model.

Three inputs and three outputs indicate that there are 3×3=9 transfer functions, MIMO transfer function concept of chapter 2 is equally applicable here. Get one In1 and one Out1 blocks in the model file of figure 3.5(d). Copy and paste each two times to get In2, In3, Out2, and Out3 in the model file. Understandably In1, In2, and In3 for $u_1(t)$, $u_2(t)$, and $u_3(t)$ respectively. Similar block naming follows for the output too. Connect all input-output blocks to respective ports like figure 3.6(b), save the file by name **test**, and keep the **test** opened.

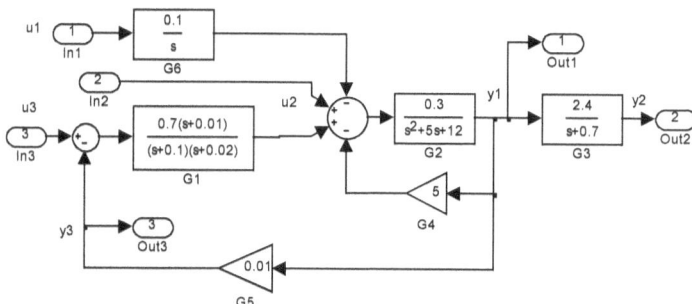

Figure 3.6(b) SIMULINK model of figure 3.5(d) with input and output ports

Calling similar to example 1 then takes place by:
>>[A,B,C,D]=linmod('test'); ↵
>>T=tf(ss(A,B,C,D)); ↵

This time we assigned all transfer functions to some user-chosen variable T which is a 3×3 matrix for obvious reason. According to MATLAB matrix indexing T(2,3) indicates transfer function between $y_2(t)$ and $u_3(t)$ i.e. $T(2,3) = \dfrac{Y_2(s)}{U_3(s)}$. Similar indexing applies to the other 8 component transfer functions of **T**. Just to view T(2,3) content, call it by:
>>T(2,3) ↵
Transfer function:

```
         0.504 s + 0.00504
-----------------------------------------
s^5 + 5.82 s^4 + 17.69 s^3 + 11.5 s^2 + 1.169 s + 0.01891
```

Clearly we have $T_{23} = \dfrac{Y_2(s)}{U_3(s)} = \dfrac{0.504s + 0.00504}{s^5 + 5.82s^4 + 17.69s^3 + 11.5s^2 + 1.169s + 0.01891}$ from above return. In a similar fashion you may call the rest 8 components of T.

Along with this example we close the chapter.

Exercises

1. Use SIMULINK predesigned block to model each of the following SISO control systems:

   (a) $H(s) = \dfrac{12}{32s^2 + 5}$
   (b) $H(s) = \dfrac{3.4}{s}$
   (c) $H(s) = \dfrac{4}{s(s+1)}$

   (d) $H(s) = \dfrac{-3s^2 + 12}{8s^5 + 32s^2 - 240s - 98}$
   (e) $H(s) = \dfrac{45}{7s(s+5)(s+4)}$

   (f) $H(s) = \dfrac{4(3s+1)}{s(2s+5+j6)(2s+5-j6)}$
   (g) $H(s) = \begin{Bmatrix} zeroes: -1, -2, -3 \\ poles: 0, j2, -j2, 2 \\ gain: 3.2 \end{Bmatrix}$

   (h) $H(s) = \{A, B, C, D\}$ where $A = \begin{bmatrix} -2 & -1 & 0 \\ 2 & 3 & 2 \\ 1 & 2 & 5 \end{bmatrix}$, $B = \begin{bmatrix} 1 \\ -2 \\ 3 \end{bmatrix}$, $C = [1 \ -2 \ 1]$, and $D = [2]$.

2. Model each of the following SISO control systems in SIMULINK: (a) $H(s) = \dfrac{s^2 + 42}{(2s^3 + 3s + 40)(s+1)(3s+1)}$ (b) $H(s) = \dfrac{s^2 + 4}{(2s+1)^4}$ (c) $H(s) =$ a second order system with $\omega_n = 2.5 \ rad/\sec$ and $\zeta = 0.77$ (d) $H(s) = \dfrac{s^2 + 42}{(2s^3 + 3s + 40)(s+1)(3s+1)} + \dfrac{s^2 + 4}{(2s+1)^4}$ (e) $H(s) = \dfrac{s^2 + 42}{(2s^3 + 3s + 40)(s+1)(3s+1)} \times \dfrac{s^2 + 4}{(2s+1)^4}$.

3. The $H(s)$ is a MIMO system which has one input and four outputs where

   $H(s) = \begin{bmatrix} H_1(s) \\ H_2(s) \\ H_3(s) \\ H_4(s) \end{bmatrix}$ and $H_1(s) = \dfrac{2}{s^3 + 1}$, $H_2(s) = \dfrac{2s+3}{s^3 + 1}$, $H_3(s) = \dfrac{s^3}{s^3 + 1}$, and $H_4(s) = \dfrac{s+8}{s^3 + 1}$. Model the $H(s)$ in SIMULINK.

4. Model each of the following MIMO control systems in SIMULINK: (a) $A = \begin{bmatrix} -1 & 1 & 2 \\ 2 & 1 & 3 \\ 3 & 8 & 6 \end{bmatrix}$, $B = \begin{bmatrix} -1 & 4 & 4 \\ -2 & -4 & 1 \\ 3 & 2 & 7 \end{bmatrix}$, $C = [-1 \ 4 \ 1]$, and $D = [-1 \ 7 \ 3]$ (b) $A = \begin{bmatrix} -1 & 1 \\ 2 & 1 \end{bmatrix}$, $B = \begin{bmatrix} -1 \\ 4 \end{bmatrix}$, $C = \begin{bmatrix} -1 & 1 \\ 3 & 0 \\ 1 & 2 \\ 3 & 5 \end{bmatrix}$, and $D = \begin{bmatrix} -1 \\ 2 \\ 3 \\ 4 \end{bmatrix}$ (c) $A = \begin{bmatrix} -1 & 1 \\ 2 & 1 \end{bmatrix}$, $B = \begin{bmatrix} -1 & 4 & -2 \\ -4 & 1 & 1 \end{bmatrix}$, $C = \begin{bmatrix} -1 & 1 \\ 2 & 3 \\ 0 & 4 \\ 5 & 6 \end{bmatrix}$, and $D = \begin{bmatrix} -1 & 4 & 3 \\ 2 & 0 & 2 \\ 3 & 5 & 1 \\ 4 & 0 & 2 \end{bmatrix}$.

5. Model series three systems of figure 2.3(b) in SIMULINK where $H_1(s) = 0.0004$, $H_2(s) = \dfrac{100(s^2+7)}{(3s+1)^3}$, and $H_3(s) = \dfrac{2}{s}$.

6. Model the parallel three systems of figure 2.3(d) in SIMULINK where $H_1(s)$, $H_2(s)$, and $H_3(s)$ are taken from problem (5).

7. Model the negative feedback system of figure 2.4(a) in SIMULINK where
$G(s) = \{A, B, C, D\}$ with $A = \begin{bmatrix} -2 & -1 & 0 \\ 2 & 3 & 2 \\ 1 & 2 & 5 \end{bmatrix}$, $B = \begin{bmatrix} 1 \\ -2 \\ 3 \end{bmatrix}$, $C = [1\ -2\ \ 1]$, and $D = [2]$ and $H(s) = 0.01\,s$.

8. In problem (7) now for the positive feedback like figure 2.4(b).

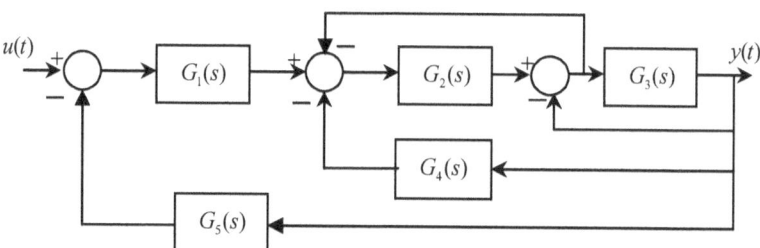

Figure E.3(1) A SISO control system

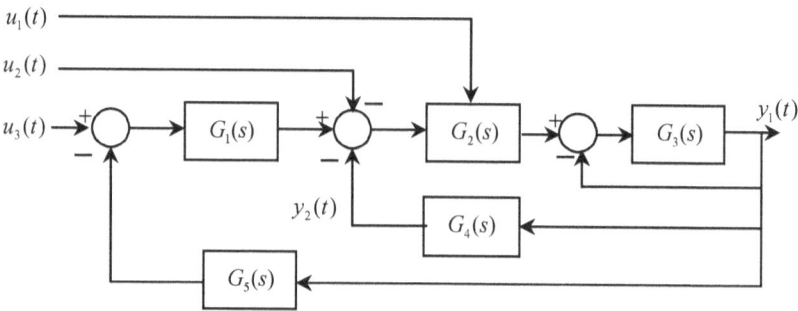

Figure E.3(2) A MIMO control system

9. Model the single input single output control system of figure E.3(1) in SIMULINK with component system functions $G_1(s) = \dfrac{s}{s^3+2}$, $G_2(s) = \dfrac{0.7s}{s^2+9.75s+11}$, $G_3(s) = \dfrac{1}{(s+0.3)(s+0.2)}$, $G_4(s) = \{A, B, C, D\}$ where $A = \begin{bmatrix} -2 & -1 & 0 \\ 2 & 3 & 2 \\ 1 & 2 & 5 \end{bmatrix}$, $B = \begin{bmatrix} 1 \\ -2 \\ 3 \end{bmatrix}$, $C = [1\ -2\ \ 1]$, and $D = [2]$, and $G_5(s) = 10^{-5}$.

10. Model the three input two output control system of figure E.3(2) in SIMULINK with component system functions $G_1(s) = \dfrac{1}{s^2+2}$, $G_2(s) = \{A, B, C, D\}$ where $A = \begin{bmatrix} -2 & -1 & 0 \\ 2 & 3 & 2 \\ 1 & 2 & 5 \end{bmatrix}$, $B = \begin{bmatrix} 1 & 0 \\ -2 & 1 \\ 3 & 9 \end{bmatrix}$, $C = \begin{bmatrix} 1 & -2 & 1 \end{bmatrix}$, and $D = \begin{bmatrix} 2 & 5 \end{bmatrix}$, $G_3(s) = \dfrac{0.7}{9s+11}$, $G_4(s) = 0.01$, and $G_5(s) = 10^{-3}$.

11. Starting from the SIMULINK model of SISO control system (problem 9) in figure E.3(1), connect appropriate input and output ports to the model and obtain the transfer function $\dfrac{Y(s)}{U(s)}$.

12. Starting from the SIMULINK model of MIMO control system (problem 10) in figure E.3(2), connect appropriate input and output ports to the model and obtain the matrix transfer function for the control system.

Answers:

(1) (a) **Transfer Fcn** with settings **Numerator coefficients:** 12 and **Denominator coefficients:** [32 0 5]
(b) **Transfer Fcn** with settings **Numerator coefficients:** 3.4 and **Denominator coefficients:** [1 0]
(c) **Zero-Pole** with settings **Zeros:** [ ], **Poles:** [0 -1], and **Gain:** 4
(d) **Transfer Fcn** with settings **Numerator coefficients:** [-3 0 12] and **Denominator coefficients:** [8 0 0 32 -240 -98]
(e) **Zero-Pole** with settings **Zeros:** [ ], **Poles:** [0 -4 -5], and **Gain:** 45/7
(f) **Zero-Pole** with settings **Zeros:** -1/3, **Poles:** [0 -2.5+3i -2.5-3i], and **Gain:** 3
(g) **Zero-Pole** with settings **Zeros:** [-1 -2 -3], **Poles:** [0 2i -2i 2], and **Gain:** 3.2
(h) **State-Space** with settings A: [-2 -1 0;2 3 2;1 2 5], B: [1;-2;3], C: [1 -2 1], and D: 2
hint: section 3.1

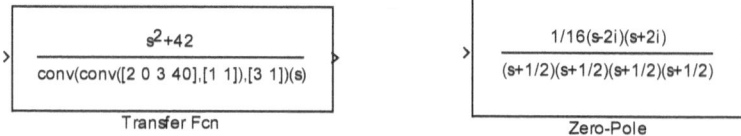

Figure E.3(3) SIMULINK model of problem 2(a)

Figure E.3(4) SIMULINK model of problem 2(b)

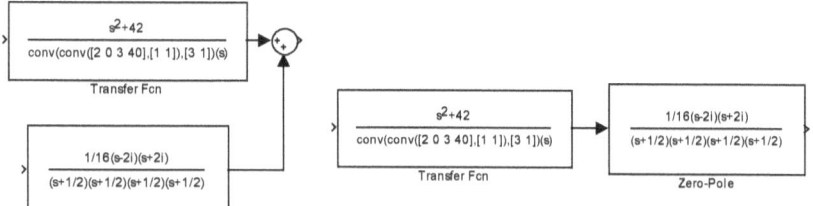

Figure E.3(5) SIMULINK model of problem 2(d)

Figure E.3(6) SIMULINK model of problem 2(e)

(2) (a) Figure E.3(3) (b) Figure E.3(4)
  (c) At MATLAB command prompt first execute [N,D]=ord2(2.5,0.77); and then enter settings in a **Transfer Fcn** block as **Numerator coefficients:** N and **Denominator coefficients:** D. See section 2.1 for ord2 details.
  (d) This transfer function is basically the summation of those in parts (a) and (b). You can add the two component systems by a **Sum** block, figure E.3(5) depicts the modeling.
  (e) This transfer function is basically the product of those in parts (a) and (b). You can seriesly connect the two component systems for the implementation. Figure E.3(6) presents the modeling.
hint: section 3.1

(3) Figure E.3(7). **Transfer Fcn** with settings **Numerator coefficients: [0 0 0 2;0 0 2 3;1 0 0 0;0 0 1 8]** and **Denominator coefficients: [1 0 0 1]**
hint: section 3.2

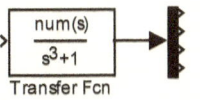

Figure E.3(7) Model of single input four output MIMO

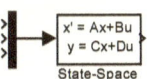

Figure E.3(8) Model of three input single output MIMO

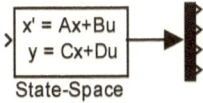

Figure E.3(9) Model of single input four output MIMO

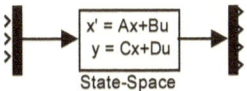

Figure E.3(10) Model of three input four output MIMO

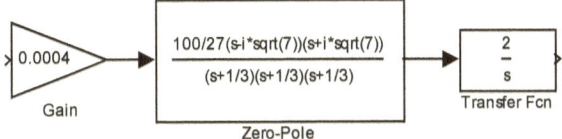

Figure E.3(11) Model of three control systems in series

(4) (a) Figure E.3(8) (b) Figure E.3(9) (c) Figure E.3(10)
hint: section 3.2

(5) Figure E.3(11). The **Gain**, **Zero-Pole**, and **Transfer Fcn** for $H_1(s)$, $H_2(s)$, and $H_3(s)$ respectively. The **Zero-Pole** setting is **Zeros: [i*sqrt(7) − i*sqrt(7)]**, **Poles: [-1/3 -1/3 -1/3]**, and **Gain: 100/27**.
hint: section 3.3

(6) Figure E.3(12)
hint: section 3.3

(7) Figure E.3(13)   hint: section 3.4
(8) Figure E.3(14)   hint: section 3.4
(9) Figure E.3(15)   hint: section 3.5

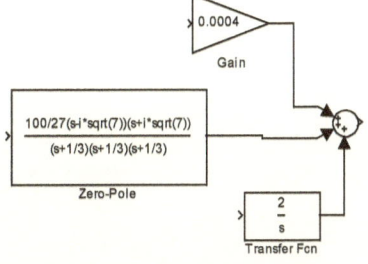

Figure E.3(12) Model of three control systems in parallel

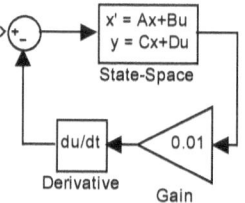

Figure E.3(13) Model of negative feedback system

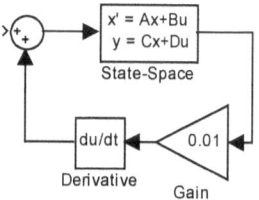

Figure E.3(14) Model of positive feedback system

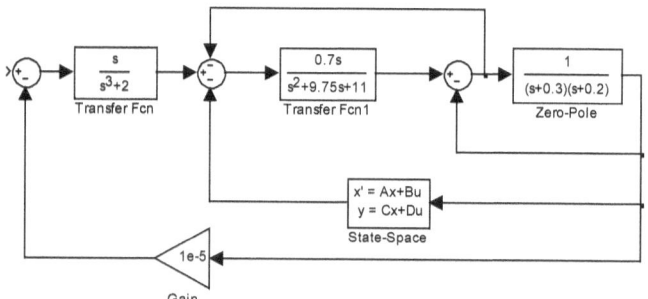

Figure E.3(15) Model of the SISO control system

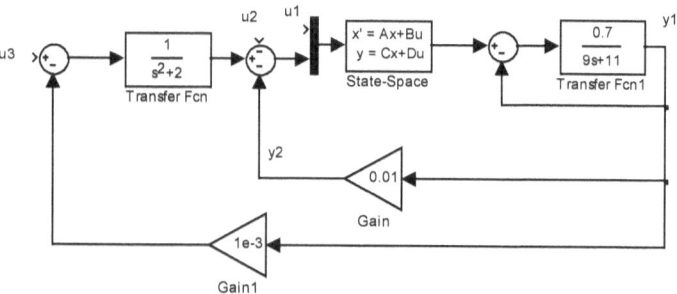

Figure E.3(16) Model of the MIMO control system

(10) Figure E.3(16), $G_2(s)$ is a two input one output MIMO which needs a **Mux** block for the split inputs.
hint: section 3.5

(11) Figure E.3(17)

$$T(s) = \frac{Y(s)}{U(s)} = \frac{0.7s^5 - 4.2s^4 - 2.1s^3 + 9.8s^2}{s^{10} + 4.95s^9 - 51.42s^8 - 103.3s^7 + 24.94s^6 - 10.63s^5 - 52.67s^4 + \overline{193.3s^3 + 184.4s^2 + 315.8s + 326.5}}$$

hint: section 3.6

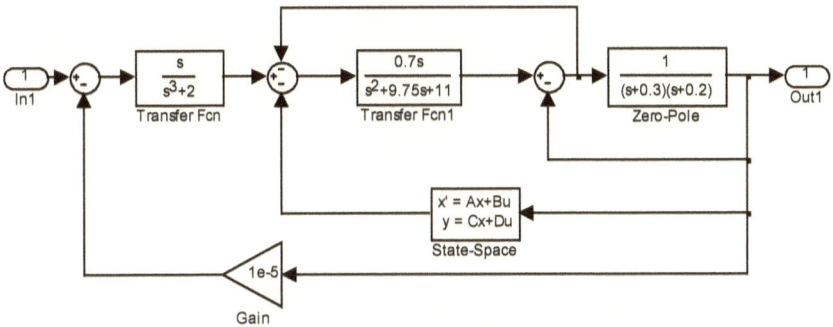

Figure E.3(17) SISO model with input and output ports

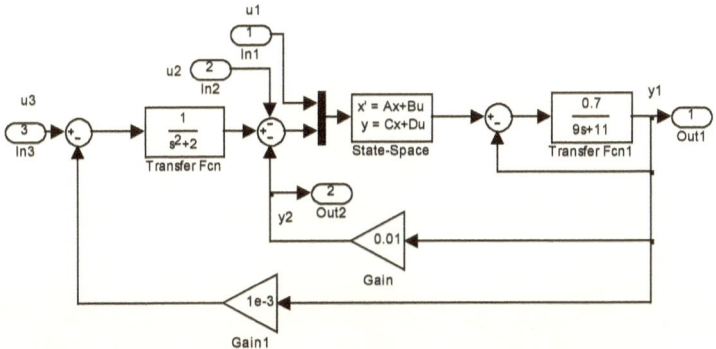

Figure E.3(18) MIMO model with input and output ports

(12) Figure E.3(18)

$$T(s) = \begin{bmatrix} T_{11} & T_{12} & T_{13} \\ T_{21} & T_{22} & T_{23} \end{bmatrix} \text{ where}$$

$$T_{11}(s) = \frac{0.1556s^5 - 0.3111s^4 - 3.267s^3 - 3.889s^2 - 7.156s - 6.533}{s^6 - 4.696s^5 - 8.818s^4 + 0.6669s^3 - 3.459s^2 + 20.11s + 36.35}$$

$$T_{21}(s) = \frac{0.001556s^5 - 0.003111s^4 - 0.03267s^3 - 0.03889s^2 - 0.07156s - 0.06533}{s^6 - 4.696s^5 - 8.818s^4 + 0.6669s^3 - 3.459s^2 + 20.11s + 36.35}$$

$$T_{12}(s) = \frac{-0.3889s^5 + 1.789s^4 + 3.344s^3 + 5.756s^2 + 8.244s + 4.356}{s^6 - 4.696s^5 - 8.818s^4 + 0.6669s^3 - 3.459s^2 + 20.11s + 36.35}$$

$$T_{22}(s) = \frac{-0.003889s^5 + 0.01789s^4 + 0.03344s^3 + 0.05756s^2 + 0.08244s + 0.04356}{s^6 - 4.696s^5 - 8.818s^4 + 0.6669s^3 - 3.459s^2 + 20.11s + 36.35}$$

$$T_{13}(s) = \frac{0.3889s^3 - 1.789s^2 - 4.122s - 2.178}{s^6 - 4.696s^5 - 8.818s^4 + 0.6669s^3 - 3.459s^2 + 20.11s + 36.35}$$

$$T_{23}(s) = \frac{0.003889s^3 - 0.01789s^2 - 0.04122s - 0.02178}{s^6 - 4.696s^5 - 8.818s^4 + 0.6669s^3 - 3.459s^2 + 20.11s + 36.35}$$

hint: section 3.6

Chapter 4

Time Domain Control System in MATLAB

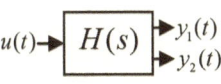

Having defined a control system in MATLAB, how one generates the system input and output samples is the subject matter of this chapter. Chapter 2 addressing is for the most part on control system formation. Pertinent code generates control system input samples which are applied to the system as defined in accordance with chapter 2. Control output from the system is facilely realized by embedded function for whose account we focus on the following:

- ❖ ❖ Code writing for control system input – periodic/nonperiodic
- ❖ ❖ SISO/MIMO simulator of control systems – data/graphics
- ❖ ❖ Step and impulse responses on control systems – SISO/MIMO
- ❖ ❖ Control system performance and error performance indices

4.1 Control system input in MATLAB

A control system needs input signal which can be voltage, current, force, displacement, or other. Realistic control signals are always continuous but a computer never generates a continuous signal due to finite memory reason. What we exercise is consider some samples of control system input and try to determine the output signal due to the input. In the sequel we explain how control system input samples are generated in MATLAB.

MATLAB does not have any knowledge of your control system input. All MATLAB can do is respond to some written command which provides convenience of generation by programming statements and pre-written function files. Suppose voltage versus time (say $u(t)$ versus t) is a control system input. Following examples illustrate how we generate a control system input i.e. all about $u(t)$ and t data generation.

✦ Control system input from mathematical expressions
Control system input is frequently described in terms of mathematical functions within some give interval. Involvement of mathematical expression requires that we write the code for the expression in MATLAB.

Suppose we intend to generate $u(t)$ versus t samples. First we must select some step size for the t, generate the t vector (subsection 1.1.2) based on chosen step size and given interval, and then write the scalar code of $u(t)$ according to appendix A, examples of which are in the sequel.

🗗 Example 1
Generate the ramp control input $u(t) = \frac{4}{3}t - 6$ over the interval $0 \le t \le 3$ sec.

Since we have to select the t step size, let it be $\Delta t = 0.1$ sec. Based on that the input sample generation is as follows:
>>t=0:0.1:3; ↵ ← t holds the t variation as a row matrix
>>u=4/3*t-6; ↵ ← Workspace variable u holds the input as a row matrix

The t and u are user-chosen variable names and both are identical size row matrix.

🗗 Example 2
Generate the control input $u(t) = 2V$ over the interval $0 \le t \le 3$ sec.

We can not generate this $u(t)$ by example 1 mentioned technique. The main problem here if the t changes, we can not relate this change to u by writing functional code. If we write u=2, only one value will be assigned to u whereas the t vector of example 1 contains many t point values. Which element value of the row matrix t does this u correspond to? Obviously there is no answer.

Let us choose the t step size as 0.2secs from 0 to 3secs then t assumes the values 0, 0.2, 0.4, ⋯ 3secs. At each of these values, the $u(t)$ must be 2. We generate the t as a row vector by writing the code t=0:0.2:3;. Let MATLAB find the number of elements in the vector t by the command length(t). Then we generate the number of ones equal to the number of elements in t and multiply each element by 2 by using the command 2*ones(1,length(t)) (appendix D.7 for ones) however the complete statement is as follows:
>>t=0:0.2:3; ↵   ← t holds the t variation as a row matrix
>>u=2*ones(1,length(t)); ↵ ← Workspace u holds input as a row matrix

Control System Analysis & Design in MATLAB and SIMULINK

⌂ Example 3

To generate the parabolic input $u(t) = 1 - \frac{2}{5}t + t^2$ N with a step size 0.1sec over the interval $0 \leq t \leq 3$ sec, we carry out the following:

>>t=0:0.1:3; ↵ ← t holds the t variation as a row matrix
>>u=1-2/5*t+t.^2; ↵ ← Workspace u holds input as a row matrix

Table 4.A Control system input data generation from mathematical expression

Control input to be generated	Command we need
$u(t) = 5e^{-t}$ for $0 \leq t \leq 0.1$ sec with a step 1 m sec	>>t=0:1e-3:0.1; ↵ >>u=5*exp(-t); ↵
$u(t) = 10(1 - e^{-3t})$ for $0 \leq t \leq 0.1$ sec with a step 1 m sec	>>t=0:1e-3:0.1; ↵ >>u=10*(1-exp(-3*t)); ↵
$u(t) = |t|$ for $-2 \leq t \leq 2$ sec with a step 0.1sec	>>t=-2:0.1:2; ↵ >>u=abs(t); ↵
$u(t) = \frac{3-|t|}{6}$ for $-2 \leq t \leq 2$ sec with a step 0.1sec	>>t=-2:0.1:2; ↵ >>u=(3-abs(t))/6; ↵
$u(t) = e^{-t^2}$ for $-2 \leq t \leq 2$ sec with a step 0.1sec (called Gaussian)	>>t=-2:0.1:2; ↵ >>u=exp(-t.^2); ↵
$u(t) = \frac{1}{\sqrt{2\pi}} e^{-\left(\frac{t-4}{3}\right)^2}$ for $-2 \leq t \leq 2$ sec with a step 0.1sec (Gaussian with different mean and scale)	>>t=-2:0.1:2; ↵ >>u=exp(-((t-4)/3).^2)/sqrt(2*pi); ↵
$u(t) = \frac{\sin \pi t}{\pi t}$ for $-2 \leq t \leq 2$ sec with a step 0.1sec (called sinc function)	>>t=-2:0.1:2; ↵ >>u=sinc(t); ↵ (sinc is MATLAB built-in with the same definition)
$u(t) = \frac{\sin t}{t}$ for $-2 \leq t \leq 2$ sec with a step 0.1sec (also called sinc function, different text defines differently)	>>t=-2:0.1:2; ↵ >>u=sinc(t/pi); ↵ (Using MATLAB built-in sinc)
$u(t) = 3e^{-t} \sin 5t$ for $0 \leq t \leq 5$ sec with a step 0.1sec (called damped sine signal)	>>t=0:0.1:5; ↵ >>u=3*exp(-t).*sin(5*t); ↵
$u(t) = \begin{cases} 1 & \text{for } t > 0 \\ -1 & \text{for } t < 0 \end{cases}$ for $-2 \leq t \leq 2$ sec with a step 0.1sec (called signum function)	>>t=-2:0.1:2; ↵ >>u=sign(t); ↵ (sign is MATLAB built-in with the left side functional definition)

⌂ Example 4

Generate the displacement input $u(t) = 0.3 \sin 2\pi f t$ m over the interval 0 to 20 $m$sec where the frequency $f = 200$ Hz.

Let us choose the time step as 0.1 $m$sec and the last time point 20 $m$sec is written as 20×10^{-3} sec whose MATLAB code is **20e-3** (1e-3 is equivalent to 10^{-3} not the natural number). The function $0.3 \sin 2\pi f t$ is coded as **0.3*sin(2*pi*200*t)** and the implementation is as follows:

>>t=0:0.1e-3:20e-3; ↵ ← t holds the t variation as a row matrix

```
>>u=0.3*sin(2*pi*200*t); ↵  ← The u holds the input as a row matrix
```

If the sinusoidal input had some phase for example $u(t) = 0.3\sin(2\pi f t - 60°)$, the command would be `u=0.3*sin(2*pi*200*t-pi/3);` because the sine input argument must be in radian.

⊟ Miscellaneous examples

In last four examples we introduced the idea behind expression based control input sample generation. We presented a collection of concise control input codes in table 4.A maintaining ongoing symbology and function.

4.2 Periodic control input in MATLAB

A periodic input repeats its wave shape every after certain period. If the input $u(t)$ is periodic over period T, the relationship $u(t) = u(t+T)$ exists at every point t and the input has the frequency $f = \dfrac{1}{T}$. To generate a periodic wave, the definition of the input in one period is enough. The $u(t)$ may quantify any physical control input like force, voltage, etc. There are many defining forms of periodic inputs, few of which are addressed in the following.

⊟ Example 1

Generate the control input wave of figure 4.1(a).

Sometimes our ability to pick the information from a given graph is important. Referring to the figure, the time period and amplitude of the wave are 1 m sec and 0.2 V respectively. The frequency of the sine wave is then $f = \dfrac{1}{T} = 1000$ Hz hence equation of the wave is $u(t) = 0.2 \sin 2\pi 1000 t$. To generate the wave by choosing a step 0.01 m sec, we exercise (just like last section) the following commands:

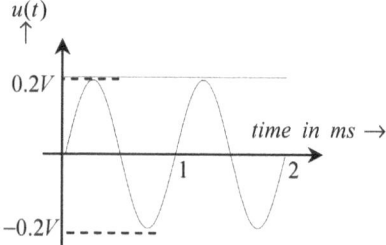

Figure 4.1(a) Plot of a two cycle sine wave

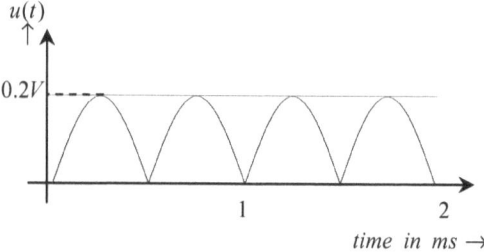

Figure 4.1(b) A full rectified sine wave

```
>>t=0:0.01e-3:2e-3; ↵     ← t holds t variation 0≤t≤2 m sec as a row
                            matrix with a step 0.01 m sec

>>u=0.2*sin(2*pi*1000*t); ↵
```

Workspace **u** holds the input $u(t)$ samples as a row matrix where the **u** is a user-chosen variable.

⌸ Example 2

Generate the full rectified sine input of figure 4.1(b).

It is exactly the wave of figure 4.1(a) but the negative halves are turned to positive. The command **abs** turns the negative value to a positive one. In example 1 we need to write **u=abs(0.2*sin(2*pi*1000*t))**; to generate the wave assuming the same step size.

⌸ Example 3

Generate the half rectified sine input of figure 4.1(c).

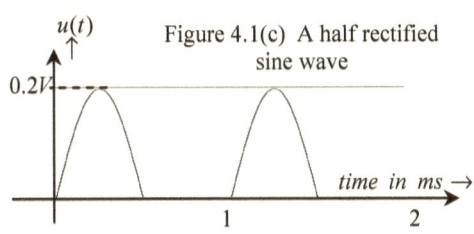

Figure 4.1(c) A half rectified sine wave

This wave is also derived from the wave of figure 4.1(a) in which the negative part of the wave is turned to zero. In order to generate this input, first we generate the input of figure 4.1(a) as in example 1, find the indexes of the t vector at which the input is less than zero by using the **find** (appendix D.6), and set the input values corresponding to these indexes to zero as follows:

>>t=0:0.01e-3:2e-3; ↵   ← t holds t variation $0 \le t \le 2$ msec as a row matrix with a step 0.01 msec
>>u=0.2*sin(2*pi*1000*t); ↵ ← u holds the input of figure 4.1(a)
>>r=find(u<0); ↵   ← r holds the t indexes when the input has negative value where r is user-chosen variable
>>u(r)=0; ↵   ← setting the u elements of negative indexes to 0, u holds the input as a row matrix

⌸ Example 4

In one period T, a square wave $u(t)$ is defined as $u(t) = \begin{cases} A & \text{for} \quad 0 \le t \le D \\ -A & \text{for} \quad D < t \le T \end{cases}$ where D is the duty cycle of the wave as a percentage of T. Figure 4.1(d) depicts the plot of one cycle square wave in which the duty cycle varies from 0 to 100%.

MATLAB format for sample generation of the wave is **square($2\pi ft, D$)**

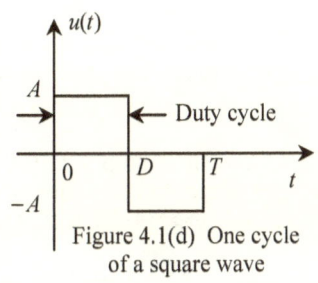

Figure 4.1(d) One cycle of a square wave

where $f = \frac{1}{T}$ is the frequency of the square wave, t is the desired time interval vector in the form of a row matrix over which we wish to see the square wave, and the **square** is a built-in function.

Let us generate a 10 Hz square wave samples with amplitude variation $\pm 1V$, with duty cycle 50%, and over the interval $-0.3 \le t \le 0.3$ sec.

Again the selection of the step size is mandatory as well as user-supplied and let it be 0.01sec. The wave sample generation is as follows:
>>t=-0.3:0.01:0.3; ↲   ← t holds t variation −0.3 ≤ t ≤ 0.3 sec as a row matrix with a step 0.01sec
>>u=square(2*pi*10*t,50); ↲ ← The workspace variable u holds the square wave as a row matrix

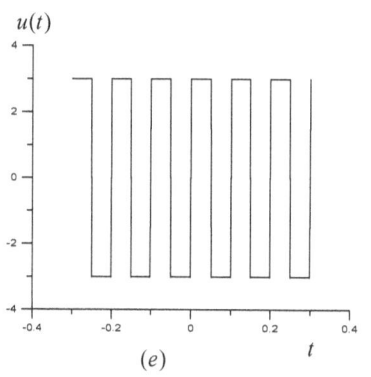
(e)

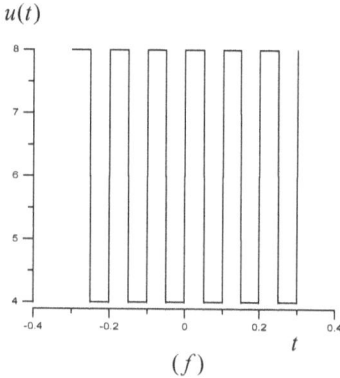
(f)

Figures 4.1(e)-(g) Different kinds of square waves: (e) amplitude ±3 V, frequency 10 Hz, duty cycle 50%, (f) amplitude swing 4 V to 8 V, frequency 10 Hz, duty cycle 50%, and (g) amplitude swing 4 V to 8 V, frequency 10 Hz, duty cycle 80%

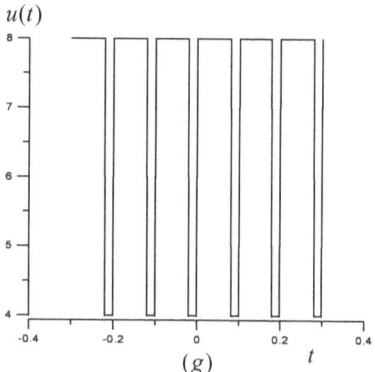
(g)

Figures 4.1(e)-(g) show the square inputs of different characteristics. Each of the inputs has the time period (duration of one cycle) $T = 0.1$ sec and frequency $f = \frac{1}{T} = 10\ Hz$ and exists over the interval $-0.3 \le t \le 0.3$ sec. Choosing a step size 0.01sec, commands for the generation of square inputs are presented as follows:
>>t=-0.3:0.01:0.3; ↲   ← t holds t variation −0.3 ≤ t ≤ 0.3 sec as a row matrix with a step 0.01sec
>>u=3*square(2*pi*10*t,50); ↲   ← Command for the figure 4.1(e)
>>u=6+2*square(2*pi*10*t,50); ↲ ← Command for the figure 4.1(f)
>>u=6+2*square(2*pi*10*t,80); ↲ ← Command for the figure 4.1(g)

In each case the workspace variable u holds the input as a row matrix. The default swing of the **square** is ±1 so just multiplying by 3 achieves the required swing of the wave in figure 4.1(e).

In figure 4.1(f) a linear mapping $y = mx + c$ is necessary to make the **square** sweep from 4 to 8 where x and y correspond to the former and latter input values respectively. The related parameters m and c are to be found from the specification of the given input. When the value of **square** is -1 (means $x = -1$), the input value of the figure 4.1(f) should be 4 (means $y = 4$) so $4 = -m + c$. Again if the value of **square** is 1 (means $x = 1$), the input value of the figure 4.1(f) should be 8 (means $y = 8$) on that $8 = m + c$. Solving the two equations, one obtains $m = 2$ and $c = 6$. Treating the **square** as x, equation of the wave in figure 4.1(f) should be $y = 2x + 6$ or **u=6+ 2*square(2*pi*10*t,50)** – that is how we determined the command.

This kind of linear transformation is often necessary for the same type of wave shape for instance square to square, triangular to triangular, sine to sine, Gaussian to Gaussian, etc. In the periodic wave generation the notion of the cycle and period must be transparent. With the time period 0.1sec and duration 0.6sec for each wave, there must be 0.6/0.1=6 cycles in the generated input.

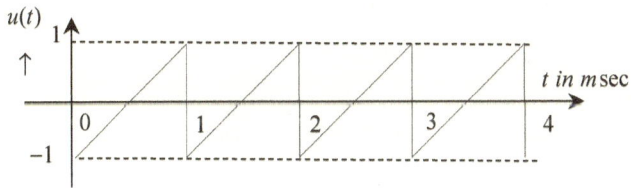

Figure 4.2(a) A sawtooth wave

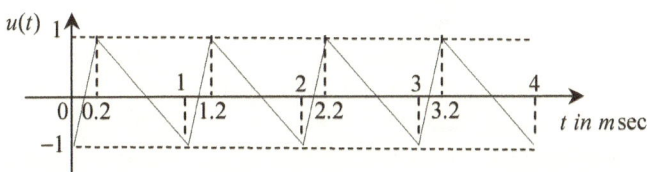

Figures 4.2(b) Periodic triangular wave of different characteristic

Example 5

A sawtooth wave is generated by the built-in function **sawtooth**. The general format for the wave generation is **sawtooth($2\pi f t$)** where T is time period of the sawtooth wave, $f = \dfrac{1}{T}$ is the frequency of the wave, and t is the desired time interval as a vector over which we wish to see the wave. The default swing of **sawtooth** is from -1 to 1.

We intend to generate the sawtooth wave of figure 4.2(a). Looking into the figure, time period T of the wave is 1 msec therefore frequency $f = 1000\,Hz$. The wave exists over the interval $0 \le t \le 4$ msec. The step size selection comes from the user and let us say it is 0.001 msec. The implementation is as follows:

```
>>t=0:0.001e-3:4e-3; ↵
```
Above **t** holds the t variation $0 \leq t \leq 4$ msec as a row matrix with a step 0.001 msec.
```
>>u=sawtooth(2*pi*1000*t); ↵
```
Above workspace variable **u** holds the input samples as a row matrix.

🗗 **Example 6**

The wave of figure 4.2(b) is a variant of wave in figure 4.2(a). In the figure 4.2(a) the maximum 1 of the triangular wave is occurring at 100% period point on the t axis (for example at 1 msec, 2 msec, etc). On the contrary the maximum 1 of figure 4.2(b) is occurring at 20% period of the wave for example at 0.2 msec, 1.2 msec, etc. The function **sawtooth** is still effective to generate the input of figure 4.2(b) as follows (assuming the step size of example 5):
```
>>t=0:0.001e-3:4e-3; ↵
>>u=sawtooth(2*pi*1000*t,0.2); ↵
```
Now there are two input arguments in the function **sawtooth** (another syntax of **sawtooth**), the first and second of which are the $2\pi ft$ description and the peak point occurrence point in terms of 0-1 of the period respectively.

🗗 **Point to remember**

The step size Δt must be much less than the period T in all above sample generations. A reasonable choice can be $\Delta t = T/100$ for one cycle.

4.3 Control system output in MATLAB

A control system we form by chapter 2 mentioned functions or techniques. Section 4.1 elucidates the control system input generation. What about the output from a control system? This section is all about the output from a control system. Of coarse the output is in sample form. The embedded function **lsim** helps us obtain the control system output samples provided that the input samples and control system are available in MATLAB. A variety of problems may appear depending upon what we intend to simulate. Following examples demonstrate some computing or graphing regarding control system time domain output.

🗗 **Example 1: Input-output samples on** $H(s)$

A control system is defined by $H(s) = \dfrac{2}{s+1}$. An input $u(t) = 2$ over $0 \leq t \leq 1\text{sec}$ is applied to the system. Determine the output $y(t)$ samples subject to $\Delta t = 0.2\text{sec}$.

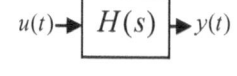

Figure 4.3(a) A control system with input and output

Figure 4.3(a) shows input-output strategy of the control system. First we form the system by **tf** of chapter 2 as follows:
```
>>H=tf(2,[1 1]); ↵          ← H⇔ H(s), H is user-chosen
```
Then required t samples we generate by:

```
>>t=0:0.2:1; ↵          ← t is user-chosen and holds t samples
```
After that the input $u(t)$ samples (example 2 of section 4.1) we generate by:
```
>>u=2*ones(1,length(t)); ↵ ← u is user-chosen and holds u(t) samples
```
The lsim has one syntax: user-supplied output variable for $y(t)$ =lsim(control system representing variable, input $u(t)$ samples, corresponding t samples) so execute the following at the command prompt:
```
>>y=lsim(H,u,t); ↵      ← y is user-chosen and holds y(t) samples
```
The y samples are in a column matrix but the t is a row one. If we wish to view the two values side by side, execute the following:
```
>>[t' y] ↵              ← Chapter 1 for ' operator
```

ans =

      0   0
  0.2000 0.7251
  0.4000 1.3187
  0.6000 1.8048
  0.8000 2.2027
  1.0000 2.5285
    ↑   ↑
    t $y(t)$

For instance the last value indicates $y(t)$=2.5285 at t=1sec in above output.

▣ Example 2: Control system input-output samples on state-space form

Suppose the $H(s)$ of figure 4.3(a) is given as $H(s)=\{A,B,C,D\}$ where $A=\begin{bmatrix}-2 & -1 & 0\\ 2 & 3 & 2\\ 1 & 2 & 5\end{bmatrix}$, $B=\begin{bmatrix}1\\-2\\3\end{bmatrix}$, $C=[1\ -2\ 1]$, and $D=[2]$. We wish to obtain $y(t)$ samples subject to example 1 $u(t)$.

When a control system is given in terms of state-space form, we still apply the lsim. The only point is the $H(s)$ is defined by **ss** of section 2.1 and do so by:
```
>>A=[-2 -1 0;2 3 2;1 2 5]; B=[1;-2;3]; C=[1 -2 1]; D=2; ↵
>>H=ss(A,B,C,D); ↵
```
The rest commands are taken from the example 1:
```
>>t=0:0.2:1; u=2*ones(1,length(t)); y=lsim(H,u,t); ↵
>>[t' y] ↵
```

ans =

      0   4.0000
  0.2000   7.5114
  0.4000   11.4027
  0.6000   14.1696
  0.8000   9.8017
  1.0000 -23.3043
    ↑   ↑
    t $y(t)$

⊟ **Example 3: Graph of control system input-output**

The lsim keeps also provision for graphing control system input and output together. Consider the example 1. We can graph the $u(t)$ and $y(t)$ versus t just by excluding the output argument:
>>lsim(H,u,t) ↵

The result is the figure 4.3(b). The constant line and curve are for the $u(t)$ and $y(t)$ respectively. Clearly the horizontal axis is over $0 \le t \le 1\sec$ as entered.

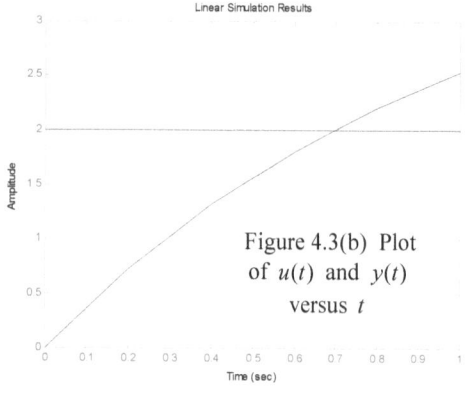

Figure 4.3(b) Plot of $u(t)$ and $y(t)$ versus t

4.4 MIMO control system output in MATLAB

The reader needs the background of MIMO control system which is explained in section 2.2. The lsim of last section is also applicable for MIMO control systems. Following five examples demonstrate computing or graphing about MIMO control systems.

⊟ **Example 1: Single input multi-output samples on $H(s)$**

Figure 2.2(a) depicts single input two output MIMO. Suppose the MIMO has $H_1(s) = \dfrac{4}{3s^2+7}$ and $H_2(s) = \dfrac{4s+5}{3s^2+7}$ and $H(s) = \begin{bmatrix} H_1(s) \\ H_2(s) \end{bmatrix}$. The $H_1(s)$ and $H_2(s)$ refer to y_1 to u and y_2 to u respectively. We wish to obtain y_1 and y_2 samples subject to $u(t) = 2V$ on $\Delta t = 0.1\sec$ over $0 \le t \le 0.5\sec$.

Let us define the MIMO observing section 2.2 symbology:
>>N1=4; D1=[3 0 7]; N2=[4 5]; D2=[3 0 7]; ↵
>>H=tf({N1;N2},{D1;D2}); ↵ ← Forming the MIMO $H(s)$, H holds $H(s)$

The $u(t)$ samples we generate (example 2 in section 4.1) by:
>>t=0:0.1:0.5; u=2*ones(1,length(t)); ↵ ← u holds $u(t)$ samples as a row matrix

The lsim syntax of last section we apply here too with identical symbology:
>>y=lsim(H,u,t); [t' y] ↵

```
ans =
         0         0         0
    0.1000    0.0133    0.2823
    0.2000    0.0529    0.5912
    0.3000    0.1179    0.9197
    0.4000    0.2068    1.2600
    0.5000    0.3174    1.6042
       ↑         ↑         ↑
       t        y₁        y₂
```

The point is return from lsim is a two column matrix, the first and second of which are for y_1 and y_2 respectively. Should you intend to separate the output samples, exercise y(:,1) and y(:,2) for y_1 and y_2 respectively.

In a similar fashion we exercise single input three output MIMO. The $H(s)$ requires N3 and D3 then and the y will be a three column matrix for the three output MIMO.

Example 2: Multi input single output samples on $H(s)$

Figure 2.2(b) depicts two input single output MIMO. With $H_1(s) = \frac{4}{3s^2+7}$ and $H_2(s) = \frac{4s+5}{3s^2+7}$, the MIMO has $H(s) = [H_1(s) \quad H_2(s)]$ where $H_1(s)$ and $H_2(s)$ refer to y to u_1 and y to u_2 respectively. We wish to obtain y samples subject to $u_1(t) = 2V$ and $u_2(t) = t^2/7 \, V$ on $\Delta t = 0.1$ sec over $0 \le t \le 0.5$ sec.

We define the MIMO according to section 2.2 symbology by:
>>N1=4; D1=[3 0 7]; N2=[4 5]; D2=[3 0 7]; ↵
>>H=tf({N1,N2},{D1,D2}); ↵   ← Forming $H(s)$, H holds $H(s)$

Now we have two different inputs and their sample generation must take place at identical t points. The samples of $u_1(t)$ and $u_2(t)$ must be as a column matrix, so must be the samples of t and do so by:
>>t=[0:0.1:0.5]'; ↵   ← See chapter 1 for column matrix entering
>>u1=2*ones(length(t),1); ↵   ← u1 holds $u_1(t)$ samples as a column matrix
>>u2=t.^2/7; ↵   ← u2 holds $u_2(t)$ samples as a column matrix

The u1 and u2 are user-chosen variables from which a two column matrix is formed as follows:
>>u=[u1 u2]; ↵   ← u holds $u_1(t)$ and $u_2(t)$ samples as a single quantity

Calling similar to example 1 takes place by:
>>y=lsim(H,u,t); [t y] ↵

```
ans =
         0        0
    0.1000   0.0134
    0.2000   0.0535
    0.3000   0.1199
    0.4000   0.2114
    0.5000   0.3266
       ↑        ↑
       t        y
```

Similarly we exercise three input single output MIMO. Then $H(s)$ requires N3 and D3 and the u will be a three column matrix for a three input MIMO with additional $u_3(t)$.

⊟ **Example 3: Multi input multi output samples on** $H(s)$

Figure 2.2(c) presents two input three output MIMO. Usually MIMO systems have common denominator transfer functions. The MIMO $H(s) = \begin{bmatrix} H_{11}(s) & H_{12}(s) \\ H_{21}(s) & H_{22}(s) \\ H_{31}(s) & H_{32}(s) \end{bmatrix}$ is consisted of the following: $H_{11}(s) = \frac{7}{s^2+5}$, $H_{12}(s) = \frac{5s}{s^2+5}$, $H_{21}(s) = \frac{1}{s^2+5}$, $H_{22}(s) = \frac{2s+3}{s^2+5}$, $H_{31}(s) = \frac{1}{s^2+5}$, and $H_{32}(s) = \frac{s+1}{s^2+5}$. See section 2.2 for input-output link. Our objective is to find y_1, y_2, and y_3 samples subject to $u_1(t)=2V$ and $u_2(t)=t^2/7$ V on $\Delta t=0.1\sec$ over $0 \le t \le 0.5\sec$.

Exercising the symbology of section 2.2, let us enter the numerator-denominator coefficients:

```
>>N11=7; N12=[5 0]; ↵    ← Numerators of H₁₁(s) and H₁₂(s)
>>N21=1; N22=[2 3]; ↵    ← Numerators of H₂₁(s) and H₂₂(s)
>>N31=1; N32=[1 1]; ↵    ← Numerators of H₃₁(s) and H₃₂(s)
>>D11=[1 0 5]; ↵         ← Denominator of H₁₁(s)
```

Since the denominator coefficients are same, so just assign **D11** to others:
```
>>D12=D11; D21=D11; D22=D11; D31=D11; D32=D11; ↵
```
After that form the $H(s)$ by:
```
>>H=tf({N11,N12;N21,N22;N31,N32},{D11,D12;D21,D22;D31,D32}); ↵
```
Input entering is exactly the one as in example 2:
```
>>t=[0:0.1:0.5]'; ↵
>>u1=2*ones(length(t),1); ↵
>>u2=t.^2/7; ↵
>>u=[u1 u2]; ↵    ← u holds u₁(t) and u₂(t) samples as a single quantity
```
Finally call the simulator by:
```
>>y=lsim(H,u,t); [t y] ↵

ans =
       0        0        0        0
  0.1000   0.0701   0.0101   0.0100
  0.2000   0.2775   0.0403   0.0398
  0.3000   0.6133   0.0896   0.0881
  0.4000   1.0624   0.1566   0.1529
  0.5000   1.6036   0.2386   0.2314
     ↑        ↑        ↑        ↑
     t       y₁       y₂       y₃
```

The y_1, y_2, and y_3 samples one can access by y(:,1), y(:,2), and y(:,3) respectively.

⊟ **Example 4: Multi input multi output samples on state space form of** $H(s)$

State space form is better in the sense that matrix dimension routinely takes care of input-output numbers. Suppose a three input four output MIMO is given by $H(s)=\{A,B,C,D\}$ where $A = \begin{bmatrix} -2 & 1 \\ 2 & 1 \end{bmatrix}$, $B =$

$\begin{bmatrix} -1 & 4 & -2 \\ -4 & 1 & 1 \end{bmatrix}$, $C = \begin{bmatrix} -1 & 1 \\ 2 & 3 \\ 0 & 4 \\ 5 & 6 \end{bmatrix}$, and $D = \begin{bmatrix} -1 & 4 & 3 \\ 2 & 0 & 2 \\ 3 & 5 & 1 \\ 4 & 0 & 2 \end{bmatrix}$. Say the inputs and outputs are labeled as $\{u_1(t), u_2(t), u_3(t)\}$ and $\{y_1(t), y_2(t), y_3(t), y_4(t)\}$ respectively. Determine the output samples subject to $u_1(t) = 2t - 2$, $u_2(t) = e^{-2t}$, and $u_3(t) = \sin t$ on $\Delta t = 0.2$sec over $0 \le t \le 0.8$sec.

First enter the state space matrices by:
>>A=[-2 1;2 1]; B=[-1 4 -2;-4 1 1]; C=[-1 1;2 3;0 4;5 6]; ↵
>>D=[-1 4 3;2 0 2;3 5 1;4 0 2]; ↵

Then form the $H(s)$ by:
>>H=ss(A,B,C,D); ↵

After that generate all inputs first generating t samples as a column matrix:
>>t=[0:0.2:0.8]'; ↵
>>u1=2*t-2; u2=exp(-2*t); u3=sin(t); ↵

Combined input as a three column matrix is obtained as:
>>u=[u1 u2 u3]; ↵

Eventually the simulator calling is conducted by:
>>y=lsim(H,u,t); [t y] ↵

```
ans =
         0    6.0000   -4.0000   -1.0000   -8.0000
    0.2000    5.9330    5.2416    6.8748   11.0613
    0.4000    6.9778   15.2845   17.0602   31.4840
    0.6000    8.7979   26.7243   29.5853   54.6989
    0.8000   11.2653   40.2637   44.8790   82.2751
       ↑        ↑        ↑        ↑         ↑
       t       y₁       y₂       y₃        y₄
```

$\quad\quad\quad\quad t \quad\quad\quad y_1 \quad\quad\quad y_2 \quad\quad\quad y_3 \quad\quad\quad y_4$

You may exercise y(:,1), y(:,2), y(:,3), and y(:,4) for y_1, y_2, y_3, and y_4 respectively.

⌸ Example 5: Graphing on MIMO

In example 4 we demonstrated just sample obtaining. What if we wish to see the four outputs as a graph? Exercise lsim(H,u,t) at the command prompt in order to see the graph. Figure 4.4(a) shows the graph. In the figure there are four plots stacked atop one another. The label in vertical axis To: Out(1) means y_1, so does the others. The horizontal axis of coarse is the t variation. The MIMO plotter automatically includes some input. For example the To: Out(1) or y_1 shows some other plot along with it. Bring mouse pointer on the curve and leftclick the mouse, a box will show Input: In(1) meaning $u_1(t)$ in addition the same box shows the value of t and $u_1(t)$, all of which you find at the upperleft corner in the figure. For instance the shown value is $t = 0.0691$ and $u_1(t) = -1.86$.

There is one demerit of this plot. For many outputs the figure may appear clumsy. The reader might say, I wish to get the data and plot it in my

own. In this regard appendix E mentioned **plot** might be useful. But you need to reduce the step size say 0.01 and reexecute the **t** related statements:
>>t=[0:0.01:0.8]';u1=2*t-2;u2=exp(-2*t); u3=sin(t);u=[u1 u2 u3]; ↵
Now there is no need to display the samples so just execute:
>>y=lsim(H,u,t); ↵
Suppose we wish to plot $y_1(t)$ versus t and get so by **plot(t,y(:,1))**. The graph is not shown for space reason.

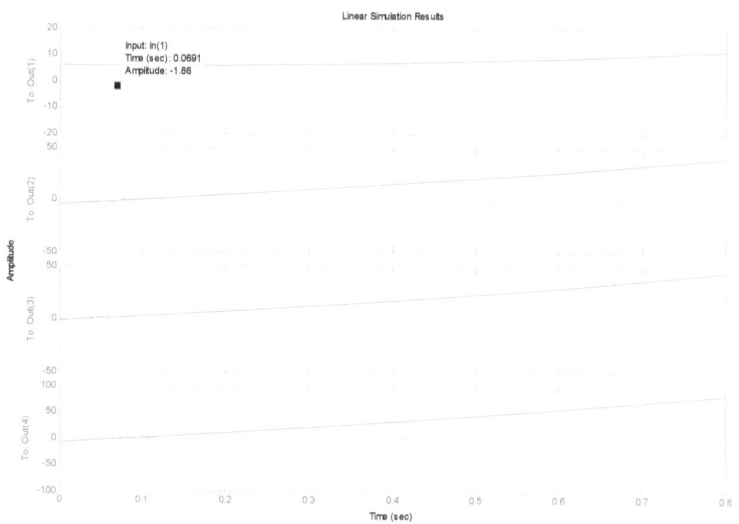

Figure 4.4(a) Plot of MIMO outputs as a single plot

The graphing technique so illustrated is on state space basis. The **lsim** evenly works on transfer function defined MIMO.

4.5 Step and impulse responses of a control system

Step response of a control system has meaningful implication. Most dynamics of a control system are based on input-output. Step response predominantly manifests how output of the control system behaves in the presence of one unit input. That one unit may quantify any physical input like force, voltage, displacement, etc.

On the contrary impulse response indicates frequency correlated behavior. Impulse input signifies all frequencies i.e. 0 to infinity. The control output subject to impulse input manifests dominant frequency presence in the system output.

However MATLAB holds embedded function **step** and **impulse** for computing or graphing on step and impulse response related problems respectively, few of which are illustrated now.

The **step** or **impulse** has many syntaxes depending on user-requirements. We need to define control system $H(s)$ by section 2.1 mentioned **tf**, **zpk**, or **ss** regardless of step and impulse responses.

🔁 **Example 1: Handling step response samples**

In this sort of problem we have a control system $H(s)$ and look for output $y(t)$ samples subject to unit step function which is defined by $u(t) = \begin{cases} 0 & \text{for } t < 0 \\ 1 & \text{for } t \geq 0 \end{cases}$. Figure 4.4(b) presents the strategy. There is no need to enter the $u(t)$ information, MATLAB does it for you.

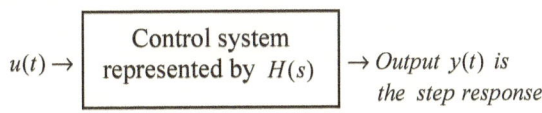

Figure 4.4(b) Concept behind the step response

Suppose step response samples for $H(s) = \dfrac{3}{5s+7}$ are to be found on $\Delta t = 0.2$ sec over $0 \leq t \leq 1$ sec.

First enter the control system to **H** by **tf**:
>>H=tf(3,[5 7]); ↵

Then we need to generate the t samples as a column matrix (chapter 1) by:
>>t=[0:0.2:1]'; ↵

The **step** has a syntax; user-supplied variable for output samples= **step**(system, t samples as a column matrix) so execute the following:
>>y=step(H,t); ↵   ← y holds $y(t)$ samples, y is user-chosen

The sample return is also as a column matrix. If you wish to see the two samples side by side, do so by:
>>[t y] ↵

```
ans =
        0         0
   0.2000    0.1047
   0.4000    0.1838
   0.6000    0.2436
   0.8000    0.2887
   1.0000    0.3229
      ↑         ↑
      t        y(t)
```

For example we read out $y(t) = 0.2887$ at $t = 0.8$ sec from above return.

🔁 **Example 2: Handling step response graph**

In example 1 if you wish to plot the step

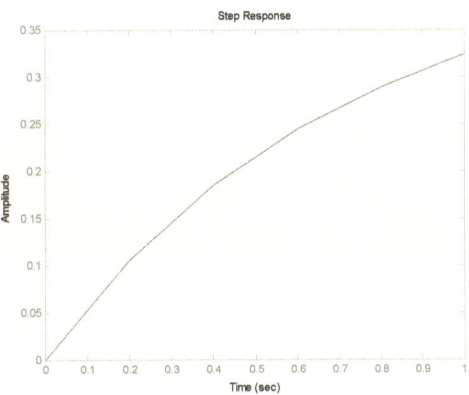

Figure 4.4(c) Step response of the control system

response i.e. $y(t)$ versus t, exclude the output argument and just call **step(H,t)**. Figure 4.4(c) presents the step response from MATLAB in which the **Amplitude** and **Time** refer to $y(t)$ and t respectively.

Example 3: Handling impulse response samples

Figure 4.4(d) shows the impulse response strategy. In MATLAB we do not enter $\delta(t)$ information. The syntax of **impulse** is similar to that of the **step**.

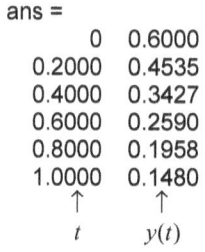

Figure 4.4(d) Concept behind the impulse response

In example 1 now we need the impulse response samples for the same control system and interval and do so by:

```
>>y=impulse(H,t); ↵
>>[t y] ↵
```

```
ans =
         0    0.6000
    0.2000    0.4535
    0.4000    0.3427
    0.6000    0.2590
    0.8000    0.1958
    1.0000    0.1480
```

$\quad t \quad\quad y(t)$

Example 4: Handling impulse response graph

All you need is exercise **impulse(H,t)** at the command prompt certainly the graph will be different from the one in figure 4.4(c) which we did not include for the space reason.

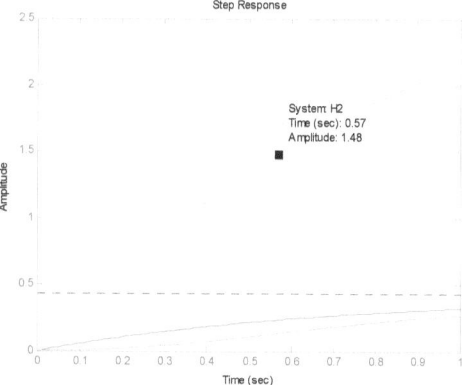

Figure 4.4(e) Step responses of multiple systems together

Example 5: Graphing step/impulse responses for multiple systems

Suppose we have three control systems which are $H_1(s) = \dfrac{3}{5s+7}$, $H_2(s) = \dfrac{3.4}{s+1}$, and $H_3(s) = \dfrac{1}{s^2+7}$ and intend to obtain the step response graph for the three control systems over $0 \le t \le 1 \sec$.

Enter the three systems one by one by tf as follows:

```
>>H1=tf(3,[5 7]); ↵        ← H1 holds H₁(s), H1 is user-chosen
>>H2=tf(3.4,[1 1]); ↵      ← H2 holds H₂(s), H2 is user-chosen
>>H3=tf(1,[1 0 7]); ↵      ← H3 holds H₃(s), H3 is user-chosen
```

The **step** has another syntax for multiple response graphing which is **step**(system 1, system 2, system 3, and so on, last bound of the interval) therefore carry out the following:
>>step(H1,H2,H3,1) ↵

Figure 4.4(e) shows the step responses of the three control systems together. The plots are in a color form. If you wish to identify the system, bring the mouse pointer on any trajectory and leftclick the mouse. The graphics window responds with a box indicating the system for example in figure 4.4(e) we see the graph for $H_2(s)$.

Should the reader need impulse responses, just execution of **impulse(H1,H2,H3,1)** is required at the command prompt.

Note: Sample values for multiple systems are not returned by the **step**, nor by the **impulse**. Starting value of t must be from $t=0$ while using **step/impulse**.

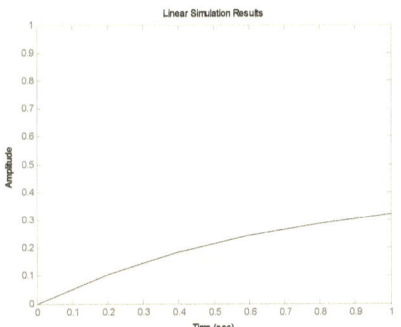

Figure 4.5(a) Plot of control system output over $0 \le t \le 1$ sec

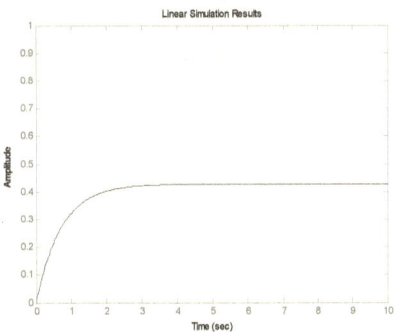

Figure 4.5(b) Plot of control system output over $0 \le t \le 10$ sec

4.6 First order control system performance

Usually in a first order system the output from a control system is exponentially settled type from which we may require time constant and steady state value. You can obtain the graph from an experiment or simulation. The important point here is how to find the two quantities from a given graph. We concentrate first on steady state value and then on the time constant.

Consider the single input single output control system of figure 4.3(a). If we have $u(t)=1 V$ and $H(s) = \dfrac{3}{5s+7}$, the graph of input $u(t)$ and output $y(t)$ versus t subject to $\Delta t = 0.2$sec over $0 \le t \le 1$sec is obtained by section 4.3 quoted technique:

>>H=tf(3,[5 7]); ↵   ← H⇔ $H(s)$, H is user-chosen

-103-

```
>>t=0:0.2:1; ↵           ← t is user-chosen and holds t samples
>>u=ones(1,length(t)); ↵ ← u is user-chosen and holds u(t) samples
>>lsim(H,u,t) ↵
```
Last command line brings about the outcome as in figure 4.5(a). The output $y(t)$ appears to be exponentially settled but we are not sure unless we increase the time interval. Say $0 \le t \le 10\,\text{sec}$ with $\Delta t = 0.02\,\text{sec}$ is chosen therefore reexecute the following:
```
>>t=0:0.02:10; u=ones(1,length(t)); lsim(H,u,t) ↵
```

The reduction of Δt is for the time constant reason which will be explained later. Anyhow figure 4.5(b) is the result of the last commands. By inspection the output is plausibly exponentially settled type. If we need the steady state value, just pick the last sample of $y(t)$. Since we need sample, reexecute the following:
```
>>y=lsim(H,u,t); ↵
```
The samples of $y(t)$ are available in y so the last sample we obtain by:
```
>>y(end) ↵

ans =
    0.4286
```
Above return says that the steady state value of the $y(t)$ is 0.4286 or symbolically $y_{ss} = 0.4286\,V$.

Given the approximate equation of output as $y(t) = y_{ss}(1 - e^{-\frac{t}{\tau}})$, the time constant τ is obtained by $\tau = \dfrac{-t_0}{\ln\left(\dfrac{y(2t_0)}{y(t_0)} - 1\right)}$ where the t_0 is user-chosen.

Let us turn our attention to ongoing control system. You may choose figure 4.5(a) or 4.5(b) output for τ finding. Since the workspace holds the later, let us work on that. Over the interval $0 \le t \le 10\,\text{secs}$ we may choose any t_0 besides the t_0 should be multiple of $\Delta t = 0.02$ sec. Say $t_0 = 1$ sec then we have $2t_0 = 2$ secs. We use the **find** function (appendix D.6) to determine the position index of these two samples and do so by:
```
>>r1=find(t==1); ↵   ← r1 is user-chosen, holds integer position index for t₀=1sec
>>r2=find(t==2); ↵   ← r2 is user-chosen, holds integer position index for 2t₀=2sec
>>-1/log(y(r2)/y(r1)-1) ↵   ← y(r2)⇔ y(2t₀) and y(r1)⇔ y(t₀)

ans =
    0.7143
```
The last command line is for the computing of τ which indicates $\tau = 0.7143$ secs.

Above technique you can apply to any experimental or simulated first order control system to find y_{ss} and τ even if it is a quasi one.

4.7 Second order control system performance

A second order control system receives special importance in control systems theory. The underlying reason is any higher order system can be decomposed into first and second order systems. The second order system has more performance parameters than the first order counterpart. Design specifications are linked with transient or steady state parameters of a second order system, momentous of which are the rise time T_r, peak time T_p, percent overshoot P.O., settling time T_s, and steady state value y_{ss}. We explain MATLAB's way of dealing these parameters in the sequel.

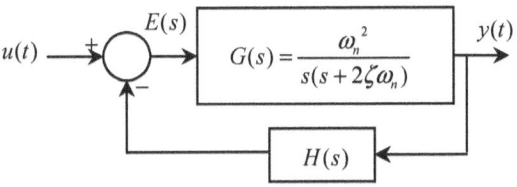

Figure 4.6(a) Prototype second order system

Figure 4.6(a) depicts a prototype second order control system where the forward path function is $G(s) = \dfrac{\omega_n^2}{s(s+2\zeta\omega_n)}$, ζ is the damping ration, and ω_n is the natural frequency.

When ζ is between 0 and 1, the system is called underdamped and above mentioned parameters are highly correlated with the system.

There may be two approaches to determine the performance, the first and second of which are the readymade MATLAB and user-written solutions respectively. Both approaches are addressed now.

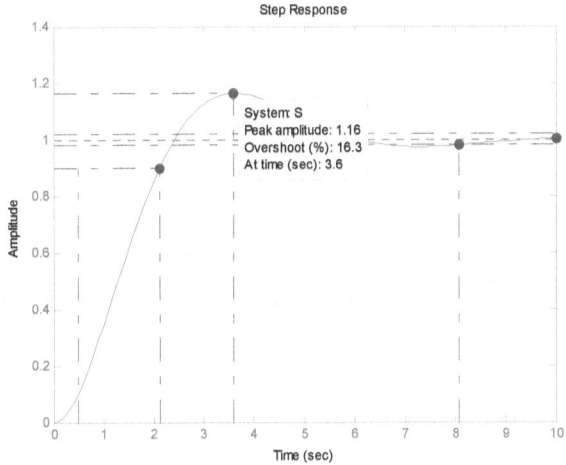

Figure 4.6(b) Step response of the prototype second order system

🗎 Readymade MATLAB solution

This approach is better if you seek for a single solution from graphical interface. Consider the prototype second order system in figure 4.6(a) with $\zeta = 0.5$, $\omega_n = 1\ rad/sec$, and $H(s) = 1$. We wish to determine the T_r, T_p, P.O., T_s, and y_{ss} subject to $u(t) = 1\ V$ over $0 \leq t \leq 10\ \sec$.

In section 4.5 we demonstrated step response both in computing and graphing forms. Since $u(t) = 1$, we may apply the **step** in order to graph the response but that will be applied to the feedback equivalent (section 2.4) and do so by:

>>z=0.5; w=1; ↵   ← z⇔ζ, w⇔ω_n, z,w are user-chosen
>>G=tf(w^2,[1 2*z*w 0]); ↵   ← G⇔$G(s)$, G is user-chosen
>>H=1; ↵   ← H⇔$H(s)$, H is user-chosen
>>S=feedback(G,H,-1); ↵   ← S is user-chosen, holds the equivalent

From the last execution we have the equivalent available in **S** on which the **step** is applied:

>>step(S,10) ↵

The response is the figure 4.6(b). Bring your mouse pointer on the graph and rightclick the mouse, you must find **Characteristics** in the popup window. Again under the **Characteristics**, you find four options namely **Peak Response, Settling Time, Rise Time**, and **Steady State**. Check each characteristic one at a time by the mouse, the result of which is the figure 4.6(b). There are four bold dots in the figure, each of which indicates one of the four characteristics. Click each bold dot to see its corresponding characteristic. For instance the shown one is for the P.O. What can we read out of the figure? The answer is

    P.O. = 16.3% which occurs at $t = 3.6$ secs i.e. $T_p = 3.6$ secs and $y(t)|_{\max} = 1.16\ V$.

Similarly clicking the other three bold dots, we get:

    $T_r = 1.64$ secs,
    $T_s = 8.08$ secs, and
    $y_{ss} = 1\ V$.

The default setting for settling time T_s is based on 2% criterion. If you wish to change that, bring mouse pointer on the figure, rightclick the mouse, find **Properties** in a popup window, find again **Characteristic** option under **Properties**, and change according to the requirement. For example with 2.7% criterion we get $T_s = 5.59$ secs.

    Again the default rise time T_r is based on 10%-90% transition. Similar to T_s, you can change the **Characteristic** for T_r as well. For example 20%-80% transition provides $T_r = 1.16$ secs. The indicatory box may turn the figure clumsy. You can remove the box by **Delete** followed by rightclick.

Control System Analysis & Design in MATLAB and SIMULINK

⚙ User-defined programming solution

The obstacle with the built-in **step** is if the input other than 1, the graphics does not suit the problem. Besides for multiple systems we can not obtain the second order responses. All second order responses we find now by programming means.

Consider the prototype second order system in figure 4.6(a) with $\zeta = 0.4$, $\omega_n = 2\ rad/\sec$, and $H(s) = 0.5$. We wish to determine the T_r, T_p, P.O., T_s, and y_{ss} subject to $u(t) = 6V$ over $0 \le t \le 5 \sec$.

Form the system as we did before:
>>z=0.4; w=2; G=tf(w^2,[1 2*z*w 0]); H=0.5; S=feedback(G,H,-1); ↵

Choose a $\Delta t = 0.01$ sec and generate the input samples as a row matrix:
>>t=0:0.01:5; u=6*ones(1,length(t)); ↵

Find the output (section 4.3) samples:
>>y=lsim(S,u,t); ↵

Finding the y_{ss}:

You can view the $y(t)$ versus t by exercising the command **plot(t,y)**. The output may or may not seem to have reached steady state. For the example at hand, the output did not. If not, increase the t bound say 20 and reexecute t=0:0.01:20; u=6*ones(1,length(t)); y=lsim(S,u,t);. View again the output by **plot(t,y)** which indicates steady state's occurrence. The y_{ss} is the last sample so get it by:

>>y_ss=y(end) ↵   ← y_ss⇔ y_{ss}, y_ss is user-chosen

y_ss =
        12.0000

Finding the peak amplitude and P.O.:

The peak amplitude is basically the maximum of $y(t)$ i.e. $y(t)|_{\max}$. We obtain that by **max** of appendix D.5:

>>y_max=max(y) ↵   ← y_max⇔ $y(t)|_{\max}$, y_max is user-chosen

y_max =
        13.3907

When T_p is needed, time corresponding to $y(t)|_{\max}$ is to be found. We need the index of maximum occurring so find it by two output arguments on **max**:

>>[y_max,I]=max(y); ↵   ← I is user-chosen, holds index for T_p
>>Tp=t(I) ↵   ← Tp⇔ T_p, Tp is user-chosen

Tp =
        2.6900

For the P.O. we use the expression $(y(t)|_{\max} - y_{ss})/y_{ss} \times 100$ whose code is **(y_max-y_ss)/y_ss*100** and which returns 11.5888%.

Finding the T_r:

Rise time T_r requires time for particular value crossing on a function for which author written **c_cross** can be applied (appendix D.8). The user has to decide percentage crossing of steady state value for example 20%-80%. We determine the times for 20% and 80% on steady state value (say t_1 and t_2 respectively) and subtraction of the two time values provide T_r i.e. $T_r = t_2 - t_1$.

For the ongoing example the two times followed by rise time we determine by:

```
>>t1=c_cross(t,y,0.2*y_ss); ↵    ← t1⇔ t₁, t1 is user-chosen
>>t2=c_cross(t,y,0.8*y_ss); ↵    ← t2⇔ t₂, t2 is user-chosen
>>Tr=t2-t1 ↵                     ← Tr⇔ T_r, Tr is user-chosen

ans =
     0.8900
```

Finding the T_s:

The settling time T_s needs some percentile criterion say 2%. It means we look for $0.98\,y_{ss}$ or $1.02\,y_{ss}$ whichever occurs first and pick the time corresponding to it. We have to exercise again the **c_cross** for time point determining:

```
>>ts1=c_cross(t,y,0.98*y_ss) ↵    ← Time for 0.98 y_ss crossing

ts1 =
    1.8050
>>ts2=c_cross(t,y,1.02*y_ss) ↵    ← Time for 1.02 y_ss crossing

ts2 =
    1.9250   4.1550
```

In above executions the **ts1** or **ts2** is user-chosen variable. The settling phenomenon becomes clear if we graph the response by appendix E mentioned plot:

```
>>plot(t,y,t,1.02*y_ss,t,0.98*y_ss) ↵
```

Figure 4.6(c) presents the step response of the prototype system for nonunity input and angular frequency. An arrow in the figure also indicates the settling point. The last time value in **ts2** corresponds to settling point which we extract by the command

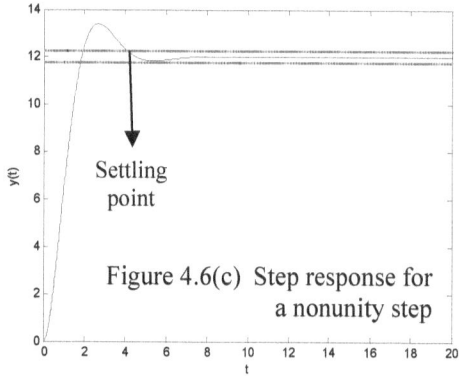

Figure 4.6(c) Step response for a nonunity step

ts2(end) i.e.:
>>Ts=ts2(end) ↵

Ts =
        4.1550

Finally we have T_s=4.155 secs. It is important to point out that irregular or nonstandard situations need computing and graphing hand in hand.

4.8 Error performance indices

Error performance of a control system requires the error signal be available in time domain. The error signal $e(t)$ we can have from s domain expression i.e. $E(s)$ or from time domain samples. We talk first about error signal sample obtaining from $E(s)$ and then about the performance indices in this section.

Let us consider the error signal on prototype second order system in figure 4.6(a). The signal is labeled by $E(s)$ in the figure. At this point there are two types of error signal; actuating error signal ($E(s)$) and system or overall error signal ($E_o(s)$) which are given by:

$$E_o(s) = \frac{[1+G(s)H(s)-G(s)]R(s)}{1+G(s)H(s)} \text{ and}$$

$$E(s) = \frac{R(s)}{1+G(s)H(s)}.$$

Think about the $E(s)$. Basically we have transfer function $\frac{1}{1+G(s)H(s)}$ considering $R(s)$ as input and $E(s)$ as output.

Choosing ζ =0.4, ω_n=2 $rad/\sec$, and $H(s) = 0.5$, let us form $G(s)$ and $H(s)$ as follows:

>>z=0.4; w=2; G=tf(w^2,[1 2*z*w 0]); H=0.5; ↵ ← Last section for symbology

Now any summation and multiplication of transfer functions we implement by **parallel** and **series** of section 2.3 respectively and exercise so by:

>>T1=parallel(1,series(G,H)); ↵ ← T1 is user-chosen, holds $1+G(s)H(s)$

The $\frac{1}{1+G(s)H(s)}$ can be written as reciprocal transfer function of $1+G(s)H(s)$ which we form (appendix D.9) by:

>>[n,d]=tfdata(T1,'v'); ↵   ← n,d are user-chosen, n holds numerator, d holds denominator of $1+G(s)H(s)$

>>T=tf(d,n); ↵   ← T is user-chosen, holds reciprocal of $1+G(s)H(s)$

What about the input ($R(s)$ is the Laplace transform of $u(t)$)? say $u(t) = 6V$ on Δt =0.01 sec over $0 \le t \le 5$ sec s. Carry out the repetition of last section command:

```
>>t=0:0.01:5; u=6*ones(1,length(t)); ↵
```
At last error signal samples we obtain by (section 4.3 for lsim):
```
>>e=lsim(T,u,t); ↵          ← e is user-chosen, holds e(t) or E(s) samples
```

If you wish to see the $e(t)$ versus t trace, exercise **plot(t,e)** at the command prompt (appendix E). The graph is not shown for space reason. From the graph one infers that the $e(t)$ does not seem to be in steady state. Increase the interval say 20 and reexecute the following:
```
>>t=0:0.01:20; u=6*ones(1,length(t)); ↵
>>e=lsim(T,u,t); ↵
```
Now execution of **plot(t,e)** indicates steady state behavior of $e(t)$. For steady state error, you may exercise **e(end)**.

Having found the error signal $e(t)$, one might be interested in error performance indices. There are four error performance indices which are defined as follows:

integral of absolute error, IAE=$\int_0^T |e(t)|\, dt$,

integral of time multiplied absolute error, ITAE=$\int_0^T t|e(t)|\, dt$,

integral of time multiplied square error, ITSE=$\int_0^T te^2(t)\, dt$, and

integral of square error, ISE=$\int_0^T e^2(t)\, dt$.

In all error indices upper limit T is user-defined. For the numerical integration we may apply **trapz** of appendix D.10 assuming that the samples of t and $e(t)$ are available (in **t** and **e** respectively).

For the last prototype second order system choosing ISE, the index is calculated as:
```
>>trapz(t,e.^2) ↵

ans =
      25.6500
```
From the above return we can say ISE=$\int_0^{20} e^2(t)\, dt$ =25.65.

▣ Comparison on error index

Sometimes we look for graphical variation of index for different damping ratio over a common time interval. In most control textbooks the comparison is for $\omega_n = 1\ rad/\sec$ and $H(s)=1$. Let us choose the damping ratio variation $0.05 \leq \zeta \leq 2$ with a step $\Delta\zeta = 0.05$ for the ongoing prototype second order system.

Suppose we wish to plot the IAE and ISE indices against $0.05 \leq \zeta \leq 2$ by choosing $T = 20$ secs. It means for every ζ, we have to conduct earlier commands and get the index out of those. The ζ variation we may exercise by a for loop (appendix D.4). In the appendix we did not include the following approach. If we have a vector and the vector follows the loop counter index, the loop counter index sequentially assumes the vector elements. For multiple statements it is better to use a script file (section 1.1.2). Regarding the index IAE or ISE accumulation, we may apply the technique of appendix D.3.

Figure 4.7(a) shows the complete code for graphical output. In the file the variables **zeta**, **ISE**, etc are user-chosen. Save the codes in a script file and run the file. After running the file, you see the output as shown in figure 4.7(b). In order to view the mark of identification, we exercise the command **legend('IAE','ISE')** at the command prompt. The two words under **legend** are user-supplied in order and under quote. The text is written in black and white form that is why the distinction is not evident in the figure, you find the

```
zeta=0.05:0.05:2;
ISE=[ ];
IAE=[ ];
for z=zeta
    w=2; G=tf(w^2,[1 2*z*w 0]); H=1;
    T1=parallel(1,series(G,H));
    [n,d]=tfdata(T1,'v');
    T=tf(d,n);
    t=0:0.01:20; u=ones(1,length(t));
    e=lsim(T,u,t);
    ISE=[ISE trapz(t,e.^2)];
    IAE=[IAE trapz(t,abs(e))];
end
    plot(zeta,IAE,zeta,ISE)
```

Figure 4.7(a) Script file for index comparison of a prototype second order system

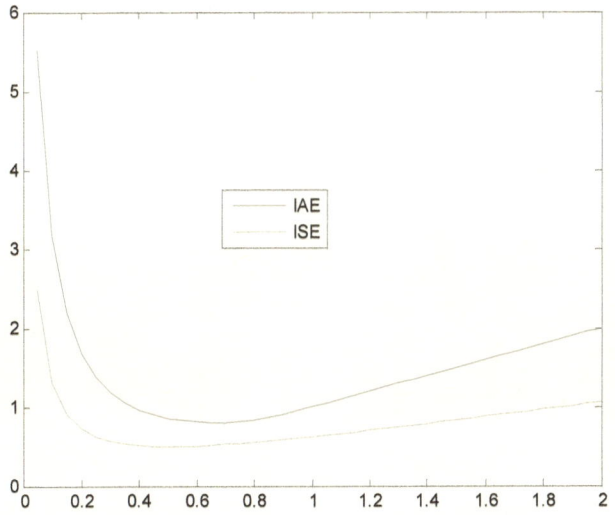

Figure 4.7(b) Comparison of IAE and ISE indices of the prototype second order system

distinction through color in MATLAB.

In last figure the horizontal and vertical axes refer to $0.05 \leq \zeta \leq 2$ and IAE or ISE index respectively.

For the indices ITAE and ITSE you need the commands trapz(t,t.*abs(e)) and trapz(t,t.*e.^2) respectively. Since scalar multiplication occurs on like dimension matrix, the t=0:0.01:20; should be replaced by t=[0:0.01:20]'; for ITAE or ITSE.

However each index manifests a minimum which might be of interest for design reason. Appendix D.5 mentioned min helps us determine the ζ corresponding to the minimum. For example the IAE gets the minimum at $\zeta = 0.65$ which we determine as follows:

>>[IAE_min,l]=min(IAE); ↵   ← IAE_min or l is user-chosen

The zeta matrix position index (i.e. integer l) also refers to $0.05 \leq \zeta \leq 2$ of figure 4.7(a) hence:

>>zeta(l) ↵

ans =
    0.6500

Similarly the ISE gets the minimum at $\zeta = 0.5$ by dint of the following commands:

>>[ISE_min,l]=min(ISE); ↵   ← ISE_min or l is user-chosen
>>zeta(l) ↵

ans =
    0.5000

Should you need the $ISE|_{min}$, just call the ISE_min:

>>ISE_min ↵

ans =
    0.6500

With the minimum index we bring an end to this chapter.

Exercises

1. Write MATLAB code for each of the following control inputs so that the samples are available in standard unit over the given interval and step size:
   (a) voltage input $u(t) = -0.7t + 6$ V with step 0.1 sec over $-1 \le t \le 3$ secs,
   (b) force input $u(t) = |-0.7t + 6|$ N with step 0.1 sec over $-1 \le t \le 3$ secs,
   (c) current input $u(t) = 4.4$ mA with step 0.1 msec over $0 \le t \le 2$ msec s,
   (d) parabolic current input $u(t) = 2 - 0.2t + 0.7t^2$ A with step 0.1 sec over $0 \le t \le 2$ secs,
   (e) damped sine input $u(t) = 3.1e^{-0.5t} \sin(1.5t + 34°)$ with step 0.1 sec over $0 \le t \le 3$ secs,
   (f) a sine input with harmonics which is composed of fundamental 60 Hz along with the 3rd and 5th harmonics and their amplitudes are 100 V, 20 V, and 1 V with step 0.0015 sec over $0 \le t \le \frac{1}{30}$ secs respectively, and
   (g) a displacement input $u(t) = 5\cos te^{-t}$ mm on $\Delta t = 1$ msec over $0 \le t \le 0.1$sec s.

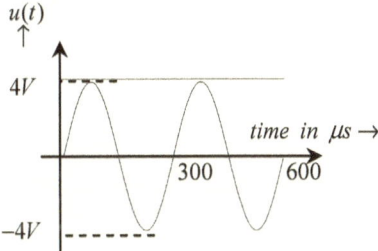

Figure E.4(1) A two cycle sine wave

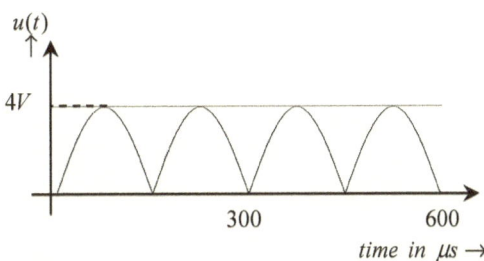

Figure E.4(2) A full rectified sine wave

2. (a) generate the samples of the two cycle sine wave in figure E.4(1) considering Δt at least one hundredth of one period (b) do the same as in part (a) for the full rectified sine wave of figure E.4(2) (c) do the same as in part (a) for the half rectified sine wave of figure E.4(3).

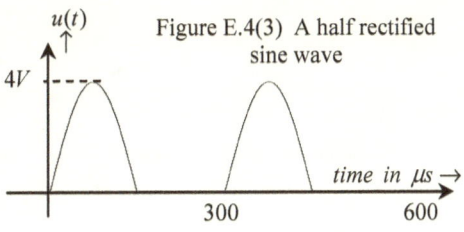

Figure E.4(3) A half rectified sine wave

3. Generate the samples of each following square wave on $\Delta t = 0.001$ sec over the interval $-0.1 \le t \le 0.1$ secs: (a) frequency 20 Hz, amplitude $\pm 2.5 V$, and duty cycle 50% (b) frequency 20 Hz, amplitude swing $2V$ to $7V$, and duty cycle 50% (c) frequency 20 Hz, amplitude swing $2V$ to $7V$, and duty cycle 70%.

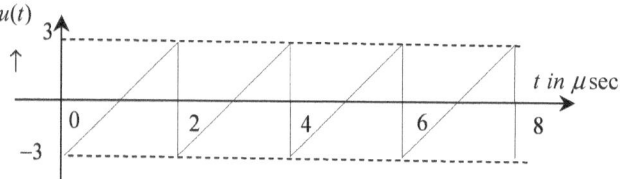

Figure E.4(4) A sawtooth wave

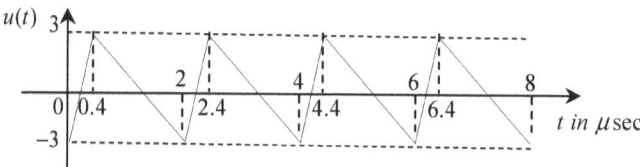

Figure E.4(5) Periodic triangular wave with peak inbetween the period

4. Generate the samples of each following triangular wave on Δt being one hundredth of the period: (a) as in figure E.4(4) (b) as in figure E.4(5).
5. A control system is given as shown in figure 4.3(a). Determine the output $y(t)$ samples for each of the following: (a) $H(s) = \dfrac{3s}{s^2+1}$ and $u(t) = 5$ on $\Delta t = 0.02$ sec over $0 \le t \le 0.1$ secs (b) $H(s) = \{A, B, C, D\}$ where $A = \begin{bmatrix} -1 & -2 & 0 \\ 2 & 3 & 2 \\ 4 & 2 & 0 \end{bmatrix}$, $B = \begin{bmatrix} 1 \\ -2 \\ 3 \end{bmatrix}$, $C = [5 \ -2 \ 3]$, and $D = [8]$ and $u(t) = t^2$ on $\Delta t = 0.5$ sec over $0 \le t \le 2$ secs. Graph $y(t)$ versus t for the samples of parts (a) and (b).
6. A MIMO control system has one input ($u(t)$) and three outputs ($y_1(t)$, $y_2(t)$, $y_3(t)$). The MIMO has the transfer function $H(s) = \begin{bmatrix} H_1(s) \\ H_2(s) \\ H_3(s) \end{bmatrix}$ where $H_1(s) = \dfrac{5}{7s^2+1}$, $H_2(s) = \dfrac{2s+5}{7s^2+1}$, and $H_3(s) = \dfrac{5s}{7s^2+1}$. Determine the output samples subject to $u(t) = 3$ on $\Delta t = 0.2$ sec over $0 \le t \le 1$ sec. Also graph the outputs.

7. A MIMO control system has three inputs ($u_1(t), u_2(t), u_3(t)$) and one output ($y(t)$). The MIMO has the transfer function $H(s) = [H_1(s) \quad H_2(s) \quad H_3(s)]$ where $H_1(s) = \dfrac{5}{7s^2+1}$, $H_2(s) = \dfrac{2s+5}{7s^2+1}$, and $H_3(s) = \dfrac{5s}{7s^2+1}$. Determine the output samples subject to $u_1(t)=3$, $u_2(t)=2t-1$, and $u_3(t)=e^{-3t}$ on $\Delta t = 0.2$ sec over $0 \le t \le 1$ sec. Also graph the output.

8. A MIMO control system has three inputs ($u_1(t), u_2(t), u_3(t)$) and two outputs ($y_1(t), y_2(t)$). The MIMO has the transfer function $H(s) = \begin{bmatrix} H_{11}(s) & H_{12}(s) & H_{13}(s) \\ H_{21}(s) & H_{22}(s) & H_{23}(s) \end{bmatrix}$ where $H_{11}(s) = \dfrac{7}{2s^2+3}$, $H_{12}(s) = \dfrac{2s}{2s^2+3}$, $H_{21}(s) = \dfrac{2}{2s^2+3}$, $H_{22}(s) = \dfrac{3s+5}{2s^2+3}$, $H_{13}(s) = \dfrac{5}{2s^2+3}$, and $H_{23}(s) = \dfrac{3s+1}{2s^2+3}$. Determine the output samples subject to $u_1(t)=3$, $u_2(t)=2t-1$, and $u_3(t)=e^{-3t}$ on $\Delta t = 0.3$ sec over $0 \le t \le 1.2$ secs. Also graph the outputs.

9. A two input three output MIMO is given by $H(s) = \{A, B, C, D\}$ where $A = \begin{bmatrix} -2 & 1 \\ 2 & 1 \end{bmatrix}$, $B = \begin{bmatrix} -1 & 4 \\ -4 & 1 \end{bmatrix}$, $C = \begin{bmatrix} -1 & 1 \\ 2 & 3 \\ 0 & 4 \end{bmatrix}$, and $D = \begin{bmatrix} -1 & 4 \\ 2 & 0 \\ 3 & 5 \end{bmatrix}$. Say the inputs and outputs are labeled as $\{u_1(t), u_2(t)\}$ and $\{y_1(t), y_2(t), y_3(t)\}$ respectively. Determine the output samples subject to $u_1(t)=3t-1$ and $u_2(t)=0.3e^{-t}$ on $\Delta t = 0.1$ sec over $0 \le t \le 0.4$ sec s. Also graph the outputs.

10. Determine the step response samples of control system $H(s) = \dfrac{3s}{s^2+1}$ on $\Delta t = 1$ sec over $0 \le t \le 3$ sec s. Also graph the response. Determine the impulse response of the system and graph the impulse response too.

11. Graph the step responses of control systems $H_1(s) = \dfrac{3s}{s^2+1}$, $H_2(s) = \dfrac{3s+1}{2s^2+1}$, and $H_3(s) = \dfrac{3s}{3s^2+1}$ together on $\Delta t = 1$ sec over $0 \le t \le 3$ sec s. Do the same but for impulse response.

12. A first order control system is given by $H(s) = \dfrac{8}{6s+9}$. Determine its steady state output value and time constant subject to $u(t) = 0.8$.

13. A second order control system as in figure 4.6(a) is given by $G(s) = \dfrac{\omega_n^2}{s(s+2\zeta\omega_n)}$ where $\zeta = 0.32$, $\omega_n = 1.3 \ rad/sec$, and $H(s) = 0.95$. Use MATLAB graphical interface to determine (a) steady state output y_{ss}, (b) 20%-70% rise time T_r, (c) peak time T_p, (d) output maximum $y(t)|_{max}$, (e) percent overshoot P.O., and (f) 2.3% settling time T_s subject to unity step input. Verify these second order performance parameters by applying code writing approach.

14. Determine the following actuating error performance indices of second order system in question (13) on $\Delta t = 0.02$ sec over $0 \le t \le 16\sec$ s: (a) ISE (b) ITSE (c) IAE (d) ITAE. Also graph the IAE and ITAE indices on $\Delta \zeta = 0.05$ over $0.05 \le \zeta \le 3.5$. In the last graph determine the ζ for each minimum index.

Answers:

(1) (a) t=-1:0.1:3; u=-0.7*t+6; (b) in part (a) u=abs(-0.7*t+6); (c) t=[0:0.1:2]*1e-3; u=4.4e-3*ones(1,length(t)); (d) t=0:0.1:2; u=2-0.2*t+0.7*t.^2; (e) t=0:0.1:3; u=3.1*exp(-0.5*t).*sin(1.5*t+pi/180*34); (f) t=0:0.0015:1/30; u=100*sin(2*pi*60*t)+20*sin(2*pi*180*t)+sin(2*pi*300*t); (g) t=0:1e-3:0.1; u=5e-3*cos(t).*exp(-t);
hint: section 4.1

(2) (a) t=0:300e-6/100:600e-6; u=4*sin(2*pi/300e-6*t); (b) in part (a) now u=abs(4*sin(2*pi/300e-6*t)); (c) r=find(u<0); u(r)=0; after part (a)
hint: section 4.2

(3) (a) t=-0.1:0.001:0.1; u=2.5*square(2*pi*20*t); (b) t=-0.1:0.001:0.1; u= 9/2+5/2*square(2*pi*20*t); (c) t=-0.1:0.001:0.1; u=9/2+5/2*square(2*pi*20*t,70);
hint: section 4.2

(4) (a) t=[0:2/100:8]*1e-6; u=3*sawtooth(2*pi/2e-6*t); (b) t=[0:2/100:8]*1e-6; u=3*sawtooth(2*pi/2e-6*t,0.4/2); hint: section 4.2

(5) Following tables for samples and figures E.4(6) and E.4(7) respectively. In figure E.4(7) the upper curve is the $y(t)$. hint: section 4.3

(a) Commands needed:
H=tf([3 0],[1 0 1]);
t=[0:0.02:0.1]';
u=5*ones(length(t),1);
y=lsim(H,u,t);
[t y]	

t	$y(t)$
0	0
0.02	0.3
0.04	0.5998
0.06	0.8995
0.08	1.1987
0.1	1.4975

(b) Commands needed: A=[-1 -2 0;2 3 2;4 2 0]; B=[1;-2;3]; C=[5 -2 3]; D=8; H=ss(A,B,C,D); t=[0:0.5:2]'; u=t.^2; y=lsim(H,u,t); [t y]	
t	$y(t)$
0	0
0.5	3.1909
1	14.1197
1.5	25.2506
2	-23.0945

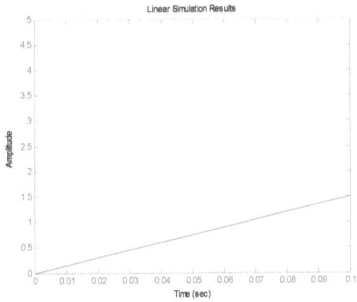

Figure E.4(6) $y(t)$ versus t for control system in problem 5(a)

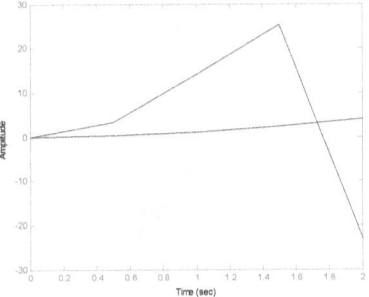

Figure E.4(7) $y(t)$ versus t for control system in problem 5(b)

(6) N1=5; D1=[7 0 1]; N2=[2 5]; D2=D1; N3=[5 0]; D3=D1; H=tf({N1;N2;N3},{D1;D2;D3}); t=0:0.2:1; u=3*ones(1,length(t)); y=lsim(H,u,t); [t' y]
Samples are in the following table:

t	$y_1(t)$	$y_2(t)$	$y_3(t)$
0	0	0	0
0.2	0.0428	0.2141	0.4282
0.4	0.1711	0.5127	0.8539
0.6	0.3841	0.8940	1.2747
0.8	0.6805	1.3558	1.6883
1	1.0587	1.8956	2.0922

Figure E.4(8) for the graph.   hint: section 4.4

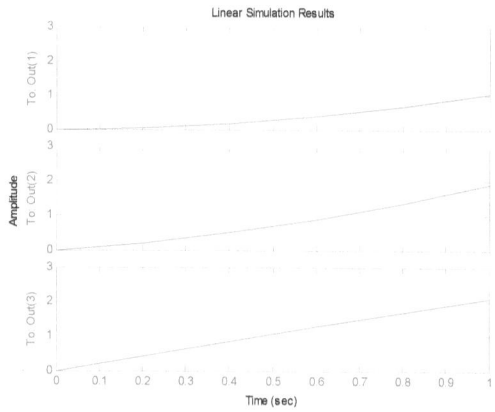

Figure E.4(8) Single input triple output MIMO outputs

(7) N1=5; D1=[7 0 1]; N2=[2 5]; D2=D1; N3=[5 0]; D3=D1; H=tf({N1,N2,N3},{D1,D2,D3}); t=[0:0.2:1]'; u1=3*ones(length(t),1); u2=2*t-1; u3=exp(-3*t); u=[u1 u2 u3]; y=lsim(H,u,t); [t y]. Figure E.4(9) for the graph.
Samples are in the following table:

t	0	0.2	0.4	0.6	0.8	1
y(t)	0	0.0953	0.2316	0.4419	0.7490	1.1691

hint: section 4.4

Figure E.4(9) Triple input single output MIMO output – right side figure

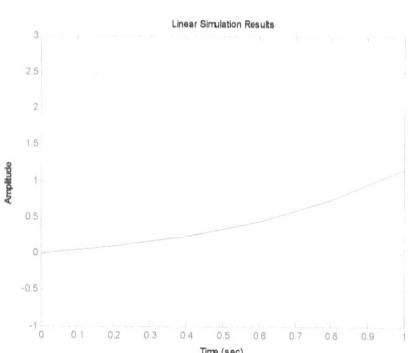

(8) N11=7; N12=[2 0]; N13=5; N21=2; N22=[3 5]; N23=[3 1]; D11=[2 0 3];
D12=D11; D13=D11; D21=D11; D22=D11; D23=D11;
H=tf({N11,N12,N13;N21,N22,N23},{D11,D12,D13;D21,D22,D23});
t=[0:0.3:1.2]'; u1=3*ones(length(t),1); u2=2*t-1; u3=exp(-3*t); u=[u1 u2 u3];
y=lsim(H,u,t); [t y]
Figure E.4(10) for the graph. Samples are in the following table:

t	0	0.3	0.6	0.9	1.2
$y_1(t)$	0	0.352	1.8731	4.2986	7.2798
$y_2(t)$	0	0.0639	0.3992	1.1926	2.5115

hint: section 4.4

Figure E.4(10)
Three input two output MIMO output – right side figure

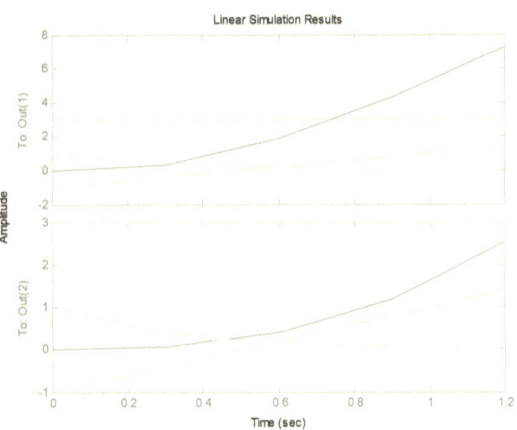

(9) A=[-2 1;2 1]; B=[-1 4;-4 1]; C=[-1 1;2 3;0 4]; D=[-1 4;2 0;3 5]; H=ss(A,B,C,D);
t=[0:0.1:0.4]'; u1=3*t-1; u2=0.3*exp(-t); u=[u1 u2]; y=lsim(H,u,t); [t y]. Figure E.4(11) for the graph.

Figure E.4(11)
Two input three output MIMO output – right side figure

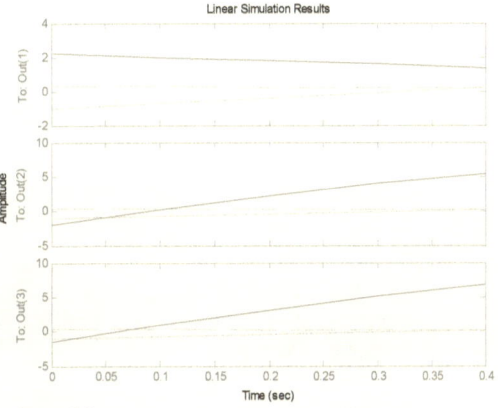

Samples are in the following table:

t	0	0.1	0.2	0.3	0.4
$y_1(t)$	2.2	1.9963	1.7937	1.5758	1.3270
$y_2(t)$	-2	0.2292	2.238	3.9949	5.4621
$y_3(t)$	-1.5	0.8974	3.1165	5.1060	6.8112

hint: section 4.4

(10) For step: H=tf([3 0],[1 0 1]); t=[0:3]'; y=step(H,t); [t y], step(H,t)
For impulse: y=impulse(H,t); [t y], impulse(H,t)
Samples are in the following table:

	t	0	1	2	3
Step	y(t)	0	2.5244	2.7279	0.4234
Impulse	y(t)	3	1.6209	-1.2484	-2.97

Figures E.4(12) and E.4(13) for the graphs respectively.
hint: section 4.5

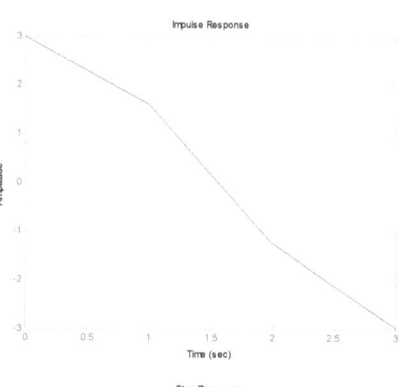

Figure E.4(12) Step response of control system in problem (10) – right side figure

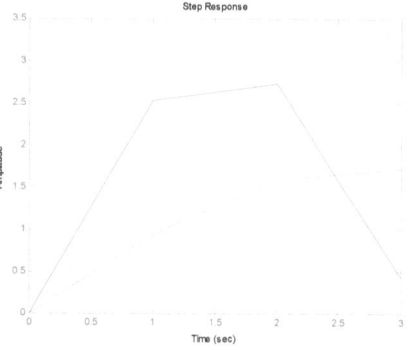

Figure E.4(13) Impulse response of control system in problem (10) – right side figure

Figure E.4(14) Step responses of three control systems in problem (11) – right side figure

Figure E.4(15) Impulse responses of three control systems in problem (11) – right side figure

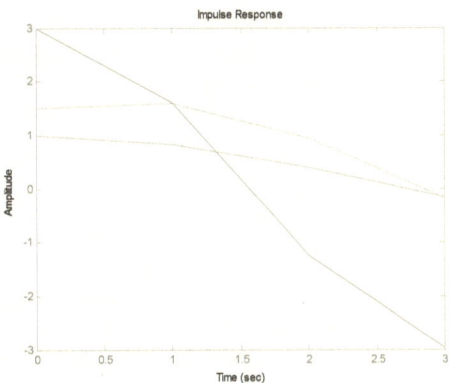

(11) Figures E.4(14) and E.4(15) respectively.
   Commands: H1=tf([3 0],[1 0 1]); H2=tf([3 1],[2 0 1]); H3=tf([3 0],[3 0 1]);
       t=[0:3]'; step(H1,H2,H3,t) or impulse(H1,H2,H3,t)
   hint: section 4.5

(12) By trial and error we chose time resolution $\Delta t = 0.01$ sec, time interval $0 \le t \le 10$ sec s, and $t_0 = 1.5$ secs.

   Commands: H=tf(8,[6 9]); t=0:0.01:10; u=0.8*ones(1,length(t));
   y=lsim(H,u,t); y_ss=y(end)
   to=1.5; r1=find(t==to); r2=find(t==2*to); tm=-to/log(y(r2)/y(r1)-1)
   $y_{ss} = 0.7111$ and $\tau = 0.6667$ secs   hint: section 4.6

(13) By choosing resolution $\Delta t = 0.01$ sec over $0 \le t \le 20$ sec s: (a) $y_{ss} = 1.0526$ (b) $T_r = 0.641$ secs (c) $T_p = 2.62$ secs (d) $y(t)|_{max} = 1.4058$ (e) P.O.=33.6% and

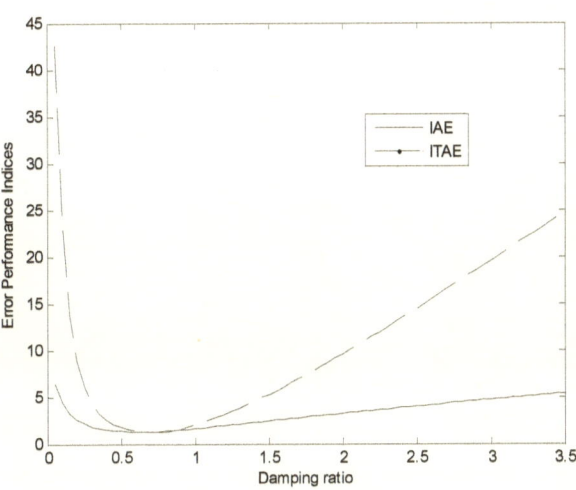

Figure E.4(16) Error performance indices versus damping ratio for IAE and ITAE

(f) T_s =8.68 secs

hint: section 4.7

(14) (a) ISE=0.8601 (b) ITSE=0.7894 (c) IAE=1.7412 (d) ITAE=3.8156

Figure E.4(16)

$IAE|_{min}$=1.2669 when ζ =0.65

$ITAE|_{min}$=1.2183 when ζ =0.75

Commands for the indices:
```
w=1.3; z=0.32; G=tf(w^2,[1 2*z*w 0]); t=[0:0.02:16]'; H=0.95;
u=ones(1,length(t)); T1=parallel(1,series(G,H)); [n,d]=tfdata(T1,'v');
T=tf(d,n); e=lsim(T,u,t); ISE=trapz(t,e.^2), IAE=trapz(t,abs(e)),
ITSE=trapz(t,t.*e.^2), ITAE=trapz(t,t.*abs(e))
```

Commands for the graph:
```
zeta=0.05:0.05:3.5; t=[0:0.02:16]'; u=ones(1,length(t)); w=1.3;
H=0.95; ITAE=[ ]; IAE=[ ];
for z=zeta
  G=tf(w^2,[1 2*z*w 0]);
  T1=parallel(1,series(G,H));
  [n,d]=tfdata(T1,'v');
  T=tf(d,n);
  e=lsim(T,u,t);
  ITAE=[ITAE trapz(t,t.*abs(e))];
  IAE=[IAE trapz(t,abs(e))];
end
plot(zeta,IAE,zeta,ITAE)
legend('IAE','ITAE')
[IAE_min,I]=min(IAE)
zeta(I)
[ITAE_min,I]=min(ITAE)
zeta(I)
```

hint: section 4.8

Chapter 5

Time Domain Control System in SIMULINK

SIMULINK blocks are so effective and well-designed that our computer virtually turns to an analog machine although behind the simulation the data processing is discrete. The hidden idea in SIMULINK block is input-output data processing, most control system topics follow so. For this reason control system modeling best be exercised in SIMULINK. Chapter 3 addresses only control system modeling. In this chapter we bring in basically modeling of input to and output from the control system. Anyhow our implementations outline the following:

- ✦ ✦ Signal modeling basics in SIMULINK including its derivative
- ✦ ✦ SISO/MIMO control system output in SIMULINK
- ✦ ✦ Control system performance parameter modeling in SIMULINK

5.1 Control input modeling in SIMULINK

In chapter 4 mostly we wrote the code of a control input and generated the input/output samples at command window of MATLAB. SIMULINK keeps provision for input generation employing modeling approach. The reader needs SIMULINK basics which are addressed in chapter 1. From a given input characteristic we pick information or its parameters and enter those parameters to pre-designed SIMULINK blocks to generate the input in continuous sense despite hidden discrete generation. In

MATLAB we view an input by command **ezplot** or **plot** contrarily the block **Scope** (subsection 1.2.5) of SIMULINK exhibits the input so generated in a model. Following sections are all about control system input modeling.

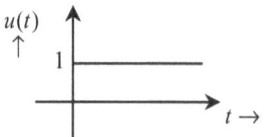

Figure 5.1(a) Unit step function

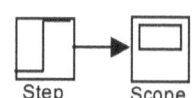

Figure 5.1(b) **Step** block connected with **Scope**

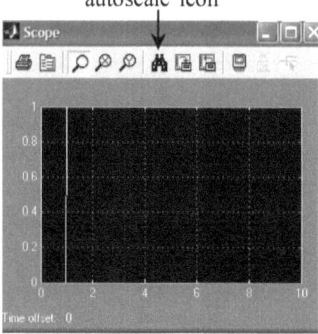

Figure 5.1(c) **Scope** output for default **Step** block

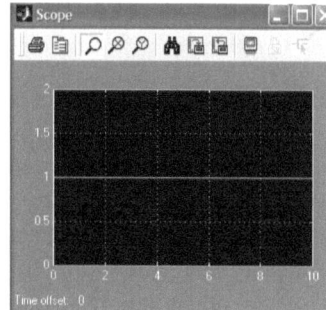

Figure 5.1(d) **Scope** output for $u(t)$

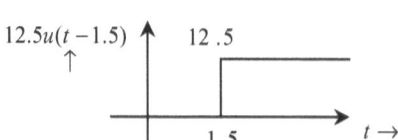

Figure 5.1(e) Unit step function shifted at 1.5 and of final value 12.5

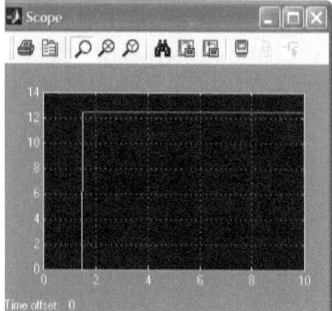

Figure 5.1(f) **Scope** output for $12.5u(t-1.5)$

5.2 Modeling unit step and its derivative inputs

The unit step or Heaviside function $u(t) = \begin{cases} 1 & \text{for } t \geq 0 \\ 0 & \text{for } t < 0 \end{cases}$ which has the plot in figure 5.1(a) is simulated by **Step** (appendix C for outlook and link) block. Open a new SIMULINK model file (subsection 1.2.2), bring one **Step** and one **Scope** blocks in the model file, connect (subsection 1.2.3) the blocks as shown in figure 5.1(b), run the model by clicking ▶ icon in the menu bar, and doubleclick the **Scope**. The autoscale icon as indicated in figure 5.1(c) has some default settings both in horizontal and vertical axes of the **Scope**. Clicking the autoscale icon adapts the default axes setting to current one. On clicking the autoscale icon we find the **Scope** output as seen in figure 5.1(c). The displayed output has the step value at 1 or functionally the default return is $u(t-1)$ but the $u(t)$ or figure 5.1(a) has the step value 0. Doubleclick the **Step** block to see its parameter window like figure 5.1(g), change the **Step time** from default 1 to 0 in the parameter window, run the model, and doubleclick the **Scope** to see the correct unit step function with autoscale setting like figure 5.1(d) i.e. figure 5.1(d) shown output refers to $u(t)$ of figure 5.1(a).

As another example the input $12.5u(t-1.5)$ has the plot in figure 5.1(e) which has final value 12.5 and is shifted at 1.5 to the right on the time axis. We need to enter the setting in the parameter

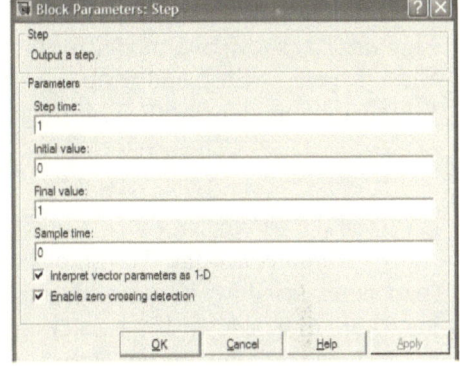

Figure 5.1(g) Block parameter window of the **Step**

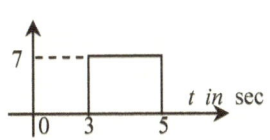

Figure 5.1(h) A finite duration pulse

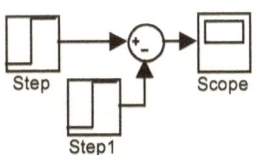

Figure 5.1(i) Two **Step** blocks model the signal in figure 5.1(h)

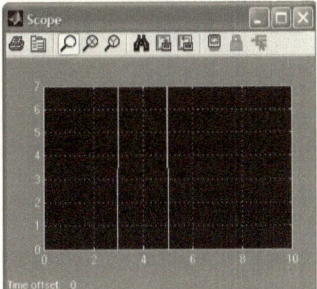

Figure 5.1(j) **Scope** output for $7u(t-3) - 7u(t-5)$

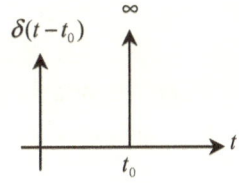

Figure 5.1(k) Ideal Dirac delta function

window of figure 5.1(g) as $\begin{Bmatrix} \text{Step value to 1.5} \\ \text{Final value to 12.5} \end{Bmatrix}$ and the **Scope** output with autoscale setting is shown in figure 5.1(f).

A finite duration constant value signal on and off is used in control system whose generation needs using two **Step** blocks. Figure 5.1(h) presents a finite duration constant signal. Mathematically we read off the signal as $7u(t-3) - 7u(t-5)$. The components $7u(t-3)$ and $7u(t-5)$ are simulated by two different **Step** blocks. The subtraction of $7u(t-5)$ from $7u(t-3)$ takes place by the **Sum** block (appendix C for outlook and link). Figure 5.1(i) shows the model for the finite pulse in figure 5.1(h). Bring two **Step** (appear as **Step** and **Step1**), one **Sum**, and one **Scope** blocks in a new SIMULINK model file. Doubleclick each of the **Step** blocks at a time and enter the settings $\begin{Bmatrix} \text{Step value to 3} \\ \text{Final value to 7} \end{Bmatrix}$ and $\begin{Bmatrix} \text{Step value to 5} \\ \text{Final value to 7} \end{Bmatrix}$ for the **Step** and **Step1** in the parameter window of figure 5.1(g) respectively.

Doubleclick the **Sum** block and find its **List of signs** as default **++** (parameter window is not shown for space reason) which indicates two inputs are to be added. If we enter one more **+** i.e. **+++** meaning three inputs are to be added, and so on. Again turning the **++** to **+-** means - connected input is subtracted from **+** connected input which is required for figure 5.1(i). Connect the blocks like figure 5.1(i), run the model, doubleclick the **Scope** block, click the autoscale icon, and view the designed pulse like figure 5.1(j) which is consistent with figure 5.1(h).

An ideal Dirac delta function $\delta(t-t_0)$ located at $t=t_0$ has zero existence time, infinite amplitude, and unity area (figure 5.1(k)). Computer never simulates this type of hypothetical function. Numerically we assume some large value with short existence which serves the purpose of Dirac delta generation. This is a special case of finite duration pulse of figure 5.1(h). As an example let us choose duration of the pulse to be 0.01sec and the function starts at 3sec i.e. $t_0=3$sec. By function we write the numerical $\delta(t-t_0)$ as $100u(t-3) - 100u(t-3.01)$ maintaining unit area of the pulse therefore modeling $\delta(t-3)$ is very similar to that of the pulse in figure 5.1(h) (i.e. $\begin{Bmatrix} \text{Step value to 3} \\ \text{Final value to 100} \end{Bmatrix}$ and $\begin{Bmatrix} \text{Step value to 3.01} \\ \text{Final value to 100} \end{Bmatrix}$ for the **Step** and **Step1** of figure 5.1(i) respectively).

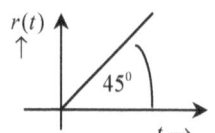

Figure 5.2(a) Ramp function

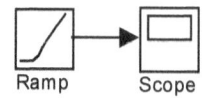

Figure 5.2(b) Ramp block connected with Scope

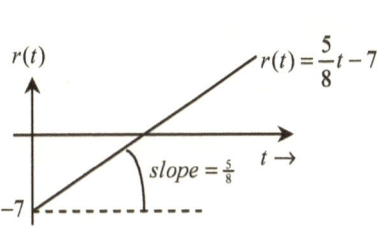
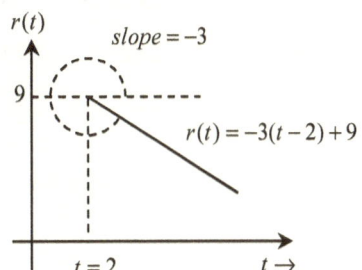

Figure 5.2(c) Ramp function with different slope and initial output

Figure 5.2(d) Ramp function $r(t) = -3(t-2)+9$ with different starting time

5.3 Modeling ramp and its derivative inputs

Ramp function graphed in figure 5.2(a) is defined as $r(t)=t$ which is a straight line with unity ($\tan 45^0 = 1$) slope and starts from $t=0\sec$. Block **Ramp** (appendix C for outlook and link) generates the $r(t)$ by default. Let us bring one **Ramp** and one **Scope** blocks in a new SIMULINK model file (subsection 1.2.2), connect (subsection 1.2.3) them according to figure 5.2(b), run the model by clicking ▶ icon in the menu bar, and doubleclick **Scope** to see the output as shown in figure 5.2(f) with autoscale setting (figure 5.1(c)). Ramp derived functions are simulated with a slight change in the parameter window data of **Ramp**.

Let us doubleclick the **Ramp** to see its parameter window like figure 5.2(e). You find there the parameters as $\begin{Bmatrix} \text{Slope} \\ \text{Start time} \\ \text{Initial output} \end{Bmatrix}$. The function $r(t)=t$ has slope 1, start time 0 (meaning beginning of the interval), and initial output 0 (meaning value of the function at the beginning bound).

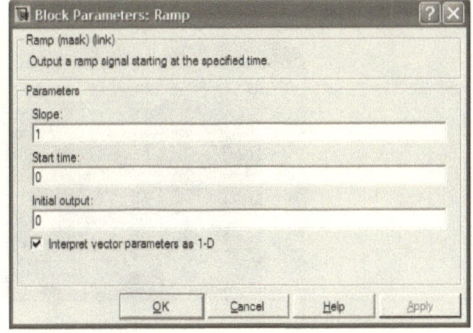
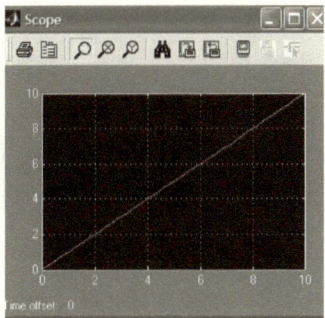

Figure 5.2(e) Parameter window of **Ramp**

Figure 5.2(f) **Scope** output for default **Ramp** i.e. $r(t)=t$

-127-

Let us see another ramp whose equation is $r(t) = \frac{5}{8}t - 7$ and its graph is the figure 5.2(c). Doubleclick the **Ramp** in model of figure 5.2(b), enter the setting of **Ramp** as $\begin{cases} \text{Slope} = 5/8 \\ \text{Start time} = 0 \\ \text{Initial output} = -7 \end{cases}$ in the parameter window, run the model, and see the output as shown in figure 5.2(g) with autoscale setting.

The function $r(t) = -3(t-2) + 9$ is also a ramp which has the parameters $\begin{cases} \text{Slope} = -3 \\ \text{Start time} = 2 \\ \text{Initial output} = 9 \end{cases}$ and which is detailed in figure 5.2(d). The corresponding **Scope** output with autoscale setting is depicted in figure 5.2(h). Referring to the figure, the block assumes the functional value for $t = 2$ over the interval $0 \le t \le 2$.

If the reader persists in being the function 0 over $0 \le t \le 2$, the functional description had better be $r(t) = [-3(t-2) + 9]u(t-2)$ and you need the model of figure 5.2(i) whose **Scope** output with autoscale setting is the

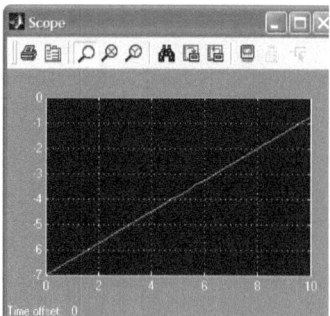

Figure 5.2(g) **Scope** output for
$r(t) = \frac{5}{8}t - 7$

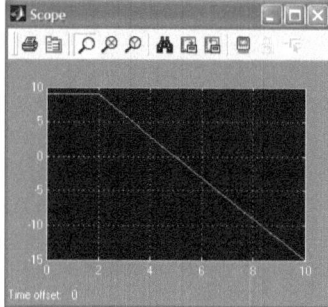

Figure 5.2(h) **Scope** output for
$r(t) = -3(t-2) + 9$

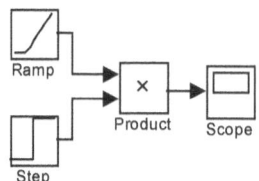

Figure 5.2(i) Model for signal
$r(t) = [-3(t-2) + 9]u(t-2)$

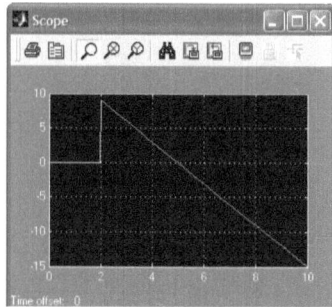

Figure 5.2(j) **Scope** output for the model of figure 5.2(i)

figure 5.2(j). The $u(t-2)$ in $r(t)$ is modeled by the **Step** block of section 5.2 with parameter setting $\begin{Bmatrix} \text{Step value to 2} \\ \text{Final value to 1} \end{Bmatrix}$. The functional multiplication of $-3(t-2)+9$ and $u(t-2)$ takes place by using **Product** block (appendix C).

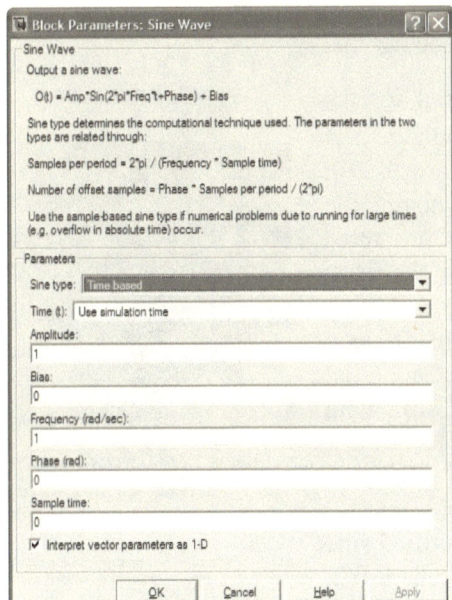

Figure 5.3(b) Block parameter window of Sine Wave

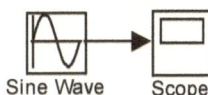

Figure 5.3(a) Model for sine wave simulation of example 1

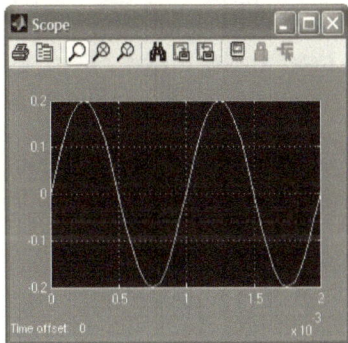

Figure 5.3(c) **Scope** output of the model for example 1

5.4 Modeling sine wave and its derivative inputs

Perhaps sine wave is the most addressed one in electrical engineering. The wave is defined as $y(t) = A\sin(2\pi ft + \theta)$ where A, f, θ, and t are amplitude, frequency, phase angle, and time interval over which the wave is to be simulated respectively. Time period of the wave is given by $T = \frac{1}{f}$. Block **Sine Wave** (appendix C for outlook and link) generates various sinusoids for which a number of examples are presented in the following.

⌸ Example 1

Generate a sine wave of frequency $1KHz$ and amplitude ± 0.2. The wave should exist for 2 milliseconds.

Amplitude of the wave can represent any physical quantity such as displacement, voltage, or current. The wave has time period $T = \frac{1}{10^3}$ secs or $1\,msec$ hence in the given time interval we expect the wave to have two cycles. The model for implementation is shown in figure 5.3(a). Bring one **Sine Wave** and one **Scope** blocks in a new SIMULINK model file (subsection 1.2.2) and connect (subsection 1.2.3) them according to figure 5.3(a). Doubleclick the **Sine Wave** to see its parameter window like figure 5.3(b) and enter the settings as $\begin{cases} \text{Amplitude}: 0.2 \\ \text{Frequency (rad/sec)}: 2*pi*1e3 \end{cases}$ keeping the others as default ($\omega = 2\pi f$ is used for the frequency, appendix A for coding). For time interval entering (subsection 1.2.6), click the menu **Simulation** down the **Configuration Parameters**, be prompted with the figure 1.7(g), and enter the **Stop time** as **2e-3** (for 2 msec duration starting from 0). Run the model by clicking ▶ icon in the menu bar and the **Scope** output should look like figure 5.3(c) with autoscale setting (figure 5.1(c)).

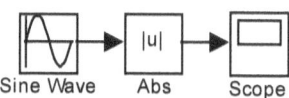

Figure 5.3(d) Model for full rectified sine wave

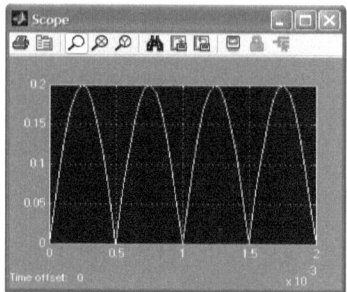

Figure 5.3(e) **Scope** return for full rectified sine wave

⊟ Example 2

Let us generate a full rectified sine wave of frequency 1*KHz*, amplitude 0-0.2, and duration 2 msec.

This is basically example 1 with the exception that the negative peaks become positive as shown in figure 4.1(b).

Figure 5.3(f) Model for half rectified sine wave

Extra simulation we do is bring one **Abs** (the block finds absolute value of its input signal, appendix C for link) and insert that between the **Sine Wave** and **Scope** like figure 5.3(d). After running the model, the reader should see the **Scope** output as shown in figure 5.3(e) with autoscale setting.

⊟ Example 3

We wish to generate a half rectified sine wave of frequency 1*KHz* and amplitude 0-0.2 and the wave should exist for 2 msec.

This wave is the wave of example 1 but negative portion of wave is set to zero like figure 4.1(c). To model the wave, we bring the **Saturation** block (appendix C for outlook and link) and connect that as shown in figure 5.3(f). The block has two saturation limits: lower and upper (parameter window is not shown for space reason). The block sets wave value to the saturation limit if the wave reaches below or above the saturation limit value. Referring to example 1, if we set the value of wave to 0 for any negative value, we have the half rectified wave but do not want to change the positive portion of the wave which happens if upper saturation limit is equal to the positive amplitude of wave. On doubleclicking the **Saturation**, we enter its settings as $\begin{cases} \text{Upper limit}: 0.2 \\ \text{Lower limit}: 0 \end{cases}$ in the parameter window and run the model. The **Scope** response with autoscale setting is shown in figure 5.3(g).

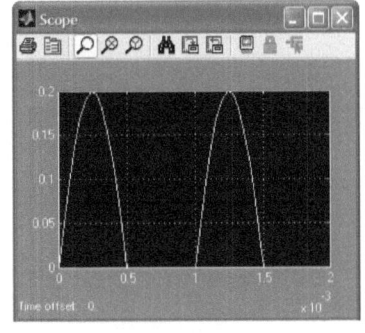

Figure 5.3(g) **Scope** output for half rectified sine wave

⊟ Example 4

Let us generate the two sinusoidal frequency wave $y(t) = 0.2\sin 2\pi ft + 0.1\sin 4\pi ft$ over $0 \leq t \leq 1ms$ where $f = 2KHz$.

The $y(t)$ has two sine components: the first with amplitude 0.2 and frequency $2KHz$ and the second one with amplitude 0.1 and frequency $4KHz$. Figures 5.3(h) and 5.4(a) show the model and its **Scope** output with autoscale setting respectively. Necessary parameter settings for **Sine Wave** and **Sine Wave1** are $\begin{cases} \text{Amplitude}: 0.2 \\ \text{Frequency (rad/sec)}: 2*pi*2e3 \end{cases}$

and $\begin{cases} \text{Amplitude}: 0.1 \\ \text{Frequency (rad/sec)}: 2*pi*4e3 \end{cases}$

respectively. Also the **Stop time** should be 1e-3.

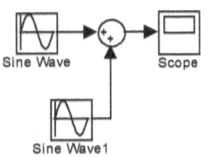

Figure 5.3(h) Model for two frequency wave

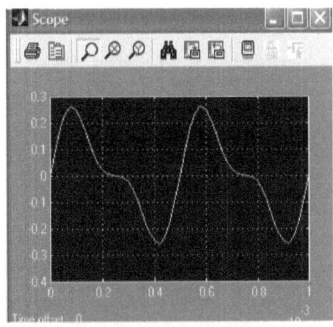

Figure 5.4(a) Two frequency wave from **Scope**

⊟ Example 5

In this example we generate three frequency wave $y(t) = 0.2\sin 2\pi ft + 0.1\sin 4\pi ft + 0.08\sin 10\pi ft$ over $0 \leq t \leq 1ms$ where $f = 2KHz$.

Obviously one needs three **Sine Wave** and a three input **Sum** (doubleclick the **Sum** block and change its **List of signs** to +++ to turn its input port number to three from default two) blocks as depicted in model 5.4(b). Settings for **Sine Wave**, **Sine Wave1**, and **Sine Wave2** blocks are $\begin{cases} \text{Amplitude}: 0.2 \\ \text{Frequency (rad/sec)}: 2*pi*2e3 \end{cases}$, $\begin{cases} \text{Amplitude}: 0.1 \\ \text{Frequency (rad/sec)}: 2*pi*4e3 \end{cases}$, and $\begin{cases} \text{Amplitude}: 0.08 \\ \text{Frequency (rad/sec)}: 2*pi*10e3 \end{cases}$ respectively. The **Stop time** of solver is **1e-3**. Simulation is similar to previous examples.

⊟ Example 6

Adding some phase in each wave which is $y(t) = 0.2\sin(2\pi ft - 60°) + 0.1\sin(4\pi ft + 10°)$ modifies the wave of example 4. In parameter windows of **Sine Waves** just append the phases as **–pi/3** and **10*pi/180** respectively.

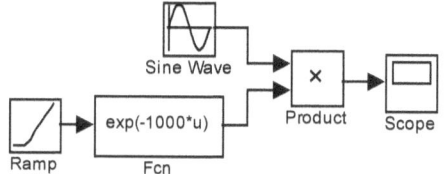

Figure 5.4(b) Model for three frequency wave

⊟ Example 7

An example of damped sine wave is $y(t) = 0.2e^{-1000t}\sin(2\pi ft - 60°)$ where $f = 2KHz$. Let us generate the wave over $0 \le t \le 5m\sec$.

One separates the $y(t)$ as e^{-1000t} and $0.2\sin(2\pi ft - 60°)$, the latter part is resembling to sine wave of example 1. The exponent part e^{-1000t} is modeled by blocks **Ramp** and **Fcn** (appendix C for outlook and link). The **Ramp** simulates t and **Fcn** performs user-defined mathematical operation on its input port signal assuming that the input variable is in terms of u.

Let us bring one **Ramp**, one **Fcn**, one **Sine Wave**, one **Product**, and one **Scope** blocks in a new SIMULINK model file. Connect the blocks as shown in figure 5.4(c). Doubleclick the **Fcn**, enter the code of e^{-1000t} as **exp(-1000*u)** considering the independent variable u in its parameter window (not shown for space reason), and enlarge the block to display its contents. Doubleclick the **Sine Wave**, enter its settings as

Figure 5.4(c) Model for damped sine wave generation

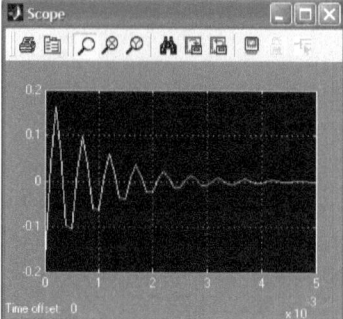

Figure 5.4(d) Damped sine wave returned by SIMULINK with adaptive setting

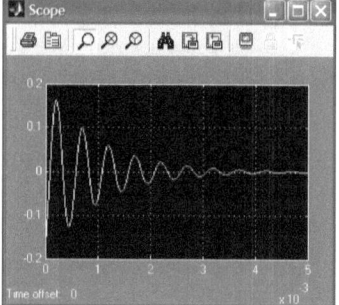

Figure 5.4(e) Damped sine wave when solver is set for fixed step

$\begin{cases} \text{Amplitude: } 0.2 \\ \text{Frequency (rad/sec): } 2*pi*2e3 \\ \text{Phase (rad): } -pi/3 \end{cases}$, change the solver **Stop time** to **5e-3** for

$0 \le t \le 5m\sec$ (figure 1.7(g)), run the model, and doubleclick the **Scope** to see the output like figure 5.4(d) with autoscale setting.

Looking into figure 5.4(d), successive maxima of the wave are not so smooth. The reason for this is SIMULINK solver adaptively selects time which is nonuniform in general. There is provision for changing t step. Since the wave existence is in millisecond range, let us choose the fixed step size as $0.01 \, m\sec$. Change the **Solver option Type** from **Variable step** to **Fixed step** in figure 1.7(g), enter the **Fixed step** size as **0.01e-3**, and run the model. The output is depicted in figure 5.4(e) with autoscale setting.

Example 8

When a sine wave is raised up or pushed down from horizontal axis by some user-defined value, the operation is called adding a bias to the sine wave. Let us consider the wave of example 1 and wish to raise the wave by 0.1 i.e. equation of the wave becomes $y(t) = 0.2\sin 2\pi ft + 0.1$ so the swing from +0.2 to –0.2 should be from +0.3 to –0.1 due to bias inclusion. However let us doubleclick the **Sine Wave** in model of figure 5.3(a) and enter the bias 0.1 in the parameter window of figure 5.3(b). The **Scope** output with autoscale adjustment is shown in figure 5.4(f).

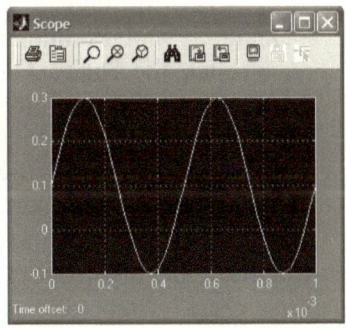

Figure 5.4(f) Sine wave of example 1 with a bias 0.1

5.5 Modeling rectangular wave and its derivative inputs

Section 4.2 addresses periodic rectangular wave generation. We illustrate some examples on SIMULINK context in the sequel.

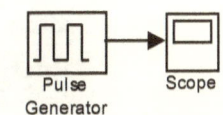

Figure 5.5(a) Model for rectangular pulse of example 1

Example 1

Let us generate a rectangular pulse of amplitude swing 0-0.3, frequency $5KHz$, duty cycle 60%, and duration $0.6 \, m\sec$. The model of figure 5.5(a) is constructed in this regard in which the block **Pulse Generator** (appendix C for outlook and link) generates the rectangular wave. Time period of the wave should be $T = \dfrac{1}{f} = \dfrac{1}{5KHz} = 0.2 \, m\sec$. Upon modeled like figure 5.5(a), we doubleclick the **Pulse Generator** and enter its settings as

$\begin{cases} \text{Amplitude}: 0.3 \\ \text{Period(secs)}: 0.2e-3 \\ \text{Pulse Width (\% of period)}: 60 \end{cases}$ in the parameter window keeping the others as default (the parameter window is not shown for space reason). For duration 0.6 msec entering, change the solver Stop time to 0.6e-3 (figure 1.7(g)). Figure 5.5(b) shows simulation output on autoscale setting (figure 5.1(c)).

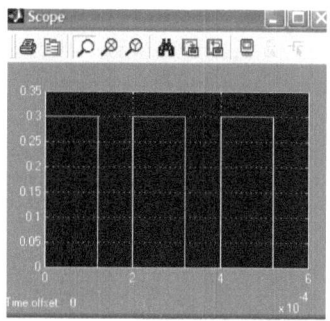

Figure 5.5(b) Scope output for model of example 1 with autoscale setting

⊟ Example 2

In last example, the amplitude swing is 0-0.3. What if we would like to have a wave of the same frequency and duration but the swing is from −0.3 to 0.3?

The difference between expected maximum and minimum is 0.6. We doubleclick the Pulse Generator of figure 5.5(a), change its amplitude to 0.6, and add a constant value of −0.3 to obtain the expected swing. The model of figure 5.5(c) depicts the implementation. In addition to blocks of model in figure 5.5(a), you need one Constant and one Sum blocks (appendix C). We enter −0.3 on doubleclicking the Constant block however the model's Scope output is presented in figure 5.5(d).

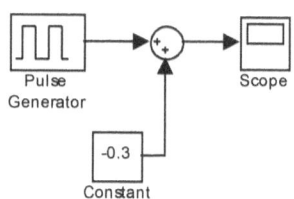

Figure 5.5(c) Model for equal positive and negative swings

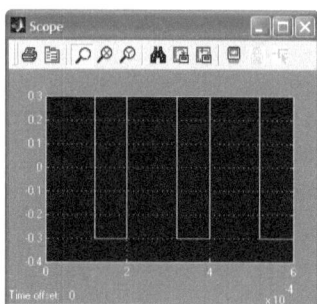

Figure 5.5(d) Scope output for model of example 2 with autoscale setting

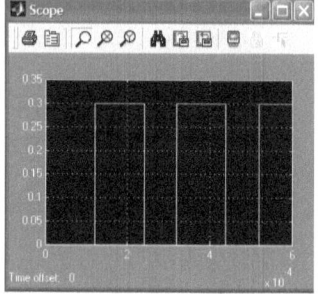

Figure 5.5(e) Shifted pulse of example 3 with autoscale setting

⊟ Example 3

In examples 1 and 2, the wave started at time $t=0$. Let us say we intend to shift the wave of example 1 to the right by one duty cycle. In absolute unit, the time shift becomes 0.12 msec. Doubleclick the Pulse Generator of figure 5.5(a) and enter the Phase delay (secs) as 0.12e-3 in the parameter window. Figure 5.5(e) is the outcome of simulation on autoscale setting.

5.6 Modeling triangular wave and its derivative signals

A triangular wave is basically a ramp function over one period with different slope, swing, and offset. Figure 5.6(a) shows a typical triangular wave whose equation is given by $f(t) = \left\{ \dfrac{At}{T} \text{ for } 0 \leq t \leq T \right\}$ excluding offset. The offset just shifts the wave up or down.

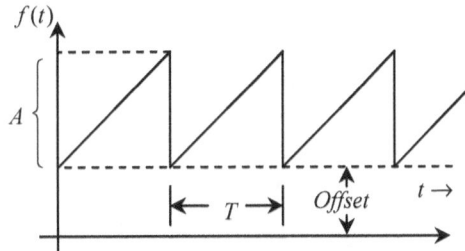

Figure 5.6(a) A triangular wave of period T seconds and amplitude swing from 0 to A

⛉ Example 1

Let us generate a triangular wave which has frequency $500\ Hz$, amplitude swing from -0.05 to 0.05, and duration 0.01secs.

We wish to introduce the **Signal Builder** block of SIMULINK which keeps provision for window interface signal design. Time period of the wave is $T = \dfrac{1}{f} = 0.002$secs

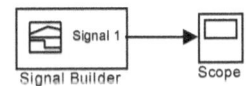

Figure 5.6(b) **Signal Builder** connected with **Scope** block

indicating five cycles in given duration.

Let us bring one **Signal Builder** (appendix C for outlook and link) and one **Scope** blocks in a new SIMULINK model file (subsection 1.2.2) and connect (subsection 1.2.3) them like figure 5.6(b). SIMULINK responds with the design window of figure 5.6(c) on doubleclicking the **Signal Builder** in which a finite

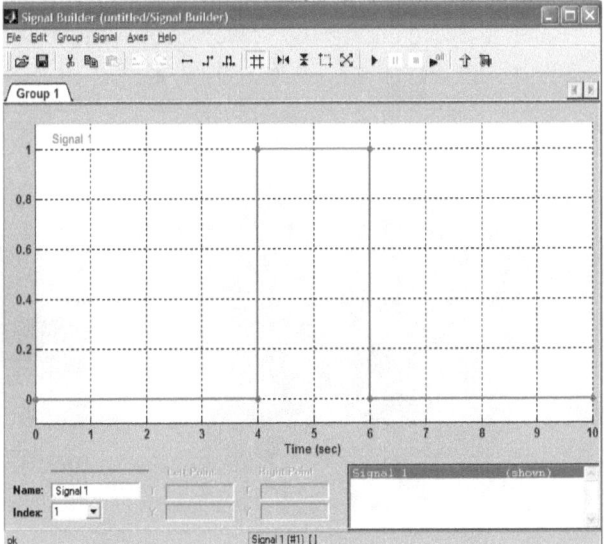

Figure 5.6(c) Design window of **Signal Builder** block

-135-

duration rectangular pulse exists as the default one. Since the wave frequency or time period and duration must be consistent in wave modeling, we first enter the wave duration in **Signal Builder** design window. To do so, click the **Axes** from the menu of the design window and find the option **Change time range** in a pulldown menu. Let us click that and enter $\begin{Bmatrix} \text{Min time}:0 \\ \text{Max time}:0.01 \end{Bmatrix}$ in the prompt window for entering the wave duration 0.01 secs (the design window is updated on account of the change). The required amplitude swing is from –0.05 to 0.05 hence we click again the **Axes** in the menu bar of the design window, click the **Set Y Display limits** in the pull down menu, and enter $\begin{Bmatrix} \text{Minimum}:-0.05 \\ \text{Maximum}:0.05 \end{Bmatrix}$ in the prompt window.

In the design window menu bar, you also find the menu for **Signal**, click the **Signal** menu, and find **Replace with** option in the pull down menu and **Triangle** in the second stage pull down menu. Click the **Triangle** and enter $\begin{Bmatrix} \text{Frequency}:500 \\ \text{Amplitude}:0.05 \\ \text{Offset}:0 \end{Bmatrix}$ for required wave in the prompt window. With that action, the updated wave appears in the design window.

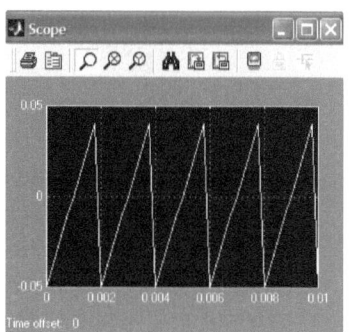

Figure 5.6(d) **Scope** output for triangular wave of example 1

We can close the **Signal Builder** window and move onto the SIMULINK model file. SIMULINK does not have any information about wave duration. Enter it by changing the solver **Stop time** to 0.01 (figure 1.7(g)). Run the model and see the **Scope** output as shown in figure 5.6(d) with autoscale setting (figure 5.1(c)). Referring to the **Scope** output, the vertical edge of each wave does not seem to be vertical. This is because of variable step or

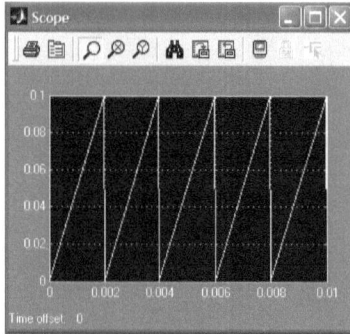

Figure 5.6(e) **Scope** output for triangular wave of example 2

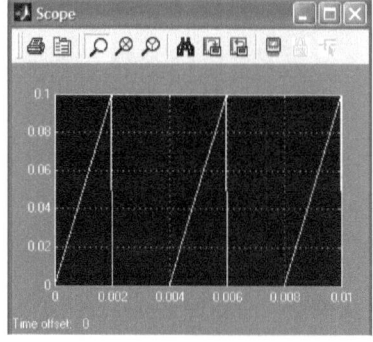

Figure 5.6(f) **Scope** output for model of figure 5.6(h)

adaptive setting of solver. If we make the solver setting fixed step and small within the duration, for sure the wave would appear as expected.

🗗 Example 2

We intend to generate a triangular wave which has frequency $500\ Hz$, amplitude swing from 0 to 0.1, and duration 0.01secs.

This is basically the wave of example 1 with little modification. The previous swing is from −0.05 to 0.05 but now we need from 0 to 0.1 so just adding an offset of 0.05 will have our simulation done. Referring to the design window of figure 5.6(c), click the **Signal** down **Replace With** down **Triangle**, enter $\begin{Bmatrix} \text{Frequency}: 500 \\ \text{Amplitude}: 0.05 \\ \text{Offset}: 0.05 \end{Bmatrix}$ in the prompt window, save the design, and run the SIMULINK model of figure 5.6(b). The output is shown in figure 5.6(e) with **Fixed Step** solver option and autoscale setting. For the **Fixed** step size 0.00001 (our chosen),

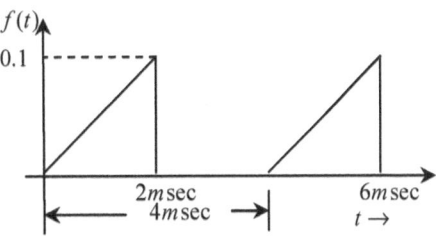

Figure 5.6(g) A triangular wave with some off interval

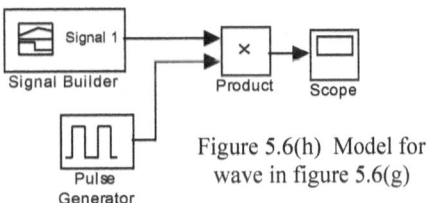

Figure 5.6(h) Model for wave in figure 5.6(g)

click the **Simulation** in model menu bar and then click **Configuration Parameters** to see the window of figure 1.7(g). As the **Solver options** type popup, select the **Fixed-step** from the popup and enter **0.00001** in the slot of **Fixed-step size** leaving other settings unchanged.

🗗 Example 3

Periodic wave of figure 5.6(g) is to be generated over $0 \le t \le 0.01$ secs.

The periodic wave has time period $4\ msec$ but the wave is off from 2 to $4\ msec$ considering the first cycle. To simulate so, we first form a triangular wave (like example 2) of swing from 0 to 0.1 and the time period $2\ msec$ (means frequency $500\ Hz$) and then set the alternate cycle to zero.

Let us imagine the first period of rectangular pulse (similar to figure 5.5(b)) whose amplitude value is 1 from 0 to $2\ msec$ and 0 from 2 to $4\ msec$ and the pulse continues this way for the other cycles.

The current wave will be accomplished if the rectangular pulse is multiplied with the triangular wave for what reason first we generate the wave of example 2 employing the **Signal Builder** and connect the **Pulse Generator** as shown in figure 5.6(h). The parameter window settings for the

Pulse Generator should be $\begin{cases} \text{Amplitude}: 1 \\ \text{Period(secs)}: 4e-3 \\ \text{Pulse Width (\% of period)}: 50 \end{cases}$. The **Scope** output is shown in figure 5.6(f) with autoscale setting.

For more examples on triangular wave the reader is referred to section 4.2 and [36].

5.7 Modeling triggered and user-defined nonperiodic signals

Any wave or signal can be turned on or off at some instant of its cycle depending on user requirement – this is termed as triggering a wave.

Let us say we have a sine wave $y = A\sin 2\pi f t$ whose frequency and amplitude are $f = 50\,Hz$ and $A = 0.8$ respectively. The wave is off in one quarter of the period and in rest three quarters of the period the wave is on. Or it can be rephrased as the wave is triggered at 25% of the period. Figure 5.7(a) illustrates triggering of the sine wave over four cycles or 0.08 seconds – which we plan to simulate.

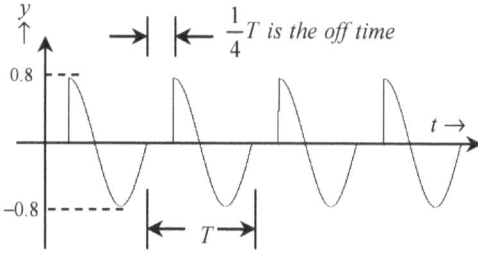

Figure 5.7(a) A sine wave is triggered at 25% of its period

To mention about SIMULINK solution, we generate a rectangular pulse of the same period as that of the sine wave but with amplitude swing from 0 to 1. The rectangular pulse (section 5.5) must have duty cycle which is exactly on time of the sine wave. Since duty cycle of the pulse starts at zero, we shift the pulse to right by exactly (1−duty cycle) where the duty cycle is expressed as percentage of time period.

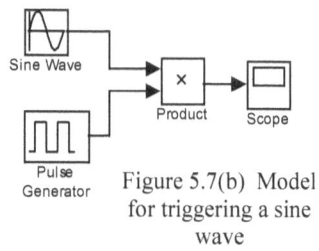

Figure 5.7(b) Model for triggering a sine wave

So to simulate the current problem, let us bring one **Sine Wave**, one **Pulse Generator**, one **Product**, and one **Scope** blocks following the link of appendix C in a new simulink model file (subsection 1.2.2) and connect (subsection 1.2.3) them as presented in figure 5.7(b). The time period of given wave is $\frac{1}{50\,Hz} = 0.02\,\text{sec}$ hence the

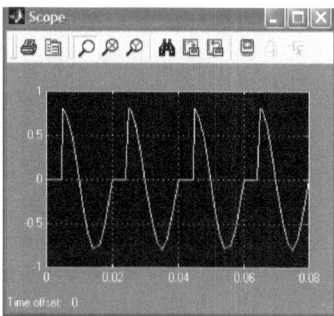

Figure 5.7(c) **Scope** output for triggered sine wave

duty cycle is either 75% of period or 0.015secs. With that the off period is

25% of period or 0.005secs. Enter the settings of **Sine Wave** and **Pulse Generator** as $\begin{Bmatrix}\text{Amplitude}: 0.8 \\ \text{Frequency(rad/sec)}: 2*pi*50\end{Bmatrix}$ and $\begin{Bmatrix}\text{Period(secs)}: 0.02 \\ \text{Pulse Width (\% of period)}: 75 \\ \text{Phase Delay (secs)}: 0.005\end{Bmatrix}$

keeping the others as default in the parameter windows respectively and set the solver **Stop time** to **0.08** (figure 1.7(g)) for four cycles. After running the model, we obtain the output from the **Scope** as depicted in figure 5.7(c) with autoscale setting (figure 5.1(c)).

This sort of triggering may take place for any other previously discussed waves.

⌗ **Nonperiodic wave**

All along we have been discussing the periodic waves of various types in previous sections. In the case of a nonperiodic wave, we just generate one cycle of a periodic wave employing techniques illustrated so far.

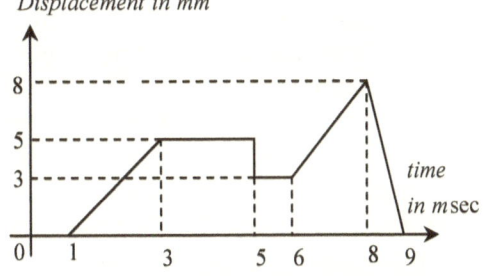

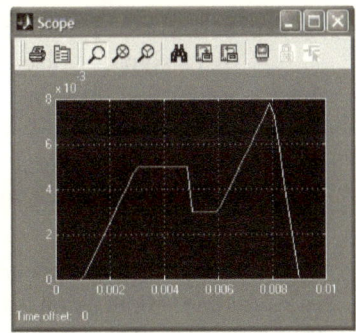

Figure 5.7(d) A displacement versus time function

If the signal is nonperiodic and defined by straight line segments and edges, **Signal Builder** is the appropriate block to generate it which we introduced in section 5.6. Just to be specific by an example, we wish to design the finite displacement versus time variation of figure 5.7(d) employing the **Signal Builder** block in SIMULINK.

Figure 5.7(e) **Scope** output for signal in figure 5.7(d) with autoscale setting

Let us bring the block in a new SIMULINK model file and doubleclick it. The design window of the figure 5.6(c) appears with default setting of a finite duration pulse. Referring to figure 5.7(d), the variation has vertical and horizontal axes ranges as 0 to 8mm and 1 to 9msec respectively but we design the function in standard units – meter and second respectively. Line segments of figure 5.7(d) have the coordinates (0,0), (1,0), (3,5), (5,5), (5,3), (6,3), (8,8), and (9,0) all in (msec,mm). Collecting consecutive time and displacement coordinates, we have [0 1 3 5 5 6 8 9]×10⁻³ sec and [0 0 5 5 3 3 8 0]×10⁻³ meter respectively.

Concerning design window of figure 5.6(c), click the **Signal** down **Replace** with down **Custom** from the menu bar, enter

$\begin{cases} \text{Time Values}: [0\ 1\ 3\ 5\ 5\ 6\ 8\ 9]*1e-3 \\ \text{Y Values}: [0\ 0\ 5\ 5\ 3\ 3\ 8\ 0]*1e-3 \end{cases}$ in the prompt window for entering horizontal and vertical coordinates in standard units, click the **Axes** down **Change time range** in menu bar, enter $\begin{cases} \text{Min time}: 0 \\ \text{Max time}: 9e-3 \end{cases}$ in the prompt window for horizontal range, and save the signal design by clicking save icon of design window menu bar. The design window displays the function of figure 5.7(d).

Just to verify, let us connect the **Signal Builder** in conjunction with the **Scope** as shown in figure 5.6(b), change the solver **Stop time** to **9e-3**, and run the model. The **Scope** output with autoscale setting as seen in figure 5.7(e) confirms the design.

The reader might say what the displacement has to do with voltage or current signal? Let us imagine a displacement sensor which produces voltage causing from the displacement. The sensor has some displacement to voltage proportionality constant often supplied by the manufacturer which you can model by using a **Gain** block. Therefore the **Signal Builder** and **Gain** together represent the sensor voltage generation.

⯑ Expression based wave

All signals discussed so far mostly employed built-in blocks. User can define function based signal too. For example a voltage signal is given by $u(t) = \cos^2 t \sin^2 t$ which we intend to generate over $0 \le t \le 2\pi$.

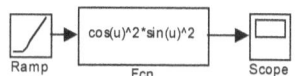

Figure 5.7(f) Expression based model of a signal

Block **Fcn** (appendix C) helps us simulate expression based signal. Recall that the **Ramp** block (section 5.3) simulates independent variable t. We just enter expressional vector code (appendix A) of given signal in **Fcn** parameter window but assuming that the independent variable is u. Therefore we enter **cos(u)^2*sin(u)^2** for $\cos^2 t \sin^2 t$ in the parameter window of **Fcn** on doubleclicking it. Figure 5.7(f) depicts the model for such signal generation. Enlarge the **Fcn** to see its contents and enter the **Stop time** as **2*pi** for interval.

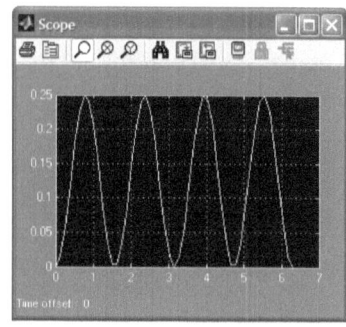

Figure 5.7(g) **Scope** output for expression based signal

On running the model we see the output for $u(t)$ in **Scope** like figure 5.7(g) on autoscale setting.

5.8 SISO and MIMO control system outputs in SIMULINK

As explained in chapter 1 control system implementation in SIMULINK requires three key elements – source, system, and sink. Modeling of sources is addressed in sections 5.1 through 5.7. SISO or MIMO control system modeling you find in chapter 3. As sink element you have to bring **Scope** or **Display** block in the model to view the output. Following examples demonstrate how to obtain model output in SIMULINK.

⊞ Example 1

An input $u(t) = \begin{cases} 1 & \text{for } t \geq 0 \\ 0 & \text{for } t < 0 \end{cases}$ is applied to the SISO fourth order control system $H(s) = \dfrac{7s^3 - 7s + 42}{s^4 - 118s^2 - 240s}$ (as in figure 5.8(a)). Obtain the output $y(t)$ wave shape over $0 \leq t \leq 0.2 \sec s$ from SIMULINK modeling.

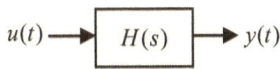

Figure 5.8(a) A SISO control system with input and output

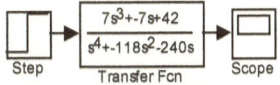

Figure 5.8(b) Model of control system in figure 5.8(a)

The reader is referred to sections 3.1 and 5.1 for modeling of $H(s)$ and $u(t)$ respectively. For the $y(t)$ wave shape viewing you just need the **Scope** block so figure 5.8(b) shows the complete model. For the interval you need to enter 0.2 as the **Stop time** (section 1.2.6). Model running yields the **Scope** output of figure 5.8(c) with autoscale setting (figure 5.1(c)).

The $H(s)$ so exercised is in numerator-denominator polynomial form in fact it can be in any other form like zero-pole-gain, state-space, etc.

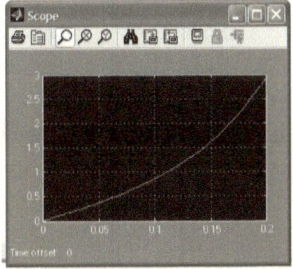

Figure 5.8(c) Scope output from model in figure 5.8(b)

⊞ Example 2

This example demonstrates an interconnected SISO control system output. In section 3.5 an interconnected system is presented in figure 3.5(a) whose model is shown in figure 3.5(b). Suppose the $u(t)$ in the control system is a ramp function defined as $u(t) = t$. Determine the output $y(t)$ wave shape over $0 \leq t \leq 5 \sec s$.

The input $u(t)$ is modeled by **Ramp** and the output viewing needs a **Scope** which you see in figure 5.8(d). Enter 5 as the **Stop time**. Figure 5.8(e) portrays the $y(t)$ wave shape with autoscale setting.

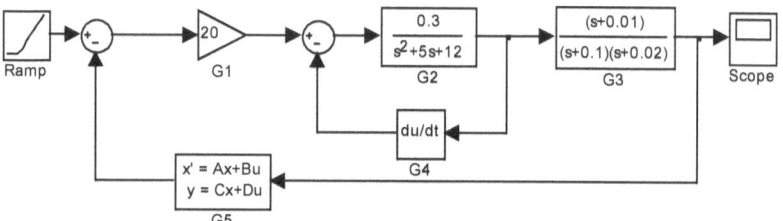

Figure 5.8(d) Model of figure 3.5(b) with input and output elements

⛶ Example 3

MIMO system of figure 3.2(b) has one input and three outputs; we wish to see the three output wave shapes for $u(t)=t$ over $0 \le t \le 5 \sec s$.

All you need in model of figure 3.2(b) is connect one **Ramp** for input and three **Scope**s for three outputs. Figure 5.8(f) shows the model. Enter the **Stop time** as **5** and the simulation is like previous examples.

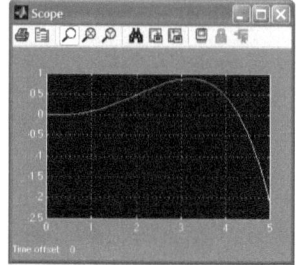

Figure 5.8(e) Output from model in figure 5.8(d)

⛶ Example 4

In the first paragraph of page 69 we have one MIMO control system defined in terms of state-space with three inputs and one output. Suppose the three inputs are $u_1(t)=t$, $u_2(t)=1$, and $u_3(t)=\sin t$; we wish to see the output $y(t)$ wave shape over $0 \le t \le 0.5 \sec s$.

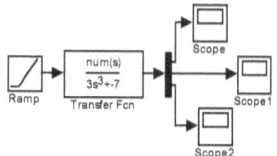

Figure 5.8(f) Single input triple output MIMO model

Clearly the three inputs are simulated by **Ramp**, **Step**, and **Sine Wave** blocks respectively. Figure 5.8(g) shows the complete model. For the **Step** you need to enter the **Step time** as 0. The **Sine Wave** generates basically $\sin t$ by default. However figure 5.8(h) presents model output on autoscale setting.

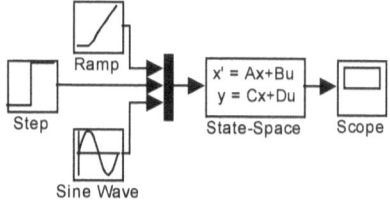

Figure 5.8(g) Model of triple input single output MIMO

⛶ Example 5

Apart from state-space if we have multiple input multiple output, connect the input blocks and **Scope**s as they are required. For example in figure 3.5(c) we have triple inputs and outputs. Suppose the inputs are the example 4 mentioned ones.

Now there is no need for **Mux** block nor for **Demux**. In the model 3.5(d) we just connect the necessary blocks as seen in figure 5.8(i). Having the model constructed, the reader can easily simulate the control system as done in previous examples.

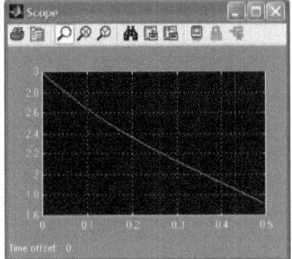

Figure 5.8(h) Model output of figure 5.8(g)

⊟ **What about step response?**

Step response is basically a special case of SISO control system which you can model similar to example 1.

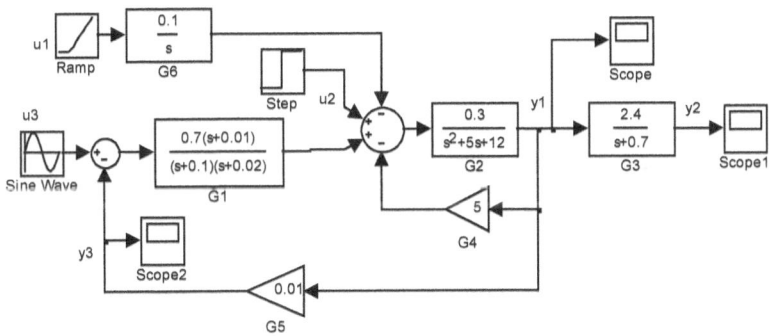

Figure 5.8(i) Model of figure 3.5(d) with input and output elements

⊟ **What about impulse response?**

We suggest you apply MATLAB technique in determining the impulse response (section 4.5). Impulse response truly speaking can never be modeled in SIMULINK as far the definition. In order to implement the response numerically, we apply a short existent high amplitude unity area pulse which serves the purpose of impulse response. Let us see the following example.

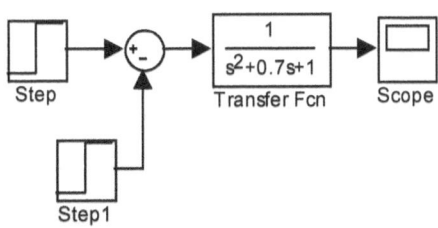

Figure 5.9(a) Modeling a practical impulse response

We wish to find impulse response for the control system $H(s) = \dfrac{1}{s^2 + 0.7s + 1}$ over $0 \le t \le 10$ sec in SIMULINK.

Section 5.2 explains modeling of a finite pulse (figure 5.1(h)). A practical impulse function is based on very short duration finite pulse.

Assume that the impulse starts at $t=0$ and the short duration is 0.01 secs (user-chosen) so the amplitude of step function should be 1/0.01 or 100 therefore the practical impulse function has expression $\delta_{practical}(t) = 100u(t) - 100u(t-0.01)$.

Figure 5.9(a) presents the model. Doubleclick the **Step** to change its **Step time** from default 1 to 0, enter its **Final Value** as 100 (for $100u(t)$), doubleclick the **Step1** to change its **Step time** from default 1 to 0.01, and enter its **Final Value** as 100 (for $100u(t-0.01)$).

5.9 Control system performance in SIMULINK

In chapter 4 we have presented MATLAB approach to determine the first and second order responses of a control system. For a first order system we suggest that the reader apply the technique of MATLAB because SIMULINK approach becomes clumsy. Nevertheless some quantities of the performance are addressed below.

◆ **Steady state value of a control output**

If the model output shows steady state behavior, you can have the value by connecting a **Display** block instead of **Scope**. If you wish to keep the **Scope**, that is also OK.

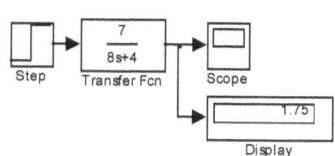

Figure 5.9(b) Steady state value of a first order system

Think about the first order control system $H(s) = \dfrac{7}{8s+4}$ connected like figure 5.8(a) with $u(t)$ step input. You can view the $y(t)$ wave shape over $0 \le t \le 10$ sec in SIMULINK by model of figure 5.9(b). A **Display** block (appendix C) is connected to the output which shows value of $y(t)$ at the last time point i.e. $t = 10$ secs so we say $y_{ss} = 1.75$.

You can apply this modeling to other order control systems like second or third.

◆ **Peak value of a control output**

If control output $y(t)$ exhibits a maximum, we obtain that employing the block **MinMax Running Resettable**. Think about the control system $H(s) = \dfrac{1}{s^2 + 0.7s + 1}$ subject to input $u(t) = 1$ over $0 \le t \le 20$ secs. We wish to find $y(t)|_{max} = 1.309$.

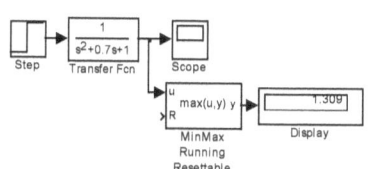

Figure 5.9(c) Model for finding peak value

Figure 5.9(c) shows the model. The step time of the **Step** block is set to 0. The **MinMax Running Resettable** is set for minimum by default. Doubleclick the block and change the setting to **max**. Enter the **Stop time** as 20 and run the model to see $y(t)|_{max}=1.309$ in the **Display**.

* **P.O. of a control output**

Having found y_{ss} and $y(t)|_{max}$, one obtains the percent overshoot (P.O.) by $(y(t)|_{max}-y_{ss})/y_{ss}\times 100$. The **MinMax Running Resettable** and **Transfer Fcn** of figure 5.9(c) provide $y(t)|_{max}$ and y_{ss} respectively. For the subtraction we use a **Subtract** block whose input ports + and – connect to $y(t)|_{max}$ and y_{ss} respectively. For the division of $y(t)|_{max}-y_{ss}$ by y_{ss}, we apply **Divide** block. Actually numerator and denominator go to × and ÷ of the **Divide** respectively. Figure 5.9(d) shows the model along with 100 multiplication conducted by a **Gain** block. As you see, the **Display** indicates the P.O. as 30.98% after running the model.

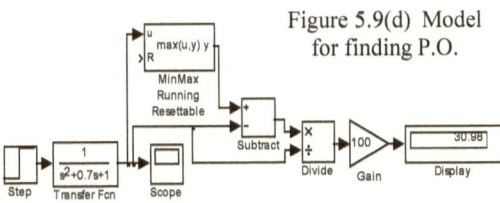

Figure 5.9(d) Model for finding P.O.

* **Peak time of a control output**

For peak time merely SIMULINK modeling becomes clumsy. The easiest way is transport the SIMULINK data into MATLAB and apply the technique of section 4.7. You need **To Workspace** block for data transport. For the ongoing second order system the model is shown in figure 5.9(e). The **To Workspace** has different data formats for saving, out of which array type is the appropriate one so doubleclick **To Workspace** and select the **Save format** as **Array**. Run the model with correct **Stop time** and the data will remain at workspace by default names **tout** and **simout** for t and $y(t)$ respectively. The T_p then we find by:

Figure 5.9(e) Transporting SIMULINK data into MATLAB

>>[y_max,l]=max(simout); ↵   ← l is user-chosen, holds index for T_p
>>Tp=tout(l) ↵   ← Tp⇔T_p, Tp is user-chosen

Tp =
    3.2967   ← i.e. $T_p=3.2967$ secs

* **Rise time of a control output**

As an example 20%-60% rise time of ongoing second order system is $T_r=0.6838$ secs which we wish to find.

Using the model of figure 5.9(e) we make the t and $y(t)$ data available to **tout** and **simout** respectively. Section 4.7 mentioned **c_cross** determines (make sure the file is available in the working path of MATLAB) the rise time at the command prompt as follows:
>>Tr=c_cross(tout,simout,0.6*simout(end))-c_cross(tout,simout,0.2*simout(end)) ↵

Tr =
   0.6838

For other percentage rise time for example 10%-90%, the command would have been **c_cross(tout,simout,0.9*simout(end))-c_cross(tout,simout, 0.1*simout(end))**. The **simout(end)** is the last value of $y(t)$ which is meant to be y_{ss}.

◆ **Settling time of a control output**

SIMULINK approach for finding settling time is somewhat clumsy. The main problem here is staying in a SIMULINK model (chapter 1), we can not enter to another model until the **Stop time** is elapsed. We suggest MATLAB-SIMULINK combined tactic for settling time.

We wish to determine the settling time for ongoing second order system.

Simulate the model of figure 5.9(e), run the model, move on to MATLAB prompt, and obtain the approximate steady state value by using the command:
>>y_ss=simout(end); ↵

The symbology of section 4.7 is also applicable here. Find the time point crossing for lower bound of settling time phenomenon:
>>ts1=c_cross(tout,simout,0.98*y_ss); ↵

Find the time point crossing for upper bound of settling time phenomenon:
>>ts2=c_cross(tout,simout,1.02*y_ss); ↵

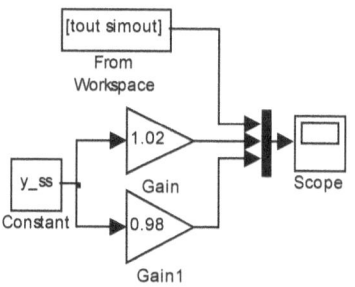

Figure 5.9(f) Model for settling time phenomenon

Figure 5.9(f) shows the model for viewing settling time phenomenon for which you need to open a new SIMULINK model file. After getting one **Constant**, two **Gains**, one **Mux**, one **From Workspace**, and one **Scope** blocks in the model file, doubleclick the **Constant**, enter **y_ss** for the steady state value in the parameter window, doubleclick the **Gain**, enter its

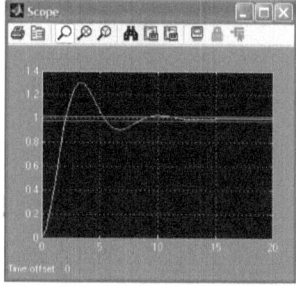

Figure 5.9(g) **Scope** output for settling time phenomenon

Gain as 1.02 for 2% upper criterion, doubleclick the Gain1, enter its Gain as 0.98 for 2% lower criterion, doubleclick the Mux, change its input number from default 2 to 3, doubleclick the From Workspace to change its Data from default simin to [tout simout], and connect the blocks like figure 5.9(f). You may need to enlarge some block to see its contents e.g. Gain.

The block From Workspace imports MATLAB data into SIMULINK but there is a specific style of data importing. The t and $y(t)$ data must be a two column matrix in which the first and second columns refer to t and $y(t)$ respectively. For this reason we entered [tout simout] in the parameter window of From Workspace. However Scope output of model in figure 5.9(g) shows the settling phenomenon on autoscale setting. Figure 5.9(g) hints that settling happened from the upper bound hence get it by:

>>Ts=ts2(end) ↵

Ts =
    11.0967

Above return says that the settling time T_s is 11.0967 secs. We suggest you follow the order of programming and modeling we mentioned while finding the settling time. If you do not do so, model might be nonoperational due to adaptive time selection of SIMULINK.

◆ **Reading off the Scope screen**

Whatever quantity we found from SIMULINK modeling can be read off the Scope screen. Take the example of just addressed settling time. Figure 5.9(g) shows the Scope output. Bring the mouse pointer at the last intersection of upper bound and output, leftclick the mouse again and again at the intersection, and find the zoomed t and $y(t)$ coordinates. From the coordinates you can easily read off the settling time.

5.10 Error performance indices in SIMULINK

We suggest the reader go through section 4.8 for error performance index finding in MATLAB. The bottomline is any index needs integration which we determine in SIMULINK by Integrator block (appendix C). Another important point is the reader has to detect the $e(t)$ signal point in the model. In this section mainly we address single index determination.

◆ **Single index finding**

Let us consider the integral time multiplied absolute error (ITAE) which is $\int_0^T t\,|e(t)|\,dt$. The $e(t)$ in fact can be from any control system. We wish to find the ITAE for the prototype second order system in figure 4.6(a) subject to $\zeta = 0.6$, $\omega_n = 1.3\,rad/\sec$, $H(s) = 0.8$, and $u(t) = 1.1\,V$ over $0 \le t \le 25 \sec s$.

The complete model you see in figure 5.10(a). The **Scope** is immaterial here. Referring to the figure the error signal $e(t)$ is before the **Transfer Fcn**. The **Ramp** and **Abs** provide t and $|e(t)|$ respectively. The **Product** block performs $t|e(t)|$ operation. The **Display** shows the ITAE index from the **Integrator** which performs $\int t|e(t)|\,dt$. In your part bring the blocks of figure 5.10(a) in a newly opened model file. Doubleclick the **Step** to change its **Step time** to 0 and **Final value** to 1.1 (for $u(t)$). Doubleclick the **Transfer Fcn** and enter its **Numerator** and **Denominator** coefficients

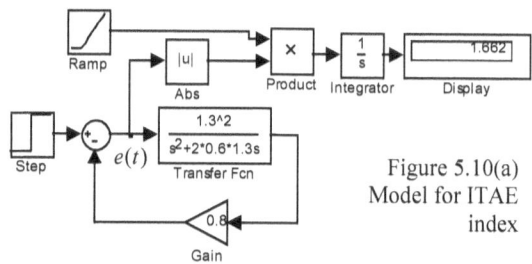

Figure 5.10(a)
Model for ITAE index

as [1.3^2] and [1 2*0.6*1.3 0] respectively. Flip the **Gain** block. The last bound of time interval which is 25 is entered as the **Stop time**. Finally connect the blocks like figure 5.10(a) and run the model to see the index 1.662 in the **Display**.

For the other indices you need slight modification in the model of figure 5.10(a). For example the model of figure 5.10(b) is applicable for the IAE index finding. Since there is no t

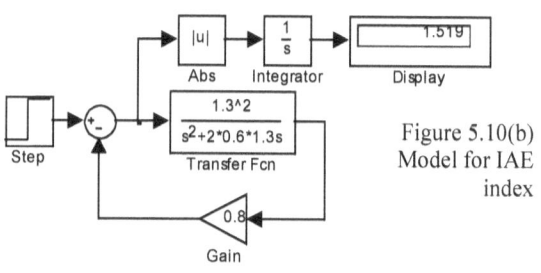

Figure 5.10(b)
Model for IAE index

multiplication in the index definition, the **Ramp** is eliminated, nor do we need a **Product** block. For the ongoing system the index is IAE= $\int_0^{25}|e(t)|\,dt$ = 1.519 which is seen in the **Display** of figure 5.10(b).

◆ **Index variation against damping ratio**
The comparison we exercised in section 4.8 can be implemented for index versus damping ratio variation.

We intend to terminate the chapter with this example.

Exercises

1. Apply SIMULINK blocks to model the following control signals (related to step function) and verify their wave shapes inspecting in **Scope**:
   (a) standard step or Heaviside function $u(t)$ (b) shifted step function $9u(t-3.5)$ (c) a finite duration pulse of amplitude $15V$ which exists over $2\sec s \le t \le 9\sec s$ (d) a practical Dirac delta function $2\delta(t-5)$ by choosing duration 0.005sec over $0 \le t \le 12\sec s$.

2. Apply SIMULINK blocks to model the following control signals (related to ramp function) and verify their wave shapes inspecting in **Scope**:
   (a) $r(t) = 0.8t$ over $0 \le t \le 5\sec s$ (b) $r(t) = 0.77t + 3.1$ over $0 \le t \le 5\sec s$ (c) $r(t) = 2.1(t-1.2) - 4$ over $0 \le t \le 5\sec s$ (d) $r(t) = [2.1(t-1.2) - 4]u(t-1.2)$ over $0 \le t \le 5\sec s$.

3. Apply SIMULINK blocks to model the following control signals (related to sine wave) and verify their wave shapes inspecting in **Scope**:
   (a) a sine wave of frequency $3KHz$ and amplitude $\pm 0.3V$ which should exist for 1 millisecond (b) a full rectified sine wave of frequency $3KHz$, amplitude $0\text{-}0.3V$, and duration $1\,m\sec$ (c) a half rectified sine wave of frequency $3KHz$, amplitude $0\text{-}0.3V$, and duration $1\,m\sec$ (d) two sinusoidal frequency wave $y(t) = 0.9\sin 2\pi ft + 0.3\sin 4\pi ft$ over $0 \le t \le 1ms$ where $f = 3KHz$ (e) three frequency wave $y(t) = 0.9\sin 2\pi ft + 0.3\sin 4\pi ft + 0.13\sin 10\pi ft$ over $0 \le t \le 1ms$ where $f = 3KHz$ (f) in part (e) add phases $10°$, $20°$, and $-30°$ to the frequency components respectively (g) damped sine wave $y(t) = 25e^{-170t}\sin(2\pi ft + 80°)$ over $0 \le t \le 2m\sec$ where $f = 3KHz$ (h) part (a) with an offset $0.2V$.

4. Apply SIMULINK blocks to model the following control signals (related to rectangular wave) and verify their wave shapes inspecting in **Scope**:
   (a) a rectangular pulse of amplitude swing $0\text{-}0.9V$, frequency $500Hz$, duty cycle 75%, and duration $4\,m\sec$ (b) in part (a) on amplitude swing $\pm 0.9V$ (c) in part (b) now the wave is shifted by ½ duty cycle to the right.

5. Apply SIMULINK blocks to model the following control signals (related to triangular wave) and verify their wave shapes inspecting in **Scope**:
   (a) triangular wave which has frequency $100\,Hz$, amplitude swing $\pm 0.9V$, and duration $30\,m\sec$ (b) in part (a) now the amplitude swing is $0\text{-}1.8V$ (c) in part (b) now every

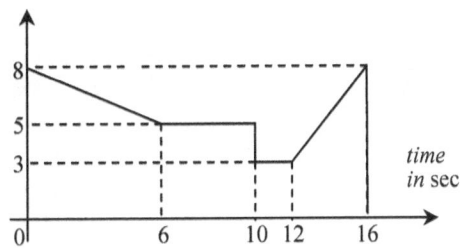

Figure E.5(1) A displacement versus time function

alternate cycle is taken out starting from the first one.

6. Apply SIMULINK blocks to model the following control signals and verify their wave shapes inspecting in Scope: (a) a sine wave whose frequency, amplitude, and duration are $f=30\ Hz$, $A=3.8\ V$, and 0.2 secs respectively is triggered at 12.5% of the period (b) displacement signal of figure E.5(1) (c) expression based signal $e^{-t}\sin 2t$ over $0 \le t \le 6 \sec$.

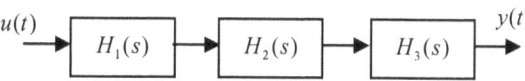

Figure E.5(2) Three control systems connected in series

7. The SISO control system in figure 5.8(a) applies $u(t)=1\ V$ and $H(s) = \dfrac{5s^2+7s+42}{s^3+8s^2+40s}$. Obtain the output $y(t)$ wave shape over $0 \le t \le 0.2 \sec s$ through SIMULINK modeling.

8. An input $u(t)=1\ V$ is applied to the series control system in figure E.5(2) with $H_1(s) = \dfrac{6.7}{(s+3)(s+5)}$, $H_2(s) = \dfrac{1}{s^2+7}$, and $H_3(s) = \{A, B, C, D\}$ where

$$A = \begin{bmatrix} -2 & -1 \\ 2 & 3 \end{bmatrix},\ B = \begin{bmatrix} -1 \\ -2 \end{bmatrix},\ C = [2\ \ 1],\ \text{and}\ D = [9].$$ Obtain the output $y(t)$ wave shape over $0 \le t \le 2 \sec s$ through SIMULINK modeling.

9. A sinusoidal input of amplitude $5\ V$ and frequency $100\ Hz$ is applied to the parallel control system in figure E.5(3) where the element transfer functions are taken from problem 8. Obtain the output $y(t)$ wave shape over $0 \le t \le 0.03 \sec s$ through SIMULINK modeling.

10. In figure 3.5(a) a SISO control system is presented. An input $u(t)=t\ V$ is applied to the system (example 1 of section 3.5). Obtain the output $y(t)$ wave shape over $0 \le t \le 2 \sec s$ through SIMULINK modeling.

11. In figure 3.5(c) a MIMO control system is shown. Three inputs $u_1(t)=2.5\ V$, $u_2(t)=0.5\ t\ V$, and $u_3(t)=2.5\ V$ are applied to the system (example 2 of section 3.5). Obtain the output $y_1(t)$, $y_2(t)$, and $y_3(t)$ wave shapes over $0 \le t \le 5 \sec s$ through SIMULINK modeling.

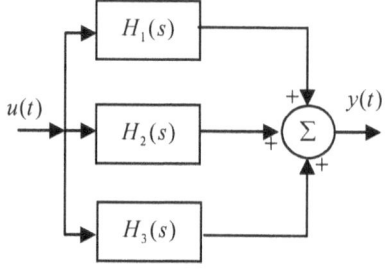

Figure E.5(3) Three control systems connected in parallel

12. Determine the approximate steady state value of the first order system $G(s) = \dfrac{7}{2s+3}$ subject to $u(t)=1.9\ V$ over $0 \le t \le 13 \sec s$ through SIMULINK modeling.

13. Exercise SIMULINK model based on second order control system in figure 5.8(a) and determine the following: (a) steady state value y_{ss} (b) peak value $y(t)|_{max}$ (c) P.O. (d) peak time T_p (e) 20%-60% rise time T_r and (f) 2.5% settling time T_s for $H(s) = \dfrac{3.8}{2s^2 + 0.9s + 3}$ subject to input $u(t) = 2.5\,V$ over $0 \le t \le 50$ secs.

14. Exercise SIMULINK model based on second order control system in figure 4.6(a) and determine the following error performance indices: (a) ISE (b) IAE (c) ITSE (d) ITAE subject to $\zeta = 0.73$, $\omega_n = 3.5\,rad/sec$, $H(s) = 0.8$, and $u(t) = 2.5\,V$ over $0 \le t \le 40$ secs.

Answers:

Since **Scope** output speaks itself that is why SIMULINK models are not provided in many problems.

(1) See section 5.2
(2) See section 5.3
(3) See section 5.4, figure E.5(4) for (a), figure E.5(5) for (b), figure E.5(6) for (c), figure E.5(7) for (d), figure E.5(8) for (e), figure E.5(9) for (f), and figure E.5(10) for (g).
(4) See section 5.5

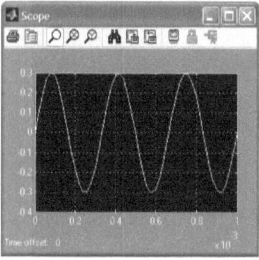
Figure E.5(4) **Scope** output of problem 3(a)

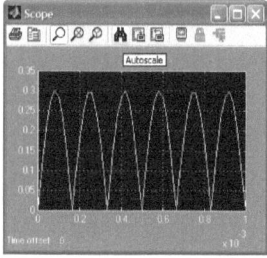
Figure E.5(5) **Scope** output of problem 3(b)

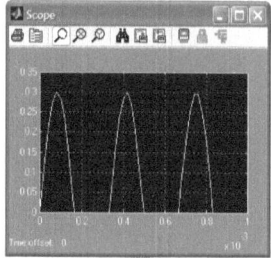
Figure E.5(6) **Scope** output of problem 3(c)

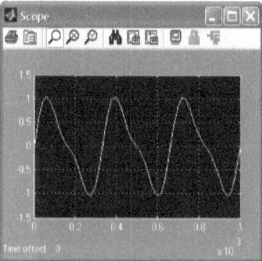
Figure E.5(7) **Scope** output of problem 3(d)

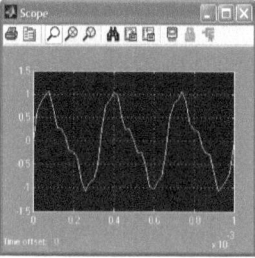
Figure E.5(8) **Scope** output of problem 3(e)

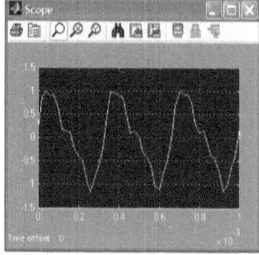
Figure E.5(9) **Scope** output of problem 3(f)

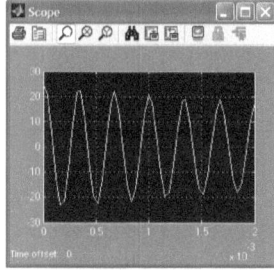
Figure E.5(10) **Scope** output of problem 3(g)

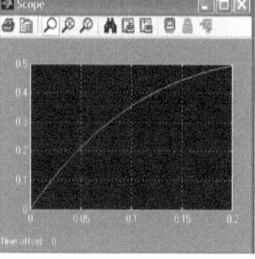
Figure E.5(11) **Scope** output of problem 7

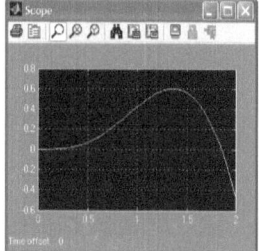
Figure E.5(12) **Scope** output of problem 8

(5) See section 5.6
(6) See section 5.7

(7) See section 5.8, figure E.5(11)
(8) See section 5.8, figure E.5(12)
(9) See section 5.8, figure E.5(13)
(10) See section 5.8, figure E.5(14)
(11) See section 5.8, figures E.5(15), E.5(16), and E.5(17) respectively

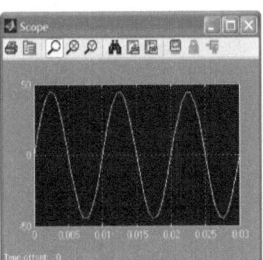

Figure E.5(13) Scope output of problem 9

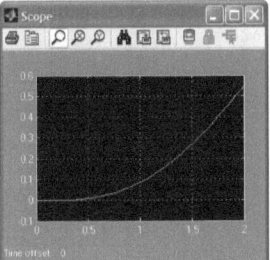

Figure E.5(14) Scope output of problem 10

(12) See section 5.9, y_{ss} =4.433 V
(13) (a) y_{ss} =3.167 V (b) $y(t)|_{max}$=4.912 V (c) P.O.=55.1% (d) T_p=2.7157 secs (e) T_r =0.4974 secs (f) T_s =13.9299 secs   Hint: section 5.9
(14) (a) ISE=2.241 (b) IAE=1.349 (c) ITSE=0.5443 (d) ITAE=0.5132
Hint: section 5.10

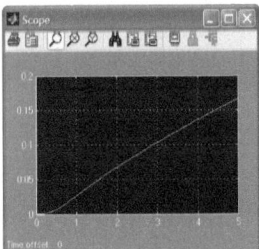

Figure E.5(15) Scope output of problem 11 for y1

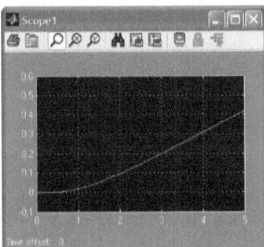

Figure E.5(16) Scope1 output of problem 11 for y2

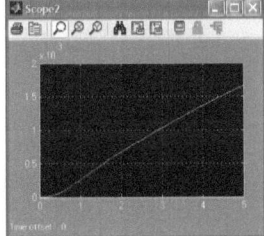

Figure E.5(17) Scope2 output of problem 11 for y3

Mohammad Nuruzzaman

Chapter 6

Control System in Frequency Domain

$\omega \rightarrow \boxed{H(j\omega)} \rightarrow \overset{Y(j\omega)}{\underset{\omega}{\uparrow}}$

Control system analysis does not employ only time domain technique, frequency based analysis plays a momentous role too. Having a control system defined in MATLAB, adequate tools and embedded functions are there to facilitate the frequency domain analysis specially in controller design. Replacing the Laplace variable s of a control system $H(s)$ by $j\omega$ yields the $H(j\omega)$ which is a complex quantity and termed as frequency spectrum. In this chapter all calculations, manipulations, analyses, and graphings are solely confined to ω and $H(j\omega)$ data. Principal frequency based topics which we focus are the following:

- ❖❖ Real-imaginary-magnitude-phase frequency response both in data and graph forms
- ❖❖ Pole-zero map and relevant damping-natural frequency
- ❖❖ Important frequency domain plots – Nyquist and Nichol

6.1 Why frequency domain analysis?

Ultimate objective of control system course is to apply automation to industry related problems or products. A controller circuit is nothing but combination of resistive-inductive-capacitive elements in case of passive circuit controller. Again some electronic elements like diode/transistor become part of the design if we seek for an active controller. A designer may design any controller circuit. The fact of the matter is every electrical system works fine at some frequency or over certain frequency band. Designed controller may fulfill design criteria but may not function at the operating frequency (s) for this reason frequency domain analysis is carried out on a controller circuit. For example power system frequency is 50 or 60 Hz globally. If you design some controller for power system whose operating frequency is 200 Hz, that will not work properly. Besides design specification is sometimes provided in frequency domain. In the following sections we explore the frequency domain tools available in MATLAB for control systems.

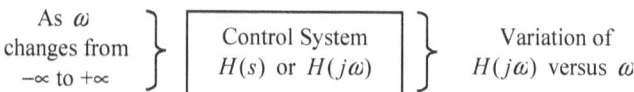

Figure 6.1(a) Frequency response of a control system

6.2 Frequency response of a control system

When we have input frequency which varies from $-\infty$ to $+\infty$, what output response of a control system should be is termed as the frequency response and the strategy is schematized in figure 6.1(a). If we have control system $H(s)$, we replace the s by $j\omega$ and work on $H(j\omega)$. In section 2.1 we addressed how a $H(s)$ is defined by **tf**, **zpk**, or **ss**. The same entering style applies here too. Built-in **freqresp** (abbreviation for frequency response) helps us compute the $H(j\omega)$ as will be explained in the following.

✦ **Value of $H(j\omega)$ at a particular angular frequency**

Consider the control system $H(s) = \frac{s}{s+1}$. For example at $\omega = 4 \, rad/sec$, the $H(j\omega)$ (i.e. $H(j\omega) = \frac{j\omega}{j\omega+1}$) becomes $0.9412 + j\,0.2353$ which we wish to compute.

The **freqresp** has a syntax **freqresp**(system, required angular frequency) which we call as follows:
```
>>H=tf([1 0],[1 1]); ↵       ← Defining H(s) as in section 2.1, H⇔ H(s)
>>R=freqresp(H,4) ↵          ← Workspace R is any user-chosen variable

R =
     0.9412 + 0.2353i        ← R holds the H(j4)
```

✦ **Values of the $H(j\omega)$ at a set of angular frequencies**

We wish to compute above $H(j\omega)$ for $\omega = 0$, 10, and 100 rad/sec which should be 0, $0.9901 + j\,0.099$, and $0.9999 + j\,0.01$ respectively.

The **freqresp** also keeps option for returning the frequency response for multiple angular frequencies. In the second input argument of **freqresp**, now we insert the required frequencies as a row matrix as follows:

>>R=freqresp(H,[0 10 100]); ↵ ← Second argument holds frequencies

At this point in MATLAB context, the return to R (user-chosen variable name) for the set of frequencies is a three dimensional array with the first two dimensions empty. Discussion of the three dimensional array is beyond the scope of text (reference 34). To remove the first two singleton dimensions from the R, we employ the command **squeeze** on the R as follows:

>>V=squeeze(R) ↵   ← The V is any user-chosen variable name

V =
   0
   0.9901 + 0.0990i
   0.9999 + 0.0100i

As the return says, computed values of $H(j\omega)$ are available as a column matrix in the workspace variable V for the three frequencies respectively. The V(1), V(2), and V(3) return the three frequencies separately respectively.

♦ **Range of $H(j\omega)$ values for a range of angular frequencies**

Very often for graphing or analysis reason we need to have $H(j\omega)$ values for a range of frequencies. For instance we intend to obtain ongoing $H(j\omega)$ over $-10 \leq \omega \leq 10\ rad/sec$ with a ω step $0.1\ rad/sec$ (i.e. $\Delta\omega = 0.1\ rad/sec$). Under this circumstance we generate the ω vector as a row matrix (chapter 1) using the colon operator by executing first w=-10:0.1:10; and then R=freqresp(H, w); V=squeeze(R); at the command prompt. Though the second input argument of **freqresp** is a row matrix, the return to V is a column matrix.

♦ **Separating various components of $H(j\omega)$**

Once we have the $H(j\omega)$ calculated, further requirements can be the separation of real, imaginary, magnitude, and phase angle components of $H(j\omega)$ i.e. $Re\{H(j\omega)\}$, $Im\{H(j\omega)\}$, $|H(j\omega)|$, and $\angle H(j\omega)$ which require the uses of built-in functions **real, imag, abs,** and **angle** respectively. For instance values stored in just mentioned V can be separated by using the commands **real(V), imag(V), abs(V),** and **angle(V)** respectively – in single or multiple frequency case. The return from the **angle** is by default in radian. If we wish to see the return in degrees, use the command **rad2deg**. We may assign the return from each of these four functions to some chosen variable for further numerical processing.

◆ **Decibel (dB) values of $H(j\omega)$ magnitudes**

Sometimes it is desirable to have decibel (dB) values of the magnitude $|H(j\omega)|$ which happens through the command **20*log10(abs(V))** on earlier V where $\log_{10} x$ has the code **log10(x)** and the decibel is defined as $20\log_{10}|H(j\omega)|$.

◆ **Horizontal frequencies in terms of Hz instead of rad/sec**

We might be interested in Hertz (Hz) frequency instead of angular one ($rad/\sec$) for what reason we employ **f=w/2/pi** where f is user-chosen and holds the Hertz frequency as column matrix (because $f = \dfrac{\omega}{2\pi}$).

◆ **Bypassing $\log_{10} 0 = -\infty$ point in dB values**

In the decibel plot it is very common that $|H(j\omega)|=0$ for some frequency. Since $\log_{10} 0 = -\infty$, the computer prints some error message. Under this type of situation, we add a negligible positive quantity epsilon to the $|H(j\omega)|$ values, which has the MATLAB code **eps**. For the ongoing system function, we should use **20*log10(eps+abs(V))**.

◆ **Suppressing high dB values of $H(j\omega)$ magnitudes**

Even though $\log_{10} 0 = -\infty$ is overcome by using the **eps**, the values of **20*log10(eps)** become too much negative like −300 dB or so which has no practical importance. In most system analysis, we restrict the lowest dB as −50 or −60 dB. If any dB value is less than −50 or −60, we force that to be −50 or −60. This sort of dB axes manipulation needs some programming technique. Before we do that, let us assign the calculated dB values of ongoing V to some variable D (any user-chosen name) as follows:

>>D=20*log10(eps+abs(V)); ↵ ← D holds the dB values as a column matrix

Let us say any dB value stored in D less than −50 will be set to −50. For this purpose the function **find** (appendix D.6) becomes useful as follows:

>>r=find(D<=-50); D(r)=-50; ↵

Concerning above implementation, the r (any user-chosen variable) holds the integer position index of column matrix D where $20\log_{10}|H(j\omega)| \le -50$ (conducted by the command **r=find(D<=-50);**). Only $20\log_{10}|H(j\omega)| \le -50$ elements in the D are set to −50 by writing the command **D(r)=-50;**. Thus the last D holds the expected $20\log_{10}|H(j\omega)|$ values as a column matrix.

♦ **Normalization of $H(j\omega)$ magnitudes with respect to maximum**

Some control system may not have $|H(j\omega)|$ values ranging 0 to 1. If dB variation is needed from 0 to some value, the $|H(j\omega)|$ must be between 0 and 1. For example the control system $H(s)=\dfrac{3s}{s+1}$ shows non 0-1 variation. Let us form the system as follows:

>>H=tf([3 0],[1 1]); ↵   ← Defining $H(s)$, H⇔ $H(s)$

We wish to find $|H(j\omega)|_{max}$ over the interval $-10 \le \omega \le 10$ rad/sec with a ω step 0.1 rad/sec. First we determine the $|H(j\omega)|$ values by applying earlier functions at indicated ω points as follows:

>>w=-10:0.1:10; R=freqresp(H,w); V=squeeze(R); ↵

We know that the last V is holding complex $H(j\omega)$ values as a column matrix. Appendix D.5 cited **max** finds the maximum of $|H(j\omega)|$ as follows:

>>M=max(abs(V)) ↵   ← M holds the $|H(j\omega)|_{max}$ value

M =
    2.9851

Normalization means finding the $\dfrac{H(j\omega)}{|H(j\omega)|_{max}}$ values so we just divide the V by the M as follows:

>>S=V/M; ↵ ← S holds the normalized complex $H(j\omega)$ values where S is user-chosen variable

Obviously the S is a column matrix as well.

6.3 Graphing frequency spectrum of a control system

Last section demonstrates frequency response of a control system for the most part on calculation context. This section is all about graphing the frequency response of a control system.

Frequency response of a control system $H(j\omega)$ in graphical form has four components namely Re{$H(j\omega)$}, Im{$H(j\omega)$}, $|H(j\omega)|$, and $\angle H(j\omega)$ versus ω over certain interval of ω. To graph the frequency response, there can be two options – either get the data and plot on your own or use the readymade function like **bode**, both of which are addressed in the following.

♦ **Getting data first and plotting the response afterwards**

In section 6.2 we have addressed computations all along on $H(s)=\dfrac{s}{s+1}$ or $H(j\omega)=\dfrac{j\omega}{j\omega+1}$. We wish to plot the $|H(j\omega)|$ versus ω over $0 \le \omega \le 10$ rad/sec by choosing a ω step 0.01 rad/sec.

First we follow the techniques of section 6.2 for obtaining the frequency spectrum data as follows:

>>H=tf([1 0],[1 1]); ↵   ← Defining the $H(s)$, H⇔ $H(s)$

```
>>w=[0:0.01:10]';  ↵   ← The w holds the $\omega$ values as a column matrix
>>R=freqresp(H,w); ↵   ← freqresp calculates $H(j\omega)$ and assigns to R
>>V=squeeze(R); ↵      ← Removing singleton dimensions, V holds $H(j\omega)$
                         complex values as a column matrix
```

After that we employ appendix E cited **plot** as follows:
```
>>plot(w,abs(V)) ↵   ← Graphing the $|H(j\omega)|$ versus $\omega$ like figure 6.1(b)
```

The **plot** has two input arguments, the first and second of which are the ω and $|H(j\omega)|$ data as a row or column matrix respectively. In a similar fashion the commands **plot(w,real(V))**, **plot(w,imag(V))**, and **plot(w,angle(V))** graph the spectra $Re\{H(j\omega)\}$ versus ω, $Im\{H(j\omega)\}$ versus ω, and $\angle H(j\omega)$ versus ω respectively (graphs are not shown for space reason). Not only that, decibel spectrum of

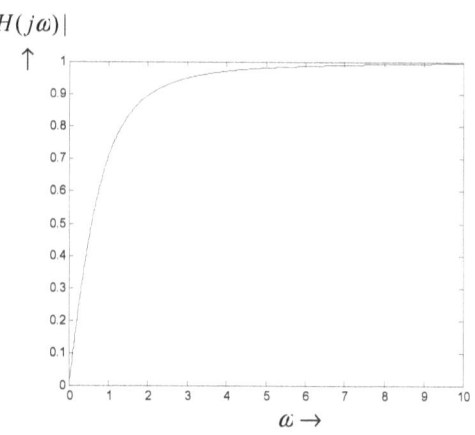

Figure 6.1(b) Plot of $|H(j\omega)|$ versus ω

section 6.2 (whose values are in **D** in the section) is graphed too by the **plot** by using **plot(w,D)** regardless of with or without dB suppression.

✦ Employing readymade bode plotter

MATLAB is so resourceful in programming and built-in function sense that we may have different options for the same type of problem. The readymade tool means we use bode plot to see the control system frequency response but mainly the magnitude and phase ones. The function we employ is **bode** which keeps provision for accepting different input-output arguments as will be explained in the following.

Graphing a system function:
We wish to plot the $H(j\omega)$ over ω for the control system $H(s) = \frac{s}{s+1}$ (or $H(j\omega) = \frac{j\omega}{j\omega+1}$) from 0.1 to 100 $rad/\sec$ by using the command **bode** and do so at the command prompt as follows:
```
>>H=tf([1 0],[1 1]); ↵ ← Defining the $H(s)$, H⇔$H(s)$
>>bode(H,{0.1,100}) ↵
```
Now the **bode** has two input arguments, the first and second of which are the system function assignee name and the ω interval

description respectively. In the ω interval description we use the second brace (not the first, nor the third brace) and input only bounds of the interval, ω step size is automatically chosen by the **bode**. Figure 6.1(c) presents the bode plot for the example $H(j\omega)$. By default the **bode** returns decibel spectrum $20\log_{10}|H(j\omega)|$ and phase spectrum $\angle H(j\omega)$ in degrees versus ω together.

There are different options hidden in bode plot of figure window 6.1(c). Rightclick on the mouse in plot area of figure 6.1(c), find the **Properties** in a popup, and click the **Properties** to see prompt window like figure 6.1(d).

To graph $|H(j\omega)|$ versus ω over the same interval, we click **Units** menu of the figure 6.1(d) and select **Absolute** in the **Magnitude** in popup.

To graph $|H(j\omega)|$ versus frequency in Hertz over the same interval, we click **Units** menu of the figure 6.1(d) and select **Hz** in the **Frequency** in popup.

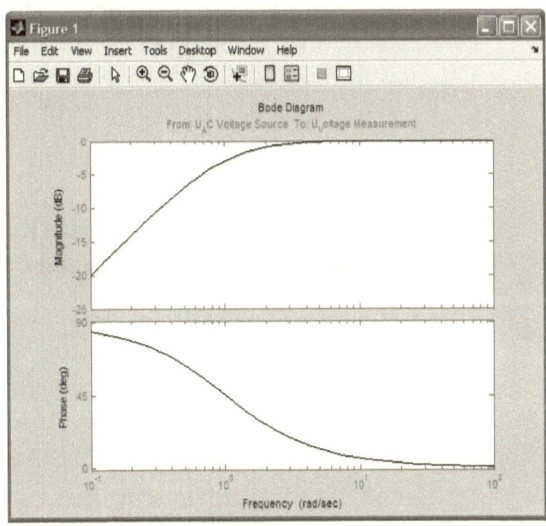

Figure 6.1(c) **Bode** plot of a control system

If you want to see only the $20\log_{10}|H(j\omega)|$ versus ω, rightclick on the mouse in plot area of figure 6.1(c), find the **Show** in a popup, and click the **Phase** under the **Show** to see only the magnitude spectrum. Similarly you can view only the phase spectrum $\angle H(j\omega)$ by clicking the **Magnitude** under the **Show**.

The default phase in the bode plot is in degree. If you

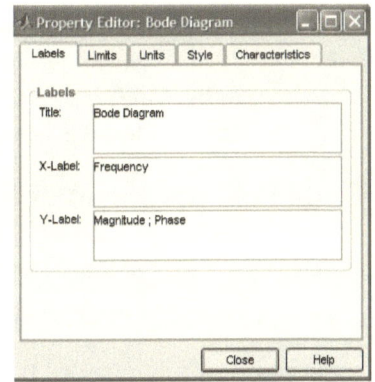

Figure 6.1(d) **Property** editor for the bode plot

want to turn that to radian, click the **Units** menu of figure 6.1(d) and select the **radians** in the **Phase in** popup.

The default title of figure 6.1(c) is **Bode Diagram**. If you intend to change that to some other for example **System Response**, click **Labels** menu of the figure 6.1(d) and type **System Response** from keyboard deleting the **Bode Diagram** under the **Title** slot of the **Labels**. In a similar fashion you can change the horizontal and vertical axes labeling of figure 6.1(c).

Mouse based access on the graph data:

Mouse pointer helps us find frequency spectrum data. For this bring the mouse pointer at any point on the **bode** drawn curve (figure 6.1(c)) and leftclick the mouse. An indicatory box appears on top of the **bode** plot in which you find the mouse point coordinates of horizontal and vertical axes quantities.

Adding grid lines to the bode graph:

Figure 6.1(c) shows only the bode graph. If you wish to include grid lines to horizontal and vertical axes, execute the command **grid** at the command prompt.

Graphing multiple control systems:

We have been exercising all along keeping the $H(j\omega)$ in the H. Suppose there is another control system available in workspace variable H1 (i.e. $H_1(j\omega)$) and we wish to plot the $H(j\omega)$ and $H_1(j\omega)$ over the same ω interval. The **bode** also keeps the provision for graphing multiple control systems. For the graphing, the command we need is **bode(H,H1,{0.1,100})** – the names of assignee holding the control systems and interval description all separated by a comma respectively. In case of three control systems, we just employ the command **bode(H,H1,H2,{0.1,100})** where H2 indicates the third control system over the same interval.

Note: The **bode** does not function for negative frequencies unlike the **freqresp** of last section and its return is always in magnitude-phase angle form.

6.4 Pole-zero map of a control system

Pole-zero map is widely used in control system analysis to test the stability of the system. The concept of pole-zero is applicable if the given control system is in rational form $\frac{P(s)}{Q(s)}$, the roots on forming equations $Q(s)$ =0 and $P(s)$=0 are called poles and zeroes of the system respectively. Given a control system $Y(s)$, MATLAB built-in function **pzmap** (abbreviation for pole zero map) locates the poles and zeroes of $Y(s)$ in the s plane which is complex in general. Denoted by symbol one writes $s = \sigma + j\omega$ indicating the real and imaginary components in s plane. The syntax we apply is **pzmap**(system) where the system is defined in accordance with chapter 2. In s plane, poles and zeroes are differentiated by the markers × and o respectively.

Let us find the pole-zero map for the control system $Y(s) = \dfrac{2(2s+1)}{s^2+1}$.

The poles (means roots of $s^2 + 1 = 0$) and zeroes (means roots of $2(2s+1) = 0$) of $Y(s)$ are $s = \pm j$ and $s = -\dfrac{1}{2}$ respectively. First we define the system by **tf** of section 2.1 as follows:

>>Y=tf([4 2],[1 0 1]); ↵

In above execution the Y is any user-chosen variable which holds $Y(s)$. Then we call the **pzmap** on entered Y as follows:

>>pzmap(Y) ↵

Response is shown in figure 6.2(a). Arrows in figure 6.2(a) indicate relative positions of the poles and zeroes of $Y(s)$.

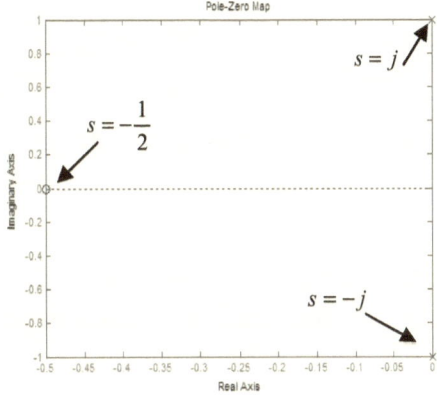

Figure 6.2(a) Pole-zero map of the system $Y(s)$

Horizontal and vertical axes of the figure 6.2(a) refer to σ and ω respectively. In the case of multiple poles or zeroes, only one is displayed. Let us go through the following relevant problems.

▣ Extracting poles and zeroes from a control system

Having the $Y(s)$ stored in Y, how do we extract poles and zeroes from the Y? The answer is exercise **pole** and **zero** commands on the Y respectively:

>>P=pole(Y) ↵

P =

    0 + 1.0000i
    0 - 1.0000i

The P is any user-chosen variable which retains the poles. Similarly you may execute Z=zero(Y) for the zeroes where the Z is a user-chosen variable. As another option you may use **pzmap** with two output arguments like [P,Z]= pzmap(Y) to avoid writing more commands.

▣ Pole/zero for series/parallel/feedback systems

You can analyze the pole-zero map for additive control systems e.g. $Y(s) = \dfrac{2(2s+1)}{s^2+1} + \dfrac{-7s^4 + 2s^3 + 24}{s^5}$ with little manipulation. When the control

-163-

systems are in addition form, they are treated as parallel system and the equivalent is obtained by the **parallel** as follows (chapter 2):

>>Y1=tf([4 2], [1 0 1]); ↵ ← Workspace Y1 holds $\frac{2(2s+1)}{s^2+1}$

>>Y2=tf([-7 2 0 0 24], [1 0 0 0 0 0]); ↵ ← Workspace Y2 holds $\frac{-7s^4+2s^3+24}{s^5}$

>>Y=parallel(Y1,Y2); ↵ ← Workspace Y holds the system $Y(s)$

The Y, Y1, and Y2 are user-chosen variables. Now execute the command **pzmap(Y)** or **[P,Z]=pzmap(Y)** for the analysis.

As an example of multiplied system consider $Y(s) = \frac{2(2s+1)}{s^2+1} \times \frac{-7s^4+2s^3+24}{s^5}$ where component systems are aforementioned ones. You just need to use Y=series(Y1,Y2); [P,Z]=pzmap(Y) for the analysis. In a similar fashion feedback system is also handled by first finding the equivalent by **feedback**.

6.5 Natural frequency and damping of a control system

Starting from a control system we may seek for the natural angular frequency (ω_n) and damping ratio (ζ) of the system. These two quantities are connected with the poles in the s plane. We know that $s = \sigma + j\omega$, $\omega = \omega_n\sqrt{1-\zeta^2}$, and $\sigma = \zeta\omega_n$. Every pole of a control system provides one set of (ω_n, ζ). The embedded function **damp** provides (ω_n, ζ) set for every pole in the system with the syntax [variable for ω_n, variable for ζ]=**damp**(control system). The control system is defined by the functions of section 2.1.

For example the control system $Y(s) = \frac{2s+1}{s^2+2}$ has the poles $\pm j\sqrt{2}$ indicating $\zeta = 0$ and $\omega_n = \pm 1.414 \, rad/\sec$ from $\sigma = 0$ and $\omega = \pm\sqrt{2}$. First we enter the system with the help of **tf** as follows:

>>Y=tf([2 1],[1 0 2]); ↵

Then call the **damp** as:

>>[wn,z]=damp(Y) ↵ ← Workspace wn and z hold ω_n and ζ respectively

wn =
    1.4142
    1.4142
z =
    0
    0

In above implementation **wn** and **z** are user-chosen variables. Since one complex conjugate poles give birth to one angular frequency, we see one value of ω_n in above return. As you see, zero (which is $-1/2$) has no role in the damping.

If a pole is located in the right half s plane, one ζ becomes negative. You may verify that for $Y(s) = \dfrac{2s+1}{s^2 - 2}$ by executing Y=tf([2 1],[1 0 -2]); [wn,z]=damp(Y).

There is another feature hidden in the **damp**, you may display the information about eigenvalue, ζ, and ω_n excluding the output argument i.e.

```
>>damp(Y) ↵
     Eigenvalue            Damping      Freq. (rad/s)
0.00e+000 + 1.41e+000i    0.00e+000     1.41e+000
0.00e+000 - 1.41e+000i    0.00e+000     1.41e+000
```

According to appendix A, the return e.g. 0.00e+000 is $0 \times 10^0 = 0$ this is just the MATLAB numeric style.

Constant damping ratio and natural angular frequency curves

Having drawn a pole-zero map by the **pzmap** of last section, we may have a look at s plane poles and zeroes for constant ζ and ω_n for design reason. The command **sgrid** without any input argument facilitates that. Reexecute the parallel system related commands of last section:

```
>>Y1=tf([4 2], [1 0 1]); Y2=tf([-7 2 0 0 24], [1 0 0 0 0 0]); Y=parallel(Y1,Y2); ↵
>>pzmap(Y) ↵
```

Afterwards we conduct the following:

```
>>sgrid ↵
```

Figure 6.2(b) is the outcome from above execution. There are two kinds of dotted traces in the figure; straight line and curve. They refer to constant damping ratio ζ and constant natural angular frequency ω_n curves respectively. For instance the arrow indicated ones in the figure correspond to $\zeta = 0.3$ and $\omega_n = 1.25$ rad/sec. Since the

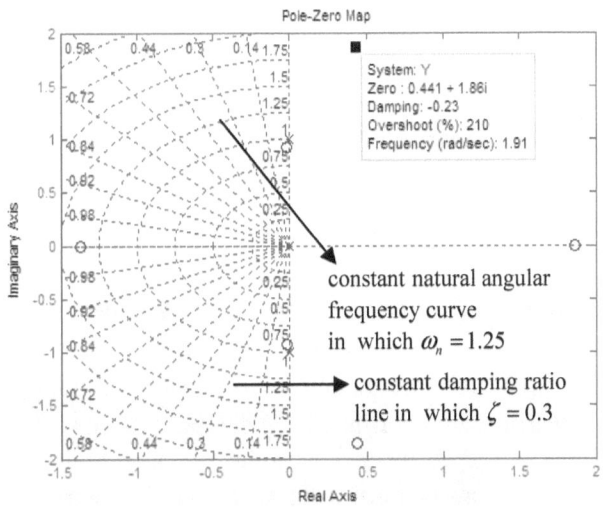

Figure 6.2(b)   Pole-zero map with s plane grid

right half s plane is the zone of control system instability, no ζ and ω_n lines are drawn in the right half. If you bring mouse pointer at any pole or zero on the pole-zero map and leftclick the mouse, you see its (ζ, ω_n) information. Every ζ line is symmetric about the $\omega = 0$ or horizontal axis and makes an

angle $\cos^{-1}\zeta$ in the lefthalf s plane. For example $\zeta=0.3$ makes an angle 72.5424^0 which we get by:
>>acosd(0.3) ↵   ← Appendix A for the trigonometric function

ans =
        72.5424

Obviously there are two such lines in the figure 6.2(b).

⊟ **User-defined constant (ζ, ω_n) on pole-zero map**

Suppose you have drawn a pole-zero map of a control system. On top of that you wish to place constant (ζ, ω_n) line/curve. The **sgrid** helps us do that but it needs two input arguments i.e. sgrid(ζ, ω_n). Say $\zeta=0.7$ and $\omega_n=1.5\ rad/sec$ then for the ongoing parallel system we have to exercise:
>>pzmap(Y) ↵
>>sgrid(0.7,1.5) ↵

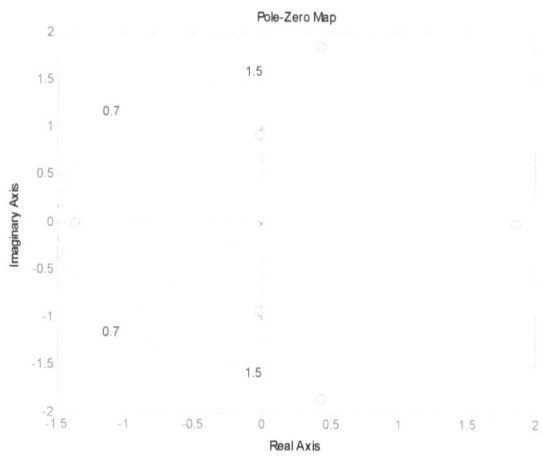

Figure 6.2(c)   Pole-zero map with user-defined (ζ, ω_n)

Figure 6.2(c) is the lucid attachment of that.

6.6 Gain and phase margins of a control system

The gain and phase margins are important when design of a control system is performed in frequency parameters. Given a control system $H(s)$, its magnitude and phase components are $|H(j\omega)|$ and $\angle H(j\omega)$ respectively. In dB scale the magnitude spectrum is $20\log_{10}|H(j\omega)|$. The gain and phase margins are interlinked and defined as follows:

gain margin: how far dB magnitude spectrum or $20\log_{10}|H(j\omega)|$ is away from 0 dB when phase spectrum $\angle H(j\omega)=-180^0$ is called the gain margin and

phase margin: how far phase spectrum or $\angle H(j\omega)$ is away from -180^0 when magnitude spectrum or $20\log_{10}|H(j\omega)|=0$ dB is called the phase margin.

Both spectra are against ω. The ω for gain and phase margins are called -180^0 crossover angular frequency and $0\ dB$ crossover angular frequency respectively. Since the magnitude spectrum is in dB, above happens also in bode plot of the control system as addressed in section 6.3. MATLAB built-in function **margin** determines the two margins provided that the control system is its input argument. The control system is defined in accordance with section 2.1. In order to test the **margin** by a control system, let us choose $Y(s) = \dfrac{234}{s(s+3)(s+1.5)}$. We intend to find the two margins for this system.

The given system is best entered by the pole-zero-gain form and do so by:
>>Y=zpk([ ],[0 -3 -1.5],234); ↵

Just call the function as:
>>margin(Y) ↵

Figure 6.3(a) is the outcome from the last command line. In the title of the graph you find indication of the two margins which clearly specifies that the gain margin is $-21.3\,dB$ which occurs at $\omega = 2.12\,rad/\sec$ and the phase margin is -48.6^0 which occurs at $\omega = 5.87\,rad/\sec$.

Should you need to access the margin values, exercise four output arguments (chapter 1) in addition to the **margin** as follows:
>>[Gm,Pm,wg,wp]=margin(Y) ↵

Gm =
    0.0865
Pm =
    -48.5663
wg =
    2.1213
wp =
    5.8654

In above execution the variables **Gm**, **Pm**, **wg**, and **wp** are user-chosen and refer to gain margin, phase margin, gain margin associated angular frequency, and phase margin associated angular frequency respectively. The gain margin return is not is dB scale instead in absolute scale that is why 0.0865 is assigned to **Gm**. For the dB margin, you may use the command 20*log10(Gm).

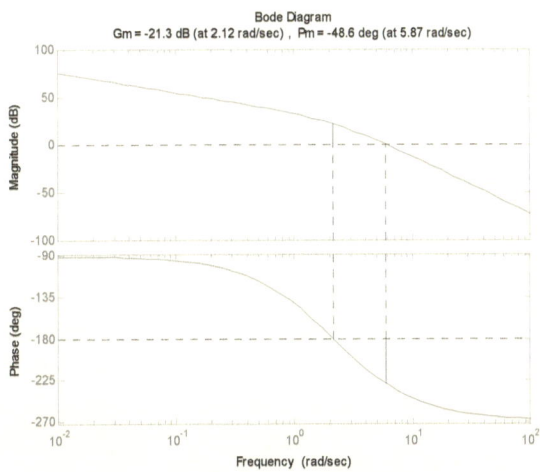

Figure 6.3(a) Gain and phase margins of a control system

Some warning might be seen on MATLAB screen because of random control system selection. Without the loss of generality, you may ignore the warning.

6.7 Nyquist plot of a control system

We know that the bode plot of a control system $H(s)$ is basically $20\log_{10}|H(j\omega)|$ and $\angle H(j\omega)$ versus ω considering the frequency spectrum $H(j\omega)$. Nyquist plot of the control system is basically the plot of Re{$H(j\omega)$} and Im{$H(j\omega)$} versus ω. In either plot it is assumed that ω varies commonly i.e. one coordinate intended for bode {$20\log_{10}|H(j\omega)|, \angle H(j\omega)$} or nyquist { Re{$H(j\omega)$}, Im{$H(j\omega)$} } for every single ω. The built-in function **nyquist** helps us obtain the Nyquist plot of a control system. Several cases of the plot are addressed in the following.

🗗 Just the Nyquist plot

Take the example of $H(s) = \dfrac{2s+1}{s^2+3s+9}$. We wish to obtain the Nyquist plot of $H(s)$.

The $H(s)$ is entered by using section 2.1 mentioned functions thereupon exercise the **tf** as follows:

>>H=tf([2 1],[1 3 9]); ↵

Call the plotter on the entered system:

>>nyquist(H) ↵

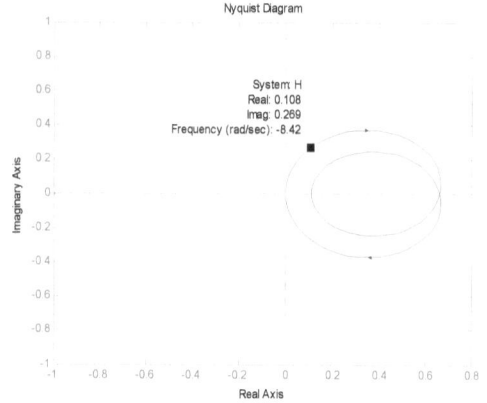

Figure 6.3(b) Nyquist plot of a control system

Figure 6.3(b) depicts the Nyquist plot in which the **Real Axis** and **Imaginary Axis** correspond to Re{$H(j\omega)$} and Im{$H(j\omega)$} respectively. The plot is a symmetric one about the real axis and as ω changes from $-\infty$ to $+\infty$ the curve becomes a contour or closed one. The arrow on the plot indicates the ω change direction. Leftclick the mouse pointer at any point on the curve and find the related coordinates for example in figure 6.3(b) we see Re{$H(j\omega)$}=0.108 and Im{$H(j\omega)$}=0.269 at $\omega = -8.42\, rad/\sec$.

🗗 Access to Nyquist plot values

Should you need to access the real and imaginary spectra samples, decide ω as a vector and use two output arguments, the first and second of which are the real and imaginary spectra values respectively. Under this circumstance no graph is drawn. For example the sample values of Re{$H(j\omega)$} and Im{$H(j\omega)$} over the ω interval $0 \leq \omega \leq 0.5\,rad/\sec$ with $\Delta\omega = 0.1\,rad/\sec$, we obtain by:

>>w=0:0.1:0.5; ↵   ← Generating ω samples
>>[R,I]=nyquist(H,w); ↵ ← R,I are user-chosen, R holds Re{$H(j\omega)$} samples, and I holds Im{$H(j\omega)$} samples

```
>>R=squeeze(R); I=squeeze(I); ↵ ← Section 6.2 for squeeze, same variable is
                                    chosen for squeeze
```

You may view the samples side by side by exercising [R I] at the command prompt.

⊟ **A segment of Nyquist plot**

A segment of the Nyquist plot you may view by exercising nyquist(H,{ ω_{min} , ω_{max} }) where ω_{min} >0. For example the plot over $0.1 \le \omega \le 5 rad/\sec$ we view by nyquist(H,{0.1,5}).

⊟ **Nyquist plot with dB magnitude lines**

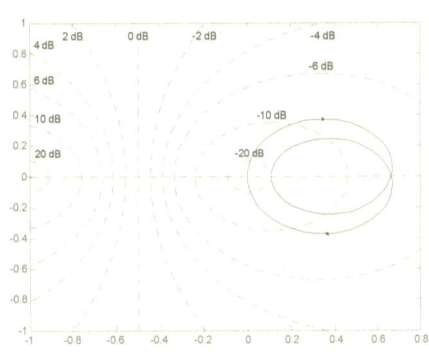

Figure 6.3(c) Inclusion of dB magnitude lines in Nyquist plot

Having a Nyquist plot obtained, you can add dB magnitude lines on the plot by the following action: rightclick the mouse on the Nyquist plot area, find the **Grid** in a popup, and click the **Grid**, the result is seen in figure 6.3(c). You can remove the dB magnitude lines by the same action.

⊟ **Identifying positive/negative angular frequencies**

Since the Nyquist plot is a closed one, it is difficult to identify which part of the curve is for the positive or negative frequencies. In order to know that, rightclick the mouse on the Nyquist plot area, find the **Show** in a popup, click the checked **Negative Frequencies**, and find only the plot for positive angular frequencies. By doing same action, you reach to the starting graph.

⊟ **Identifying the peak response**

Staying in the Nyquist plot, we can know about the peak $|H(j\omega)|$. For this rightclick the mouse on the Nyquist plot area, find the **Characteristics** in a popup, click the **Peak Response** under **Characteristics**, and find a bold dot on the Nyquist plot indicating information about the peak response.

⊟ **Zooming around (-1,0)**

Sometimes we need to zoom around (−1,0) point. This facility is also included. Rightclick the mouse on the Nyquist plot area, find the **Zoom on (-1,0)** in a popup, and click it.

6.8 Nichol's chart of a control system

By definition the Nichol's chart is $|H(j\omega)|$ versus $\angle H(j\omega)$ graph of a control system as ω varies from 0 to ∞. Embedded function **nichols** helps us obtain Nichol's chart or its related quantities. Its syntax is similar to the

nyquist of last section. Consider the last section mentioned $H(s)$. Only the graph we view by:
>>nichols(H) ⏎   ← Figure 6.3(d) shows the plot

In figure 6.3(d) the horizontal and vertical axes refer to $\angle H(j\omega)$ and $20\log_{10}|H(j\omega)|$ which you see as **Open-Loop Phase (deg)** and **Open-Loop Gain (dB)** respectively.

Click at any point on the curve of figure 6.3(d) and find its corresponding coordinate e.g. the shown one is $20\log_{10}|H(j\omega)|=-7.76\,dB$ and $\angle H(j\omega)=-57.4^{\circ}$ which occurs at $\omega=5.97\,rad/\sec$.

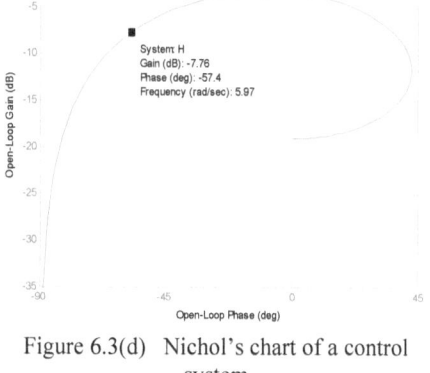

Figure 6.3(d) Nichol's chart of a control system

It is possible to include grid lines on the chart of figure 6.3(d). Rightclick the mouse on the plot area, find the **Grid** in a popup, click it, and see the figure 6.3(e) with grid lines.

The peak spectrum value we view by the following action: rightclick the mouse on the plot area, find the **Characteristics** in a popup, click the **Peak Response** under **Characteristics**, find a bold dot on the Nichol's plot, and click the dot which indicates information about the peak response.

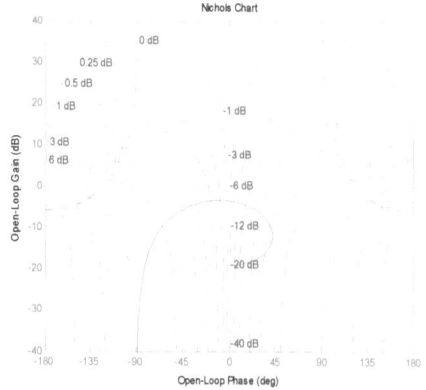

Figure 6.3(e) Nichol's chart with grid lines

A segment of the Nichol's chart you may view by exercising **nichols(H,{ ω_{min} , ω_{max} })** where $\omega_{min} > 0$. For example the plot over $0.1 \le \omega \le 5 rad/\sec$ we view by **nichols(H,{0.1,5})**.

Value accessing is achievable too e.g. obtain samples of $|H(j\omega)|$ and $\angle H(j\omega)$ with $\Delta\omega=0.1\,rad/\sec$ over ω interval $0 \le \omega \le 0.5 rad/\sec$ by:
>>w=0:0.1:0.5; ⏎   ← Generating ω samples
>>[M,P]=nichols(H,w); ⏎ ← M,P are user-chosen, M holds $|H(j\omega)|$ samples not in dB instead in absolute scale and P holds $\angle H(j\omega)$ samples in deg
>>M=squeeze(M); P=squeeze(P); ⏎ ← Section 6.2 for **squeeze**, same variable is chosen for **squeeze**

6.9 Control system DC gain and bandwidth

Here in this section we address the static or DC gain and bandwidth of a control system.

⊟ DC Gain

DC gain of a control system $H(s)$ is defined as $\underset{s \to 0}{Lt} H(s)$. This is basically the 0 frequency response of $H(j\omega)$. Built-in function **dcgain** computes the gain with syntax **dcgain(system)** where the system is defined by **tf**, **zpk**, or **ss** in accordance with section 2.1. Suppose $H(s) = \dfrac{7s^3 - 7s + 42}{s^4 - 118s^2 - 240s + 2}$ has the gain 21 which we wish to obtain.

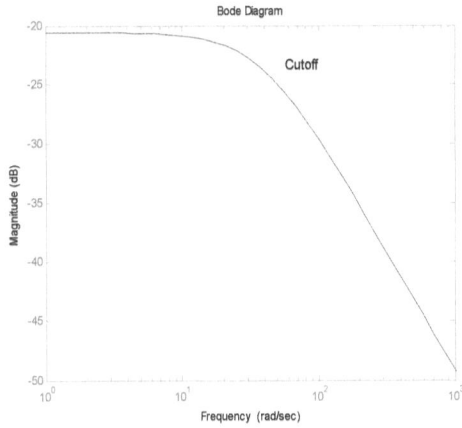

Figure 6.4(a) Lowpass behavior of a control system

Enter the system by:
>>H=tf([7 0 -7 42],[1 0 -118 -240 2]); ↵ ← H is user-chosen, holds $H(s)$
>>dcgain(H) ↵

ans =
            21

In section 2.1 a state-space system is defined. You will find the gain as 3 by exercising **dcgain(H)** for the system. Infinity gain is notified by **Inf**.

⊟ Control system bandwidth

The bandwidth concept of a control system $H(s)$ depends on the magnitude spectrum behavior of $H(j\omega)$ i.e. $|H(j\omega)|$ versus ω. As you know there are four widely known types of response – lowpass, highpass, bandstop, and bandpass. The first step of bandwidth finding is view the magnitude spectrum using **bode** of section 6.3 and be confirmed about the type of response. Bandwidth for each response

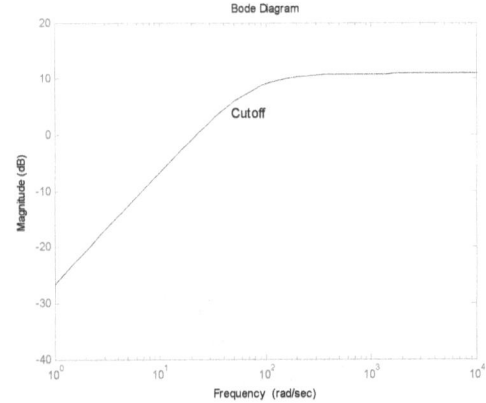

Figure 6.4(b) Highpass behavior of a control system

needs slight different technique from that of the other. We address all of them in the sequel.

Lowpass:

Example of a lowpass control system is $H(s) = \dfrac{7}{2s+75}$. Its spectrum we view by:
>>H=tf(7,[2 75]); bode(H) ↵

Graph of figure 6.4(a) certainly shows the lowpass behavior in the magnitude spectrum after running the command. For the lowpass the cutoff frequency we determine by the function **bandwidth**:
>>bandwidth(H) ↵

ans =
    37.4108

The above return is in terms of rad/sec i.e. $\omega_c = 37.4108\ rad/sec$. For this response, the $|H(j\omega)|_{max}$ occurs at $\omega = 0\ rad/sec$. Cutoff frequency is the frequency when $|H(j\omega)|$ becomes $|H(j\omega)|_{max} - 3\ dB$. If you need the Hertz frequency, you may execute **bandwidth(H)/2/pi**.

Highpass:

Example of a highpass control system is $H(s) = \dfrac{7s}{2s+150}$. View the magnitude spectrum by:
>>H=tf([7 0],[2 150]);bode(H) ↵ ← Figure 6.4(b) shows the behavior

The problem with this system is the $|H(j\omega)|_{max}$ occurs at $\omega = \infty$ or $s = \infty$ and computer never deals with infinity. The **bandwidth** does not work here instead **c_cross** is invoked (appendix D.8). In section 6.2 we explained how to get $H(j\omega)$ samples by making the use of **freqresp**. The spectrum of figure 6.4(b) indicates steady state like trajectory for large frequencies (**bode** made it from $10^0\ rad/sec$ to $10^4\ rad/sec$). Step selection here is not useful because of wide range. Appendix D.13 mentioned **linspace** can be used for specific sample number say 1000. With the symbol meaning of section 6.2, let us get the $H(j\omega)$ samples by:
>>w=linspace(10^0,10^4,1000); R=freqresp(H,w); V=squeeze(R); ↵

So V holds $H(j\omega)$ samples from which $|H(j\omega)|$ samples we get by:
>>A=abs(V); ↵   ← A is user-chosen, holds $|H(j\omega)|$ samples

Roughly the last value in A can be taken as $|H(j\omega)|$ for $\omega = \infty$ and get the value by:
>>L=A(end); ↵   ← L is user-chosen, holds the last $|H(j\omega)|$ sample

The L is basically approximate $|H(j\omega)|_{max}$. The cutoff occurs at $|H(j\omega)|_{max}/\sqrt{2}$ and call the **c_cross** accordingly:

```
>>c_cross(w,A,L/sqrt(2)) ↵
```

ans =
         76.0676   ← i.e. $\omega_c = 76.0676\ rad/\sec$

Bandpass:
Bandpass control system is different from the lowpass or highpass because of two cutoff frequencies. Another aspect of the spectrum behavior is the maximum is exhibited neither at $\omega=0$ nor at $\omega=\infty$. The two cutoff frequencies are before and after the maximum. Such example is $H(s) = \dfrac{10}{2s^2 + 0.8s + 500}$. In a similar fashion view the magnitude spectrum like figure 6.4(c) by:

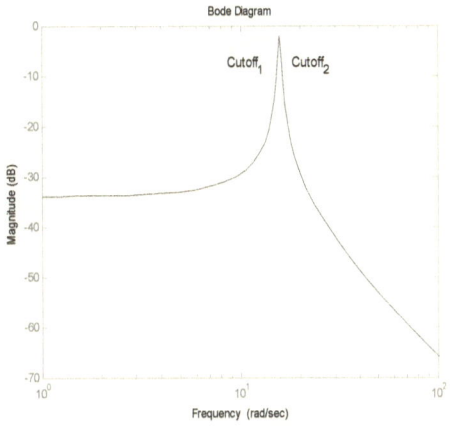

Figure 6.4(c) Bandpass behavior of a control system

```
>>H=tf(10,[2 0.8 500]);bode(H) ↵
```
The figure indicates relative positions of the two cutoffs too. Like the highpass counterpart let us select 1000 sample points for ω displayed by the **bode**:
```
>>w=linspace(10^0,10^2,1000); R=freqresp(H,w); V=squeeze(R); ↵
>>A=abs(V); ↵   ← A is user-chosen, holds $|H(j\omega)|$ samples
```
Appendix D.5 mentioned **max** helps us determine the $|H(j\omega)|_{\max}$ and its corresponding ω. This ω is termed as operating or center frequency:
```
>>[M,I]=max(A); ↵ ← M and I are user-chosen, hold $|H(j\omega)|_{\max}$ and
                      integer index related to frequency respectively
>>wo=w(I) ↵       ← wo is user-chosen, holds the center frequency
```

wo =
         15.7658   ← i.e. center frequency in $rad/\sec$

The two cutoff frequencies we find by the same **c_cross**:
```
>>wc=c_cross(w,A,M/sqrt(2)) ↵   ← wc is user-chosen
```

wc =
         15.6171   16.0135

In last execution the **wc** holds the two cutoff frequencies as a row matrix respectively.

Bandwidth is just the difference between the two frequencies:
```
>>wc(2)-wc(1) ↵
```

```
ans =
        0.3964        ← i.e. bandwidth in $rad/\sec$
```

Bandstop:

Bandstop is similar to the bandpass control system. The crucial point is whatever we computed around the maximum will be conducted now around a minimum. Another sharp contrast is cutoff is computed based on maximum. Anyhow a bandstop control system can be $H(s) = \dfrac{0.2s^2 + 100s + 800000}{0.2s^2 + 900s + 800000}$. Its

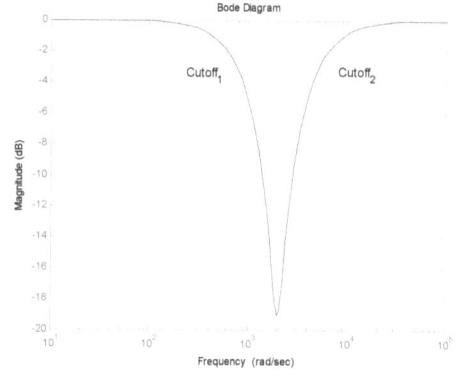

Figure 6.4(d) Bandstop behavior of a control system

related commands are the following with similar symbology:
```
>>H=tf([0.2 100 800000],[0.2 900 800000]);bode(H) ↵
```
Figure 6.4(d) is the magnitude spectrum response over $10^1 \le \omega \le 10^5$ of coarse in $rad/\sec$. The rest commands are:
```
>>w=linspace(10^1,10^5,1000); R=freqresp(H,w); V=squeeze(R); ↵
>>A=abs(V); [M,I]=min(A); ↵
>>wo=w(I) ↵
```

```
wo =
        2.0118e+003
```
The min is for the minimum. The **2.0118e+003** means 2.0118×10^3 which is the center frequency in $rad/\sec$ (appendix A). Despite the minimum, maximum is found for the cutoff by:
```
>>M1=max(A); ↵        ← M1 is user-chosen, holds $|H(j\omega)|_{max}$
>>wc=c_cross(w,A,M1/sqrt(2)) ↵
```

```
wc =
   1.0e+003 *
       0.7607   5.1646
```
Having found the two cutoffs, the bandwidth we get by:
```
>>wc(2)-wc(1) ↵
```

```
ans =
        4.4040e+003        ← i.e. $4.4040 \times 10^3$ $rad/\sec$
```

We intend to bring an end to the chapter with this.

Exercises

1. Exercise MATLAB **freqresp** function in the following questions:
   (a) determine the $H(j2)$ value for the control system $H(s) = \dfrac{9}{2s^2 + 0.01s + 2.1}$,
   (b) determine the $H(-j0.1)$, $H(j0.1)$, $H(j0.5)$, and $H(j1)$ values for the control system in part (a),
   (c) in part (a) in magnitude and phase angle form,
   (d) in part (b) in magnitude and phase angle form, and
   (e) determine the $H(j\omega)$ values of the control system in part (a) as a column matrix over $-25 \leq \omega \leq 25$ $rad/\sec$ with a ω step 0.1 $rad/\sec$.

2. In question (1) now use the state space defined control system $H(s) = \{A, B, C, D\}$ where $A = \begin{bmatrix} -2 & -1 \\ 2 & 3 \end{bmatrix}$, $B = \begin{bmatrix} -1 \\ -2 \end{bmatrix}$, $C = [2 \ 1]$, and $D = [9]$.

3. Graph all four frequency responses (magnitude, phase, real, and imaginary) of the control system in question (2) as a single plot on $\Delta\omega = 0.1$ $rad/\sec$ over $-25 \leq \omega \leq 25$ $rad/\sec$.

4. Verify that the three control systems $H_1(s) = \dfrac{9}{2s^2 + 0.01s + 2.1}$, $H_2(s) = \dfrac{4}{2.1s^2 + 0.3s + 2}$, and $H_3(s) = \dfrac{3}{3s^2 + 9.5s + 3}$ have the frequency responses (magnitude-phase angle form) over common angular frequency interval $0.01 \leq \omega \leq 100$ $rad/\sec$ as displayed in figure E.6(2).

5. Verify the frequency response behavior for each of the following control systems:
   (a) the control system $G(s) = \dfrac{1}{s + 100}$ shows a lowpass behavior,
   (b) the control system $G(s) = \dfrac{s}{s + 100}$ shows a highpass behavior, and
   (c) the control system $G(s) = \dfrac{s}{s^2 + 0.5s + 100}$ shows a narrowband behavior.

6. In the following some control system is given. Verify the pole-zero map of the system stated beside it:
   (a) $H(s) = \dfrac{5s^2 - s + 1}{s^3 - 1}$ has the pole-zero map of figure E.6(3),
   (b) $H(s) = \dfrac{3}{4.3s^5 + 8s^4 + 40s^3 - 10s^2 + 22s + 9}$ has the pole-zero map of figure E.6(4), and
   (c) $H(s) = \dfrac{1.3s^4 + 6s^3 + 7s^2 + 8s + 1}{7s^7 + 10s^3 + 2.1s^2 + 9.9s}$ has the pole-zero map of figure E.6(5).

7. Determine the gain and phase margins of each following control system:

   (a) $H(s) = \dfrac{5s^2 - s + 1}{s^3 - 1}$ (b) $H(s) = \dfrac{5(s+2)}{s(s+1)(s+3)(s+7)}$ (c) $H(s) =$
   $\{A, B, C, D\}$ where $A = \begin{bmatrix} -2 & 4 \\ 2 & 7 \end{bmatrix}$, $B = \begin{bmatrix} 1 \\ 2 \end{bmatrix}$, $C = [-9 \;\; -3]$, and $D = [-2]$.

8. Obtain Nyquist plot of the control system $H(s) = \dfrac{5s^2 - s + 1}{s^3 - 1}$. Add dB grid lines in the figure. From the Nyquist plot, find the sample values of Re$\{H(j\omega)\}$ and Im$\{H(j\omega)\}$ on $\Delta\omega = 0.1$ rad/sec over ω interval $0 \le \omega \le 0.5$ rad/sec. From the plot determine the peak response of the system. Get the plot only for positive frequencies. Obtain the Nyquist plot segment over ω interval $0.01 \le \omega \le 2$ rad/sec.

9. Obtain Nichol's plot of the control system $H(s) = \dfrac{5s^2 - s + 1}{s^3 - 1}$. Add dB grid lines in the figure. From the Nichol's plot, find the sample values of $|H(j\omega)|$ and $\angle H(j\omega)$ on $\Delta\omega = 0.1$ rad/sec over ω interval $0 \le \omega \le 0.5$ rad/sec. From the plot determine the peak response of the system. Obtain the Nichol's segment plot over ω interval $0.01 \le \omega \le 2$ rad/sec.

10. Consider the control system $H(s) = \dfrac{5s^2 - s + 1}{s^3 - 1}$. Its magnitude spectrum needs to be normalized with respect to the maximum. Obtain the normalized magnitude spectrum on $\Delta\omega = 0.01$ rad/sec over ω interval $0 \le \omega \le 16$ rad/sec.

11. In questions 6(a)-(b), determine the poles and zeroes of each control system. Do the same for the state space defined control system in problem 7(c). Add s plane grid lines to the pole-zero map of problem 7(c).

12. Consider the control system of problem 6(c). Obtain the poles and zeroes of the system. Determine damping ratio and natural frequency at every pole of the system.

13. Determine the DC gain of each following control system: (a) $H(s) = \dfrac{9s^2 - 7s + 2}{3s^5 - 118s^3 - 20s + 2}$ (b) $H(s) = \{A, B, C, D\}$ where $A = \begin{bmatrix} -2 & 4 \\ 2 & 7 \end{bmatrix}$, $B = \begin{bmatrix} 1 \\ 2 \end{bmatrix}$, $C = [-9 \;\; -5]$, and $D = [-3]$ (c) $H(s) = \dfrac{5(s+2)}{s(s+1)(s+3)(s+7)}$.

14. Identify the frequency response type (e.g. highpass) for each following control system: (a) $H(s) = \dfrac{17}{3s + 63}$ (b) $H(s) = \dfrac{4s}{s + 65}$ (c) $H(s) =$ $\dfrac{7.95}{3s^2 + 1.8s + 470}$ (d) $H(s) = \dfrac{0.3s^2 + 90s + 50000}{0.3s^2 + 700s + 50000}$ (e) $H(s) = \{A, B, C, D\}$ where $A = \begin{bmatrix} -2 & 4 \\ 2 & 7 \end{bmatrix}$, $B = \begin{bmatrix} 1 \\ 2 \end{bmatrix}$, $C = [-9 \;\; -5]$, and $D = [-3]$. For each system determine the cutoff frequency or center frequency and bandwidth whichever is pertinent.

Answers:

(1) (a) $H(j2) = -1.5254 - j\,0.0052$

(b) $H(-j0.1) = 4.3269 + j\,0.0021$, $H(j0.1) = 4.3269 - j\,0.0021$,
$H(j0.5) = 5.6249 - j\,0.0176$, and $H(j1) = 89.1089 - j\,8.9109$

(c) $H(j2) = 1.5254\angle -179.81^0$

(d) $H(-j0.1) = 4.3269\angle 0.0275^0$, $H(j0.1) = 4.3269\angle -0.0275^0$,
$H(j0.5) = 5.625\angle -0.179^0$, and $H(j1) = 89.5533\angle -5.7106^0$

(e) H=tf(9,[2 0.01 2.1]); w=-25:0.1:25; R=freqresp(H,w); V=squeeze(R);
hint: section 6.2

(2) (a) $H(j2) = 8.7647 + j\,1.0588$

(b) $H(-j0.1) = 8.0056 - j\,0.1245$, $H(j0.1) = 8.0056 + j\,0.1245$,
$H(j0.5) = 8.1263 + j\,0.5734$, and $H(j1) = 8.3846 + j\,0.9231$

(c) $H(j2) = 8.0066\angle -0.8913^0$

(d) $H(-j0.1) = 8.0066\angle -0.8913^0$, $H(j0.1) = 8.0066\angle 0.8913^0$,
$H(j0.5) = 8.1465\angle 4.036^0$, and $H(j1) = 8.4353\angle 6.2825^0$

(e) A=[-2 -1;2 3]; B=[-1;-2]; C=[2 1]; D=[9]; H=ss(A,B,C,D); w=-25:0.1:25;
R=freqresp(H,w); V=squeeze(R);
hint: section 6.2

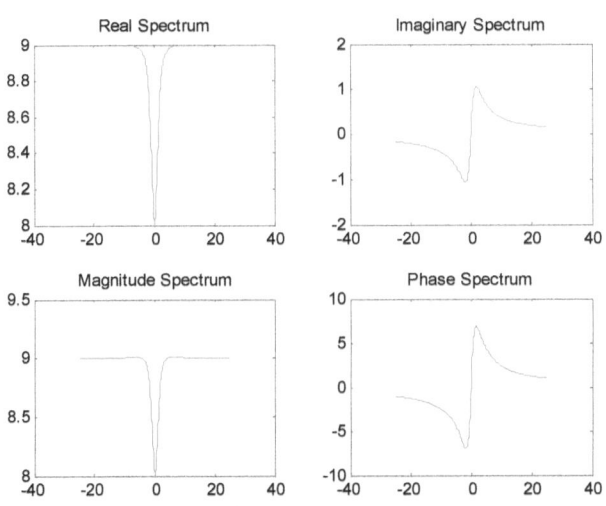

Figure E.6(1) Four frequency spectra of a control system

(3) Figure E.6(1), necessary codes are in the following:
A=[-2 -1;2 3]; B=[-1;-2]; C=[2 1]; D=[9]; H=ss(A,B,C,D); w=-25:0.1:25;
R=freqresp(H,w); V=squeeze(R);
R=real(V); I=imag(V); A=abs(V); P=rad2deg(angle(V));
subplot(221),plot(w,R),title('Real Spectrum')
subplot(222),plot(w,I),title('Imaginary Spectrum')
subplot(223),plot(w,A),title('Magnitude Spectrum')

subplot(224),plot(w,P),title('Phase Spectrum')
hint: section 6.3 and appendix E
(4) Use **bode** command.   Hint: section 6.3
(5) Exercise **bode** command for every function. Inspect each magnitude spectrum.   Hint: section 6.3
(6) Hint: section 6.4

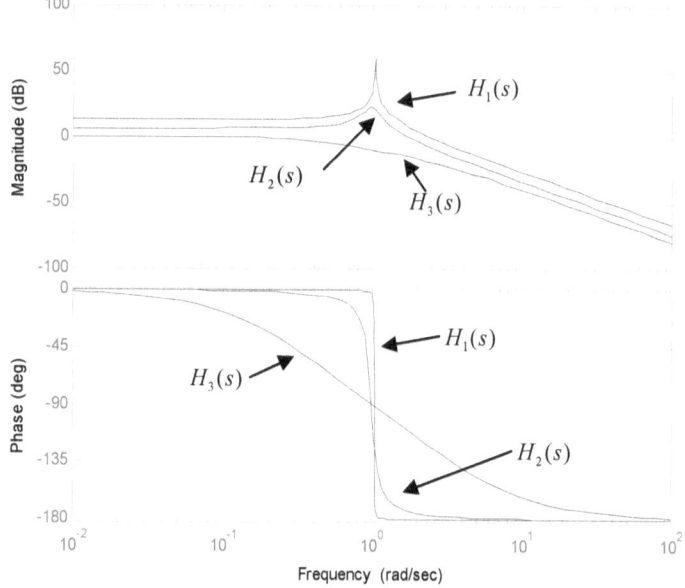

Figure E.6(2) Frequency responses of three control systems

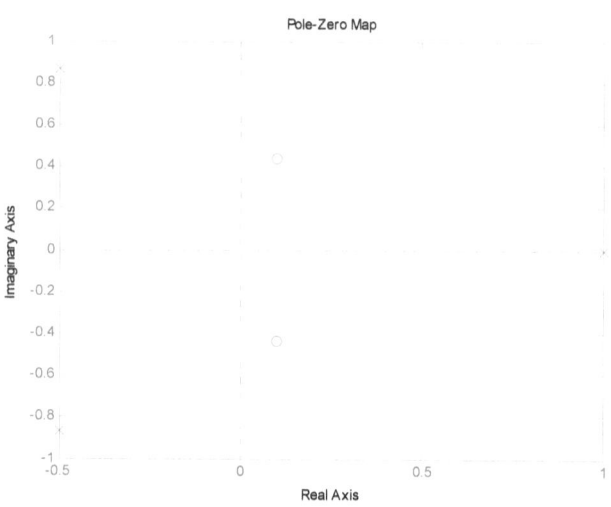

Figure E.6(3) Pole-zero map for the system of problem 6(a)

(7) (a) gain margin: 0 *dB* at $\omega=0\ rad/\sec$ phase margin: 0^0 at $\omega=0\ rad/\sec$
(b) gain margin: 27.3 *dB* at $\omega=3.52\ rad/\sec$ phase margin: 66.7^0 at $\omega=0.441\ rad/\sec$
(c) gain margin: 2.24 *dB* at $\omega=0\ rad/\sec$ phase margin: -45^0 at $\omega=2.24\ rad/\sec$
Hint: section 6.6

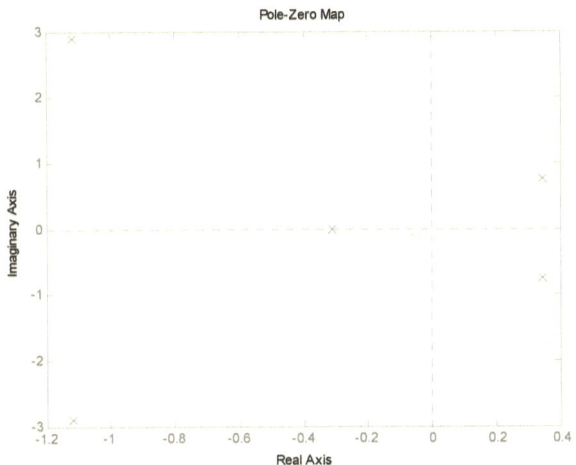

Figure E.6(4) Pole-zero map for the system of problem 6(b)

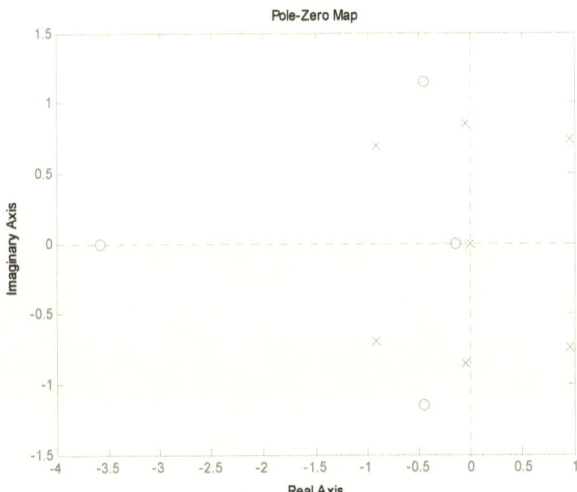

Figure E.6(5) Pole-zero map for the system of problem 6(c)

(8) Figure E.6(6). Table 6.A. In dB $|H(j\omega)|_{max}=10\ dB$ at $\omega =1.2\ rad/\sec$. Figure E.6(7). Figure E.6(8). Hint: section 6.7

(9) Figure E.6(9). Table 6.B. In dB $|H(j\omega)|_{max}=9.99\ dB$ at $\omega =1.18\ rad/\sec$. Figure E.6(10).   Hint: section 6.8

(10) Figure E.6(11).
Hint: sections 6.2 and 6.3

(11) For 6(a); poles: $-0.5+j\ 0.866$, $-0.5-j\ 0.866$, and 1 and zeroes: $0.1+j\ 0.4359$ and $0.1-j\ 0.4359$

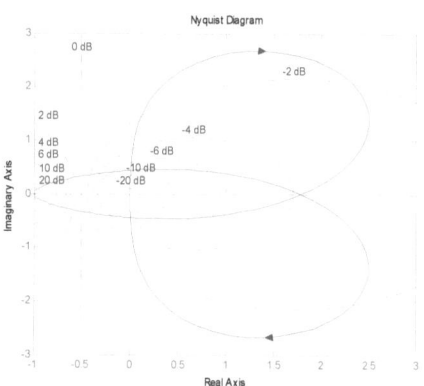

Figure E.6(6) Nyquist plot of a control system with dB grid lines

For 6(b); poles: $-1.1173+j\ 2.9001$, $-1.1173-j\ 2.9001$, $0.3432+j\ 0.7590$, $0.3432-j\ 0.7590$, and -0.3123 and zeroes: none

For 7(c); poles: -2.8151 and 7.8151 and zeroes: -4.4221 and 1.9221, figure E.6(12)
Hint: sections 6.4 and 6.5

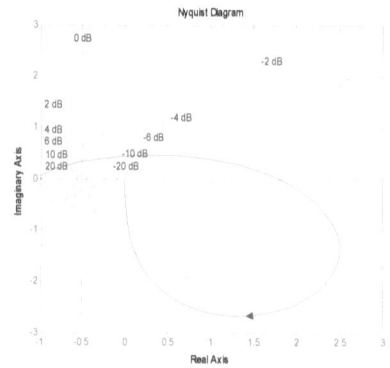

Figure E.6(7) Nyquist plot for positive frequencies

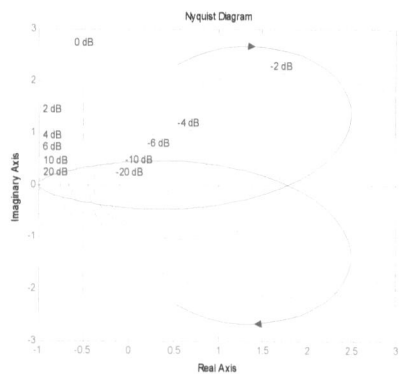

Figure E.6(8) A segment of Nyquist plot

Table 6.A Nyquist plot data

ω	Re{$H(j\omega)$}	Im{$H(j\omega)$}
0	-1	0
0.1	-0.9499	0.1009
0.2	-0.7983	0.2064
0.3	-0.5415	0.3146
0.4	-0.1737	0.4111
0.5	0.3077	0.4615

Table 6.B Nichol's plot data

ω	$|H(j\omega)|$	$\angle H(j\omega)$ in deg
0	1	180
0.1	0.9552	173.9337
0.2	0.8246	165.5054
0.3	0.6263	149.8429
0.4	0.4463	112.9031
0.5	0.5547	56.3099

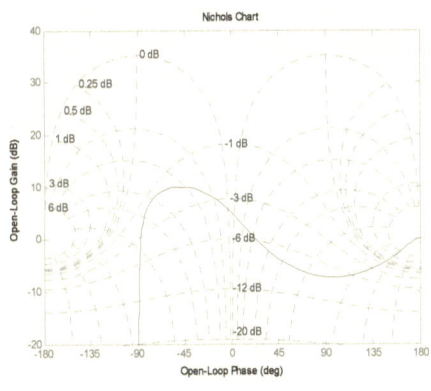

Figure E.6(9) Nichol's chart with grid lines

Figure E.6(10) A segment of Nichol's chart

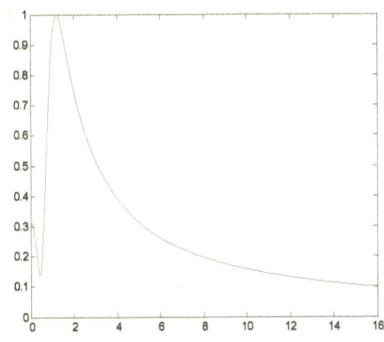

Figure E.6(11) A normalized magnitude spectrum

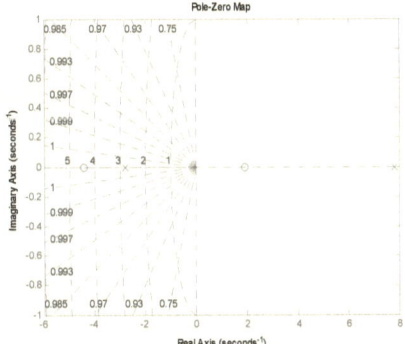

Figure E.6(12) Pole-zero map with s plane grid

(12) Table 6.C
   Hint: sections 6.4 and 6.5

Table 6.C Poles, zeroes, damping ratio, and natural frequencies of control system in problem 6(c)

poles	ζ	ω_n (rad/sec)	zeroes
0	−1	0	−3.5736
0.9619+ j 0.7384	−0.7932	1.2126	−0.4508+ j 1.1542
0.9619− j 0.7384	−0.7932	1.2126	−0.4508− j 1.1542
−0.9118+ j 0.6979	0.7941	1.1483	−0.1402
−0.9118− j 0.6979	0.7941	1.1483	
−0.0501+ j 0.8526	0.0586	0.8541	
−0.0501− j 0.8526	0.0586	0.8541	

(13) (a) 1   (b) −1.2273   (c) ∞   Hint: section 6.9
(14) (a) lowpass; ω_c =20.9501 rad / sec   (b) highpass; ω_c =66.0586 rad / sec
  (c) bandpass; center frequency: 12.4955 rad / sec and bandwidth: 0.5946 rad / sec
  (d) bandstop; center frequency: 401.3964 rad / sec and bandwidth: 2.3023×10^3 rad / sec
  (e) highpass; ω_c =6.6059 rad / sec
Hint: section 6.9

Chapter 7

Control System Root Locus and Stability

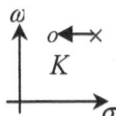

Control system design involving gain needs root locus. Gain appears in a control system invariably for example in main plant or forward path, feedback path, input, or output. Stability of a control system is of prime importance for practical reason. Engineers may design a circuit fulfilling input-output criteria but internal element behavior of the controller/system may turn the whole system unstable in the long run. Stability inspection occurs in time, frequency, or s domain. In order to elucidate the two issues we address the following in this chapter:

- ❖❖ Characteristic polynomial and Routh table implementation
- ❖❖ Root locus realization by user-defined and embedded tools
- ❖❖ Testing stability of control systems by various techniques

7.1 Characteristic equation of a control system

Figure 2.4(a) shows the basic block diagram of a control system with negative feedback. Stability of the control system depends on the roots of $1+G(s)H(s)=0$ and the equation is called the characteristic equation (C.E.). For the positive feedback control system in figure 2.4(b), the C.E. is $1-G(s)H(s)=0$. Here in this section we discuss how to form C.E. and to find its roots.

The $1+G(s)H(s)=0$ is basically a polynomial equation. The $G(s)$ or $H(s)$ is defined in accordance with section 2.1 mentioned functions. Let us go through the following examples in this regard.

◆ **Example 1: Both the $G(s)$ and $H(s)$ are in polynomial form**

Suppose we have $G(s) = \frac{s}{s^2+1}$ and $H(s) = \frac{1}{2s+1}$ from which the characteristic polynomial is $2s^3 + s^2 + 3s + 1$ which we wish to find.

Enter the two transfer functions by:
>>G=tf([1 0],[1 0 1]); ↵   ← Workspace G is user-chosen, holds $G(s)$
>>H=tf(1,[2 1]); ↵   ← Workspace H is user-chosen, holds $H(s)$

The $1+G(s)H(s)$ can be imagined as $G(s)$ and $H(s)$ are in series and the series equivalent is in parallel with 1. Section 2.3 demonstrates series-parallel control system implementation therefore we find the equivalent $1+G(s)H(s)$ and assign to user-chosen **C** by:
>>C=parallel(1,series(G,H)); ↵

Obtain the numerator and denominator polynomial coefficients of $1+G(s)H(s)$ by **tfdata** (appendix D.9):
>>[n,d]=tfdata(C,'v'); ↵   ← The n holds numerator and d holds denominator

As far as characteristic polynomial is concerned, only do we need the numerator of $1+G(s)H(s)$ hence call it by:
>>n ↵

n =
    2   1   3   1

There is an embedded function by the name **poly2str** which displays a polynomial in terms of s from the coefficients and the syntax is **poly2str**(polynomial coefficient as a row or column matrix, related variable under quote) and apply the syntax as:
>>poly2str(n,'s') ↵

ans =
    2 s^3 + s^2 + 3 s + 1

◆ **Example 2: $G(s)$ and $H(s)$ are in mixed representation**

If the control system definition is in other than polynomial like zero-pole-gain or state-space form, we need little modification. Say we have $G(s) = \{A, B, C, D\}$ where $A = \begin{bmatrix} 2 & 3 \\ 1 & 0 \end{bmatrix}$, $B = \begin{bmatrix} 1 \\ 7 \end{bmatrix}$, $C = \begin{bmatrix} 2 & 5 \end{bmatrix}$, and $D = [7]$ and $H(s) = \frac{1}{2s+1}$. The characteristic polynomial is $s^3 + 2s^2 + 7.5s - 23.5$ which we intend to obtain.

>>A=[2 3;1 0]; B=[1;7]; C=[2 5]; D=7; ↵ ← Enter state space matrices to like name variables
>>G=ss(A,B,C,D); ↵   ← Forming $G(s)$

Although $H(s)$ can be formed in terms of polynomial coefficients, pole-zero-gain form is used to see a different case:

>>H=zpk([ ],-1/2,1/2); ↵   ← Forming $H(s)$
>>C=parallel(1,series(G,H)); ↵   ← Workspace C holds $1+G(s)H(s)$

When state-space is involved in definition, resultant system is in state-space form too therefore we use **tf** on C for the conversion:

>>C=tf(C); ↵   ← Same variable is chosen again for assignment

Obviously the last C is in coefficient form hence carry out the following:

>>[n,d]=tfdata(C,'v'); ↵
>>poly2str(n,'s') ↵

ans =
     s^3 + 2 s^2 + 7.5 s - 23.5

◆ **Example 3: What about open loop $G(s)$?**

In this case you just consider the denominator polynomial coefficients of $G(s)$.

◆ **Example 4: What about the roots of a characteristic polynomial?**

First of all you have to get the polynomial coefficients like examples 1 and 2. Then apply embedded function **roots** on the polynomial. The syntax is user-chosen variable for return=**roots**(polynomial as a row matrix) and the return is a column matrix.

In example 1 the characteristic polynomial $2s^3+s^2+3s+1$ has the roots $-0.0772\pm j\,1.2003$ and -0.3456 which we intend to obtain.

The polynomial coefficients are stored in **n**, calling then takes place as:

>>r=roots(n) ↵   ← Workspace r is any user-chosen variable, holds the roots

r =
     -0.0772 + 1.2003i
     -0.0772 - 1.2003i
     -0.3456

Should you need to access each root, call r(1), r(2), etc.

◆ **Example 5: How to form a characteristic polynomial from its roots?**

Suppose we have three roots $4\pm j\,3$ and 2 and its characteristic polynomial is $s^3-10s^2+41s-50$. Our objective is to get the polynomial.

Appendix D.11 explains complex number entering. The embedded function **poly** helps us form the characteristic polynomial with the syntax user-chosen variable for return=**poly**(roots as a row matrix) and the return is a column matrix. Enter the given roots by:

>>r=[4+i*3 4-i*3 2]; ↵   ← Workspace r is any user-chosen variable, holds the roots
>>p=poly(r) ↵   ← Workspace p is any user-chosen variable, holds the polynomial

```
p =
    1   -10   41   -50
```
In order to view the s related polynomial, let us carry out the following:
```
>>poly2str(p,'s') ↵

ans =
    s^3 - 10 s^2 + 41 s - 50
```

7.2 Implementing a Routh table

Routh table helps us determine the stability of a control system unfortunately there is no specific embedded function for the table. Author written (appendix F) function **routh** as presented in figure 7.1(a) determines the Routh table.

In order to have the file, you may type the codes in a script file (chapter 1 for M or script file) and save the file by name **routh** in your working path of MATLAB or contact at page ii mentioned email for the softcopy.

Regarding the syntax, the **routh** has one input argument which is any polynomial in terms of coefficient and as a row matrix. The output of **routh** is a rectangular matrix in which the first column indicates s related power and the second through rest are the required Routh table. *Note that **routh** functions up to the 9th degree polynomial.* Following examples illustrate some table related problems.

```
function R=routh(p)
L=length(p);
A=[ ];
    for k=1:L
      A=[A;strcat('s^',num2str(k-1),'|')];
    end

if rem(length(p),2)==0
    p1=p(1:2:L);
    p2=p(2:2:L);
    H=L/2;
else
    p1=p(1:2:L);
    p2=[p(2:2:L) 0];
    H=(L+1)/2;
end
R=[p1;p2];
for k=1:L-2
    b=[ ];
    for m=2:length(p2)
      b=[b -1/p2(1)*det([p1(1) p1(m);p2(1) p2(m)])];
    end

    p1=p2;
    p2=[b zeros(1,H-length(b))];
    R=[R;p2];
end
R=[flipud([0:L-1]') R];
disp('The first column indicates')
disp(flipud(A))
```

Figure 7.1(a) Script file for Routh table

◆ **Example 1: Just the table from a characteristic polynomial**

The characteristic polynomial $s^3 - 10s^2 + 41s - 50$ has the following Routh table:

$$\begin{array}{c|cc} s^3 & 1 & 41 \\ s^2 & -10 & -50 \\ s & 36 & 0 \\ s^0 & -50 & 0 \end{array}$$

Our objective is to implement that.

Enter given polynomial coefficients to user-chosen variable **p** as follows:

>>p=[1 -10 41 -50]; ↵

Then invoke the function as:

>>A=routh(p) ↵   ← A is any user-chosen variable, holds the rectangular matrix

The first column indicates

s^3|
s^2|
s^1|
s^0|

A =

```
3   1    41
2  -10  -50
1   36   0
0  -50   0
```

As you see, what s related power is involved is illustrated at first. The first column in A is not as a part of the Routh table, this is just for mentioning the relevance. The first column you may pick by the command A(:,1), similarly second by A(:,2), and so on.

◆ **Example 2: Involvement of a variable coefficient**

Often in design problems we have one variable coefficient in the characteristic polynomial for instance $s^3 - 10s^2 + Ks - 50$. Extra attention is required here. We declare K as a variable for the polynomial by the command **syms** before using the **routh**:

>>syms k ↵   ← one space gap between **syms** and **k**

Now enter the characteristic polynomial as:

>>p=[1 -10 k -50]; ↵   ← k is any user-chosen variable, represents K

Call similar to example 1:

>>A=routh(p) ↵

The first column indicates

s^3|
s^2|
s^1|
s^0|

A =

```
+-                                        -+
|3,                    1,                 k|
|2,                   -10,              -50|
|1,                   -5+k,                0|
|0,     -1/(-5+k)*(-250+50*k),             0|
+-                                        -+
```

The first column of Routh table receives importance for stability reason which is here second in A. If you read it from the screen, you find that as
$\begin{bmatrix} 1 \\ -10 \\ -5+K \\ -\dfrac{-250+50K}{-5+K} \end{bmatrix}$. There is a function called **pretty** which displays the symbolic expression close to mathematical form. For this you need to pick the second column (done by A(:,2)) and transpose the column to row for display reason (done by .' operator):

>>C2=A(:,2); ↵   ← C2 is any user-chosen variable, holds second column of A
>>pretty(C2.') ↵

```
[                      -250 + 50 k]
[1   -10   -5 + k    - --------------- ]
[                        -5 + k   ]
```

◆ **Example 3: Involvement of two variable coefficients**
The **routh** is well suited to two variables for example the polynomial $s^3 + K_1 s^2 + K_2 s + 50$ needs the following:
>>syms k1 k2 ↵   ← k1 or k2 is any user-chosen variable, represents K_1 and K_2 respectively

The rest calling is identical with those of example 2 i.e.:
p=[1 k1 k2 50]; A=routh(p); C2=A(:,2); pretty(C2.') which results

```
[              50 - k2 k1     ]
[1   k1    - ---------------   50]  meaning
[                  k1         ]
```
$\begin{bmatrix} 1 \\ K_1 \\ -\dfrac{50 - K_1 K_2}{K_1} \\ 50 \end{bmatrix}$.

◆ **Example 4: Appearance of 0 elements**
In some polynomial e.g. $s^5 + 2s^4 + 2s^3 + 4s^2 + 11s + 10$ the first column of Routh table has 0 somewhere. If you execute p=[1 2 2 4 11 10]; A=routh(p), you find **Inf** and **NaN** as the element that means infinity and not a number.

7.3 Implementing a user-defined root locus

Root locus is basically the pole plot of a control system as gain K is changed. The gain K can be located at any point of the control system – input, output, forward path, feedback, or combined. We may use **plot** (appendix E) for the pole plot. The first deliberation is we have to find the control system equivalent considering a single gain. From the equivalent pick the denominator and find its roots or poles by using **roots** of section 7.1.

This syntax of **plot** is not explained in appendix E. When the **plot** input is a row or column matrix of complex data, it locates every complex data by a symbol which is user-supplied and as a second input argument under quote. If the poles are stored in r, we may call plot(r,'x') for graphing where x is user-chosen and indicates every pole. Let us see several examples in this regard.

✦ Example 1: Plotting the poles for a single K

The basic negative feedback control system of figure 2.4(a) incorporates a proportional controller K which you see in figure 7.1(b). Consider $G(s) = \dfrac{2s+1}{2s^2+s+2}$ and $H(s) = \dfrac{1}{s}$. We wish to locate the poles of $Y(s)/R(s)$ for $K=9$.

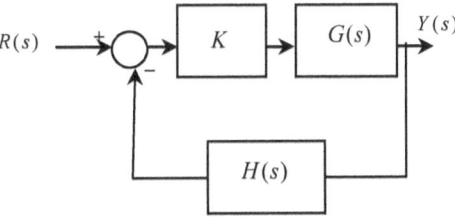

Figure 7.1(b) A negative feedback system with a proportional controller

The K we can multiply with the numerator coefficients of $G(s)$ and intend to inject K as variable that is why first assign 9 to **K**. Doing so, enter the $G(s)$ and $H(s)$ by **tf** (section 2.1):

>>K=9; ↵
>>G=tf(K*[2 1],[2 1 2]); ↵
>>H=tf(1,[1 0]); ↵

Feedback equivalent (section 2.4) we obtain by:
>>E=feedback(G,H,-1); ↵   ← E is user-chosen variable

The **E** holds $Y(s)/R(s)$ equivalent from which get the numerator-denominator polynomial coefficients (appendix D.9):
>>[n,d]=tfdata(E,'v'); ↵   ← n,d are user-chosen variable, hold numerator and denominator respectively

Denominator polynomial roots are the poles so get them by:
>>r=roots(d); ↵   ← r is user-chosen, holds the poles as a column matrix

Call the grapher to locate the poles:
>>plot(r,'x') ↵

Figure 7.1(c) is the outcome from above execution in which the horizontal and vertical axes refer to the real and imaginary axes of the poles respectively. The x marks in figure 7.1(c) correspond to poles of the system in figure 7.1(b). We could have exercised **pzmap** of section 6.4 that brings the zeroes onboard too which makes the graph congested.

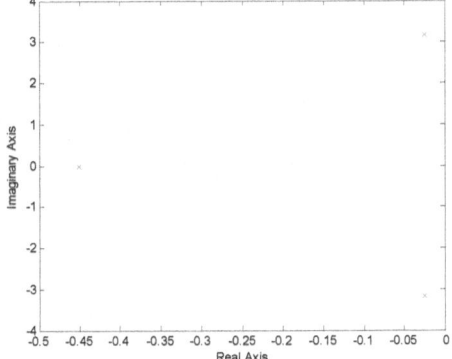

Figure 7.1(c) Plot of poles for control system in figure 7.1(b)

◆ **Example 2: Plotting the poles for a range of** K

For a range of K we change the K by a for-loop (appendix D.4). For every single K use example 1 commands. There is something crucial about the roots holding. You can accumulate the roots by tactic of appendix D.3. Suppose for example 1 mentioned control system we intend to obtain the pole plot on $\Delta K = 0.5$ over $1 \leq K \leq 10$.

As the data accumulation requires, we first assign empty matrix to r as follows:

r=[ ];

The commands of example 1 are placed within the for-loop as:

for K=1:0.5:10
    G=tf(K*[2 1],[2 1 2]);
    H=tf(1,[1 0]);
    E=feedback(G,H,-1);
    [n,d]=tfdata(E,'v');
    r=[r roots(d)];
end

The for-loop counter statement K=1:0.5:10 incorporates K variation on $\Delta K = 0.5$ over $1 \leq K \leq 10$. Poles for all Ks are available in the r as a column matrix so call the grapher to locate the poles:

plot(r,'x')

Type the codes line by line in MATLAB command window or put the codes in a script file and run the file. However figure 7.1(d) depicts the root locus of the $Y(s)/R(s)$ over $1 \leq K \leq 10$.

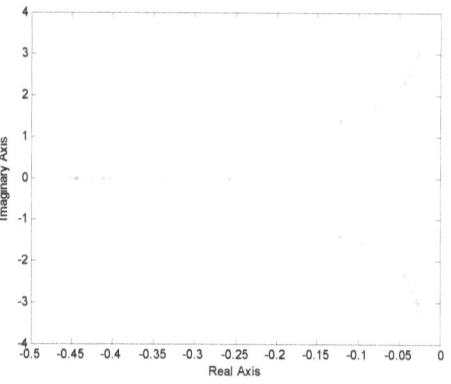

Figure 7.1(d) Plot of poles for control system in figure 7.1(b) over a K range

◆ **Example 3:** K **is in feedback path**

Not necessarily the proportional controller gain is going to be along forward path, figure 7.1(e) presents the gain K along the feedback path. Consider the previous data for $G(s)$, $H(s)$, and K.

Now the multiplication of K is with the numerator of $H(s)$ hence the complete codes for the pole plot or root locus are the following:

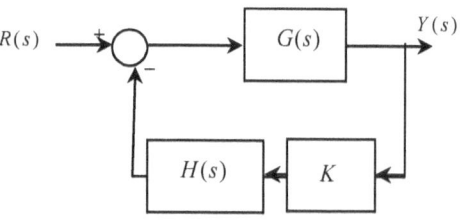

Figure 7.1(e) A proportional controller in the feedback path

```
r=[ ];
for K=1:0.5:10
        G=tf([2 1],[2 1 2]);
        H=tf(K,[1 0]);
        E=feedback(G,H,-1);
        [n,d]=tfdata(E,'v');
        r=[r roots(d)];
end
plot(r,'x')
```

The root locus of $Y(s)/R(s)$ is very similar to the one in figure 7.1(d).

7.4 Root locus by embedded function

The technique illustrated in section 7.3 in fact applies to any kind of control system. In MATLAB there is an embedded function by name **rlocus** (abbreviation for root locus) which plots the root locus of standard feedback systems. The **rlocus** takes care of figure 7.1(b) or 7.1(e) depicted control system only. We do not provide any K, machine finds itself. Let us see the following examples.

◆ **Example 1: Only the root locus plot**

In figure 7.1(b) or 7.1(e) consider $G(s) = \dfrac{2s+1}{2s^2+s+2}$ and $H(s) = \dfrac{1}{s}$. We wish to obtain the root locus plot of $Y(s)/R(s)$.

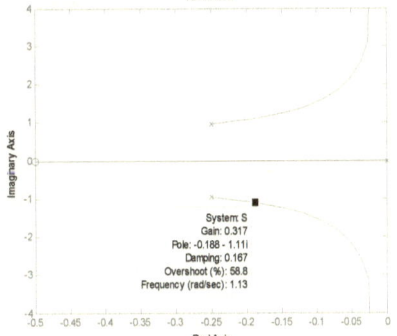

Figure 7.2(a) Standard root locus of control system in figure 7.1(b) or 7.1(e)

The syntax to plot the root locus is **rlocus**($G(s)$ $H(s)$). The $G(s)$ $H(s)$ you can find by **series** (section 2.3) once $G(s)$ and $H(s)$ are defined by section 2.1 quoted function:

>>G=tf([2 1],[2 1 2]); ⏎   ← G is user-chosen variable, holds $G(s)$
>>H=tf(1,[1 0]); ⏎   ← H is user-chosen variable, holds $H(s)$
>>S=series(G,H); ⏎   ← S is user-chosen variable, holds $G(s)$ $H(s)$
>>rlocus(S) ⏎

The result is the figure 7.2(a) that presents the standard root locus plot of the control system in figure 7.1(b) or 7.1(e). The plot is for $0 \le K < \infty$. As you know, root locus starts from pole(s) (marked by x) and ends to zeroes (marked by o) which is evident from the figure. If you click at any point on the locus, you see relevant descriptions. For example the shown one indicates $K=0.317$ for the pole $-0.188-j1.11$ moreover damping and time performance parameters are displayed too.

In last section we did it for finite K (figure 7.1(d)). Certainly there is identicalness of figure 7.1(d) with that of figure 7.2(a).

◆ **Example 2: Root locus segment for user-defined K range**
In example 1 suppose we need the root locus on $\Delta K = 0.01$ over $0 \leq K \leq 1$. To accomplish this, we need to append second argument to rlocus which is the K vector. The K as a row matrix we generate by K=0:0.01:1;. After that we call the plotter as rlocus(S,K) in order to see the root locus segment. The graph is not shown for space reason.

◆ **Example 3: Root locus for a single gain K**
Should you need the poles and zeroes or root locus for a single gain e.g. for $K=3$, just exercise rlocus(S,3).

◆ **Example 4: Poles from root locus for a range of K**
In example 2 suppose we need the poles for every single K, then we have to use an output argument i.e. P=rlocus(S,K); where P is a user-chosen variable. Since there are 101 gain values in K, so is the number of pole-sets returned. The return to P is a rectangular matrix in which every column is a pole set for a particular K obviously sequentially i.e. 1st, 2nd, etc columns for $K=0$, 0.01, etc respectively.

◆ **Example 5: Graphing multiple root locus**
Suppose there are two more control systems like figure 7.1(b) or 7.1(e) with $G_1(s) H_1(s) = \dfrac{1}{2s^3 + s + 2}$ and $G_2(s) H_2(s) = \{A, B, C, D\}$ where $A = \begin{bmatrix} 2 & 3 \\ 1 & 0 \end{bmatrix}$, $B = \begin{bmatrix} 1 \\ 7 \end{bmatrix}$, $C = [2 \quad 5]$, and $D = [7]$. Enter them to S1 and S2 respectively by:
>>S1=tf(1,[2 0 1 2]); ↵ ← S1 is user-chosen variable, holds $G_1(s) H_1(s)$
>>S2=ss([2 3;1 0],[1;7],[2 5],7); ↵ ← S2 is user-chosen variable, holds $G_2(s) H_2(s)$

There are three systems stored in S, S1, and S2 so call the plotter with three input arguments as:
>>rlocus(S,S1,S2) ↵

Figure 7.2(b) is the outcome from above execution. The text is in black and white form so you do not see the difference. MATLAB displays the plots in terms of color as red, green, and blue. Click on each plot, you find the system name like S or S1.

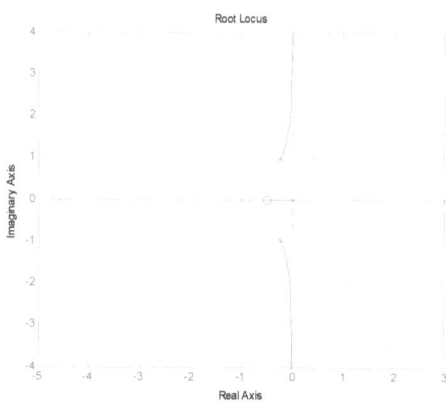

Figure 7.2(b) Root locus plots for three control systems

7.5 Stability by time domain output

You may design your control system in MATLAB or SIMULINK. If any control system output shows increasing amplitude with time, that output is unstable. At infinite time the amplitude becomes infinite which is undefined. Decreasing amplitude with time is stable. The amplitude may represent any physical quantity – voltage, current, angular speed, etc. For different control systems chapter 4 addresses lsim for MATLAB output, so does chapter 5 by SIMULINK Scope. Here we demonstrate both examples.

✦ **Example 1: Output inspection in MATLAB for open loop control system**

In section 4.3 concerning example 1 we now have $H(s) = \dfrac{2}{s-1}$ and wish to see the output $y(t)$ versus t graph over the same time interval and subject to same input.

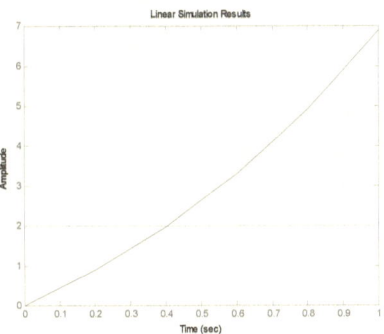

Figure 7.3(a) Output from a control system

Execute the following commands to view $y(t)$ versus t at the prompt:

>>H=tf(2,[1 -1]); t=0:0.2:1; u=2*ones(1,length(t)); lsim(H,u,t) ↵

Figure 7.3(a) shows the output whose trend is continual increase. If you change the time interval from 0-1sec to 1-100sec, you find the amplitude abruptly large. This sort of output is not realizable that is why the system is unstable. Source of the instability is pole on the right half s plane which is at $s=1$.

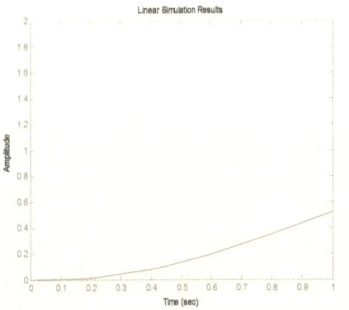
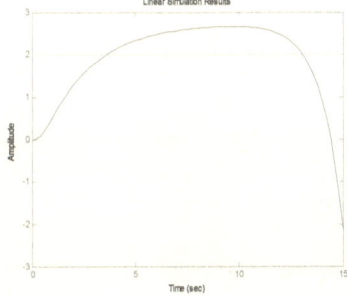

Figure 7.3(b) Output from control system in figure 2.6(a) for $0 \le t \le 1$ sec

Figure 7.3(c) Output from control system in figure 2.6(a) for $0 \le t \le 15$ sec

✦ **Example 2: Output inspection in MATLAB for interconnected system**

Figure 2.6(a) shows an interconnected SISO control system (example 4 of section 2.6). Equivalent transfer function of the system is stored in workspace OS. Let us investigate the output for different time intervals; 0-1sec, 0-15sec, 0-25sec on $u(t)=2$ and $\Delta t=0.2$sec. Reexecute the commands of section 2.6 until you get OS. After that execute the following for different intervals respectively:

>>t=0:0.2:1; u=2*ones(1,length(t)); lsim(OS,u,t) ↵
>>figure,t=0:0.2:15; u=2*ones(1,length(t)); lsim(OS,u,t) ↵
>>figure,t=0:0.2:25; u=2*ones(1,length(t)); lsim(OS,u,t) ↵

Figures 7.3(b), 7.3(c), and 7.3(d) represent the outputs respectively. The reason we used figure command is keep the previous figure and draw the recent graph in a new window. Anyhow figure 7.3(b) indicates the output increasing with no clear cut trend where it will end. In the figure 7.3(c) the output is showing a maximum then going down. In figure 7.3(d) output is drastically down to extra high negative amplitude meaning unstable system.

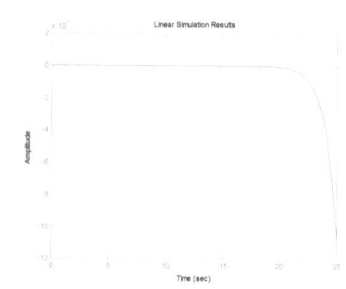

Figure 7.3(d) Output from control system in figure 2.6(a) for $0 \le t \le 25$ sec

◆ **Example 3: Output inspection in SIMULINK**

The output behavior of a control system may happen in a SIMULINK Scope return. Think about the SISO control system of example 2 in section 5.8. The output is depicted in Scope (figure 5.8(e)) over $0 \le t \le 5$ secs. If you see the output for $0 \le t \le 50$ secs by changing the Stop time of SIMULINK to 50, you find the Scope output on autoscale setting similar to figure 7.3(d) which indicates unstable result.

7.6 Stability by a pole-zero map

Pole-zero map discussion you find in sections 6.4 and 6.5. Decide which control output you are interested in and get the transfer function of it with respect to the input. Inspect the pole-zero map whether a pole is on the right half s plane.

Apparent stability:

Consider the parallel system of figure 2.3(d) whose equivalent transfer function (obviously input to output) is stored in OS (example 3 of section 2.6). You just need to exercise pzmap(OS) at the command prompt to see the pole-zero map. The map shows a pole on the right half s plane i.e. unstable output (figure is not shown for space reason). Another option can be check the pole values by the command pole whether any pole with positive real part exists. For this example execution of pole(OS) shows one of the poles is 0.8508 meaning unstable system. Other poles are not presented for space reason. Call it apparently unstable system because zero behavior is not checked.

Certainty on stability:

If some s root appears both in numerator and denominator, just looking into the poles we can not take decision on stability.

Let us investigate the pole-zero behavior of MIMO system in figure 2.6(d) (example 5 of section 2.6). The three transfer functions are stored in OS. We exercised tf(OS) for onscreen display and did not hold the result say T=tf(OS);. Now T(1) represents transfer function $\frac{Y_1(s)}{U(s)}$, so do T(2) and T(3) for the $\frac{Y_2(s)}{U(s)}$ and $\frac{Y_3(s)}{U(s)}$ respectively. Exercise pzmap(T(1)) in order to see the pole-zero map (figure is not shown for space reason). We know that symbols × and o represent pole and zero respectively. Surprisingly you find one symbol like ⊗ meaning simultaneous pole-zero at the same s point in the right half plane. How can we be sure about that? Very simple, exercise following to see all poles and zeroes:

```
>>zero(T(1)) ↵              >>pole(T(1)) ↵

ans =                        ans =
    -2.5000 + 2.3979i            -2.3792 + 2.3410i
    -2.5000 - 2.3979i            -2.3792 - 2.3410i
     1.0000                       1.0000
    -1.0000                      -0.9384
    -0.1000                      -0.4106
    -0.0200                      -0.0126
```

Above result shows one pole at $s=1$ and one zero at $s=1$. We have to get rid of the pole and zero at $s=1$ so that the system becomes stable.

Given the numerical nature of MATLAB, you find the pole/zero at 1 who knows maybe that is at 1.000001 but machine truncates the result and returns 1. There is a function called **minreal** which returns pole-zero cancelled system but it needs a user-defined tolerance with syntax user-supplied variable after cancellation=minreal(given system, user-defined tolerance). It is a good practice that you start the tolerance from 10^{-9} (which has code **1e-9**, appendix A) and increase as power of 10 i.e. then 10^{-8}, after that 10^{-7}, and so on. By doing so we found the reduced system on 10^{-5} by:

>>T1=minreal(T(1),1e-5) ↵ ← T1 is a user-chosen variable, holds pole-zero
                                                    cancelled system for $Y_1(s)/U(s)$

Transfer function:
```
20 s^5 + 122.4 s^4 + 354.4 s^3 + 281 s^2 + 29.48 s + 0.48
---------------------------------------------------------
   s^5 + 6.12 s^4 + 18.02 s^3 + 17.09 s^2 + 4.505 s + 0.054
```

In section 2.6 the transfer function is of 6th degree which is now of 5th degree from the last result. If you exercise pzmap(T1), you do not find any ⊗ symbol on the right half s plane manifesting the removal. In the left half plane still there is some ⊗ that is because of proximity on last pole-zero (at −0.0126 and −0.02 respectively). The other two transfer functions you may get reduced by T2=minreal(T(2),1e-5) and T3=minreal(T(3),1e-5) where T2 and T3 are user-chosen variables.

7.7 Stability by Routh table

Section 7.2 demonstrates Routh table implementation. In this section we address stability linking to Routh table. Two examples are elucidated in the sequel.

✦ **Example 1: Testing stability on a Routh table**

Example 1 of section 7.2 presents the Routh table for a characteristic polynomial which we stored in the workspace A. Appearance of sign change in the first column of Routh table indicates instability, A(:,2) holds the column. Any negative sign we have to check.

We can pick the second column from A by (appendix D.12):
>>C=A(:,2); ⏎ ← C is a user-chosen variable, holds the first column of table

Now check every element in C whether it is negative (appendix D.1):
>>V=C<0; ⏎ ← V is a user-chosen variable

The V is a column matrix of the same size as C in which 1 and 0 mean presence and absence of the negative sign in Routh table first column respectively. The embedded function any returns 1 and 0 for any nonzero element present and absent in a row or column matrix respectively:
>>any(V) ⏎

ans =
     1

Since the return is 1, some nonzero element is present in the V meaning unstable system. If the return were 0, that would indicate a stable system.

✦ **Example 2: Testing stability on a Routh table by different gains**

Suppose $s^3 + Ks^2 + 24s + 50 = 0$ is the characteristic equation of a control system. We wish to see stability of the system on $\Delta K = 1$ over $0 \leq K \leq 5$.

See example 2 of section 7.2 for implementation based on a variable K and do so for given polynomial by:
>>syms k, p=[1 k 24 50]; A=routh(p); C2=A(:,2); ⏎ ← Symbol meaning same as of section 7.2

Ignore the screen display from above execution importantly C2 has the first column of Routh table with variable K. You may generate K using a for-loop (appendix D.4) and get the stability result for every single K like example 1. A substitution function for K is required, which is **eval**. One of its syntaxes is **eval**(expression) but related variable must be having value assigned before calling. In example 1 **any(V)** returns stability in terms of 0/1 which needs to be held. Stability decision we may accumulate by the tactic of appendix D.3:

>>R=[ ]; for c=0:5, k=c; C=eval(C2); V=C<0; R=[R;any(V)]; end ↵
Warning: Divide by zero.
In sym.eval at 9

Above warning demonstrates that somewhere there is zero in the first column of Routh table. A small quantity epsilon (whose MATLAB code is **eps**) can be added to avoid warning:

>>R=[ ]; for c=0:5, k=c+eps; C=eval(C2); V=C<0; R=[R;any(V)]; end ↵

The for-loop counter **c** has the sequential K s. The **c**, **R**, etc are user-chosen. The **R** is a column matrix which holds the decision on stability in terms of 0/1. No gain is stored in any variable so generate it as a column matrix for display reason:

>>K=[0:5]'; ↵   ← K is a user-chosen variable, holds the gains

Side by side calling of gain and stability happens by:

>>[K R] ↵

ans =

    0   1
    1   1
    2   1
    3   0
    4   0
    5   0

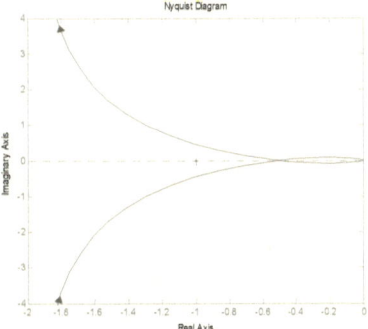

What can we infer? From $K=0$ to 2, the system is unstable and from 3 to 5 is stable. Just for illustration we chose K increment 1, finer increment can be handled this way too.

Figure 7.4(a) Nyquist plot for $K=1$

7.8 Stability by Nyquist plot

Nyquist plot is basically a frequency domain plot which is a closed loop as ω changes from $-\infty$ to $+\infty$. The product of forward and feedback path transfer functions for standard negative feedback system (figure 2.4(a)) is $G(s)H(s)$ or $GH(s)$ or $GH(j\omega)$ in frequency space. If the Nyquist plot of $GH(j\omega)$ encircles $(-1,0)$ point in GH plane, the system is unstable. Or if the loop is out of $(-1,0)$ point, the system is stable. One example we present in the following.

◆ **Example: Testing stability by Nyquist plot for different gains**

Let us consider the system transfer functions in figure 2.4(a) as $G(s) = \frac{1}{(s+1)^2}$ and $H(s) = \frac{K}{s}$. We intend to test stability of the control system subject to three gains $K=1$, $K=2$, and $K=3$.

Section 6.7 explains the Nyquist plot. Let us enter the two transfer functions (section 2.1) by:
>>H=tf(1,[1 0]); ↵   ← H is a user-chosen variable, holds $H(s)$ for $K=1$
>>G=zpk([ ],[-1 -1],1); ↵   ← G is a user-chosen variable, holds $G(s)$

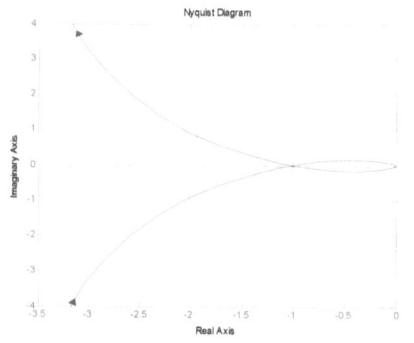 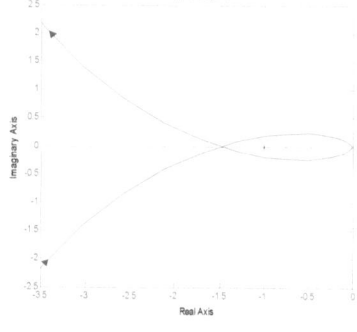

Figure 7.4(b) Nyquist plot for $K=2$ Figure 7.4(c) Nyquist plot for $K=3$

The $GH(s)$ you can form by the **series** of section 2.3:
>>GH=series(G,H); ↵   ← GH is a user-chosen variable, holds $GH(s)$
>>nyquist(GH) ↵   ← Call the grapher for response of figure 7.4(a)

Rightclick the mouse and click **Zoom on** (−1,0) in the popup for having identical response. Since the K is associated with $H(s)$, reenter it for $K=2$:
>>H=tf(2,[1 0]); ↵

Obtain $GH(s)$ accordingly:
>>GH=series(G,H); ↵

Exercise **figure** command in order to see the plot in a new window:
>>figure,nyquist(GH) ↵

Again rightclick the mouse and click the **Zoom on** (−1,0) to see figure 7.4(b). Do similar execution for the last gain:
>>H=tf(3,[1 0]); GH=series(G,H); figure,nyquist(GH) ↵

Figure 7.4(c) is the output for $K=3$ on zooming around (−1,0). What do we infer from these three gain Nyquist plots? When $K=1$, the point (−1,0) is out of the loop so the system is stable. Again for $K=2$, the point (−1,0) is exactly on the loop hence the system is marginally stable. Finally for $K=3$, the (−1,0) is inside the loop therefore the system is unstable.

7.9 Stability with delay elements

Sometimes a delay element might be present in a control system. In t and s domain that is represented by $u(t-T)$ and e^{-sT} respectively. Mathematically e^{-sT} is approximated by polynomial of certain order and we work on that approximate polynomial. Not to mention numerator and denominator are assumed to be of the same degree.

For example e^{-2s} is to be approximated on the 3rd degree. MATLAB function **pade** helps us determine that with the syntax [user-supplied variable for numerator, user-supplied variable for denominator]= **pade**(delay,degree). Simply call the function as:

>>[N,D]=pade(2,3); ↵   ← N and D are user-chosen variables, hold numerator and denominator coefficients respectively

In order to see the e^{-2s} approximated polynomial of the third order we may call the **tf** (section 2.1) on **N** and **D**:

>>tf(N,D) ↵
Transfer function:
-s^3 + 6 s^2 - 15 s + 15
----
s^3 + 6 s^2 + 15 s + 15

← Meaning $e^{-2s} \approx \dfrac{-s^3 + 6s^2 - 15s + 15}{s^3 + 6s^2 + 15s + 15}$

◆ **How to plot the e^{-sT} and its approximated version?**

Having found the approximated e^{-sT}, how one graphs the actual and approximated versions together? The answer is select some s interval say $0 \le s \le 5$ and exercise **plot** of appendix E on the two data. First we generate the s samples by choosing some step size say $\Delta s = 0.1$ and generate a row or column vector of s samples (section 1.1) by:

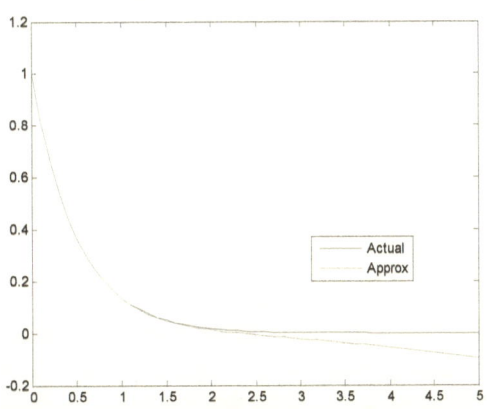

Figure 7.4(d) Graphs of e^{-sT} and approximated one together

>>s=0:0.1:5; ↵   ← **s** is a user-chosen variable, holds s samples

Then we calculate the e^{-sT} samples by the scalar code (appendix A):

>>y=exp(-2*s); ↵ ← **y** is a user-chosen variable, holds e^{-sT} samples

The approximated numerator and denominator polynomial coefficients are stored in **N** and **D** respectively. The function **polyval** evaluates a polynomial with the syntax **polyval**(polynomial coefficients as a row matrix, intended s values as a row matrix). If we use scalar division by ./, approximated transfer function will be calculated for every single s :

```
>>a=polyval(N,s)./polyval(D,s); ↵   ← a is a user-chosen variable
```
In last execution **polyval(N,s)** calculates numerator $-s^3+6s^2-15s+15$ samples, so does **polyval(D,s)** on the denominator $s^3+6s^2+15s+15$. However **a** holds all approximated polynomial samples. Next call the grapher:
```
>>plot(s,y,s,a) ↵
```
In order to attach a mark of distinction, exercise command **legend** with user-supplied texts **Actual** and **Approx** for actual and approximated respectively:
```
>>legend('Actual','Approx') ↵    ← Both texts have to be under quote
```
The result is the figure 7.4(d) in which the horizontal and vertical axes refer to the s and e^{-sT} samples respectively. Since the text is written in black-white form, you do not see the distinction. In MATLAB graphics you find that in terms of blue and green colors respectively. What do we infer from the graph of figure 7.4(d)? The approximation is roughly accurate up to $s=2$, after that deviation is explicit.

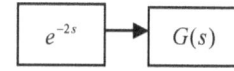

Figure 7.4(e) Delay element in series

✦ **How to apply the delay element?**

Well the approximated e^{-sT} is just another rational transfer function which may appear in series, parallel, feedback, or other form. You may form that transfer function by **T=tf(N,D)** where the **T** is a user-chosen variable. Now apply the previous section mentioned stability technique treating **T** as another transfer function.

Figure 7.4(f) Delay element of SIMULINK

Referring to figure 7.4(e), the e^{-sT} and $G(s)$ are in series. Suppose **T** and **G** hold the approximated e^{-sT} and $G(s)$ respectively then we exercise **series(T,G)** for their equivalent (section 2.3).

✦ **What about SIMULINK delay block?**

In SIMULINK (chapter 3) we also have a block by name **Transport Delay** (appendix C, figure 7.4(f)) which simulates $u(t-T)$ or e^{-sT}. If $T=$ 2secs, we doubleclick the block and enter **Time delay** as 2 in the parameter window. Having entered, **Transport Delay** of figure 7.4(f) represents $u(t-2)$ or e^{-2s}. Treat the **Transport Delay** like other SIMULINK blocks e.g. **Transfer Fcn**.

That brings an end to the chapter 7.

Exercises

1. Concerning the standard negative feedback system of figure 2.4(a), determine the characteristic polynomial for each following control system:

   (a) $G(s) = \dfrac{s}{s^2 + 0.5s + 1000}$ and $H(s) = \dfrac{9}{2s^2 + 0.01s + 2.1}$,

   (b) $G(s) = \dfrac{s}{s^2 + 0.5s + 1000}$ and $H(s) = \{A, B, C, D\}$ where $A = \begin{bmatrix} -2 & -1 \\ 2 & 3 \end{bmatrix}$,

   $B = \begin{bmatrix} -1 \\ -2 \end{bmatrix}$, $C = [2 \ 1]$, and $D = [9]$, and

   (c) $G(s) = \{A, B, C, D\}$ where $A = \begin{bmatrix} -2 & -1 \\ 2 & 3 \end{bmatrix}$, $B = \begin{bmatrix} -1 \\ -2 \end{bmatrix}$, $C = [2 \ 1]$, and

   $D = [9]$ and $H(s) = \{$pole: -2, zero: -0.1, and gain: $0.01\}$.

2. In question (1) determine roots of the characteristic equation for every feedback control system.

3. Determine the characteristic equation when its roots are the following: (a) $-2, -0.5,$ and 3 (b) $-2 \pm j\,5$ and 4 (c) $0, 0, 0,$ and $\pm j$.

4. Exercise author-written routh to determine Routh table for each following characteristic equation:
   (a) $2s^4 + 1.01s^3 + 22.15s^2 + 20.05s + 20$
   (b) $3s^7 + 4s^6 - 3s^5 + s^3 + 3.5s + 7$
   (c) $2s^4 + 3s^3 + Ks^2 + 2$
   (d) $2s^4 + s^3 + K_1 s^2 + K_2$
   (e) $s^4 + s^3 + 4s^2 + 2s + 4$

5. In figure 7.1(b) the control system functions are the following: $G(s) = \dfrac{2s^3 + 1}{s^4 + s + 7}$ and $H(s) = \dfrac{3}{2s + 9}$. Locate only the poles of $Y(s)/R(s)$ for $K = 3$.

6. In figure 7.1(e) the control system functions are the following: $G(s) = \{A, B, C, D\}$ where $A = \begin{bmatrix} -2 & -1 \\ 2 & 3 \end{bmatrix}$, $B = \begin{bmatrix} -1 \\ -2 \end{bmatrix}$, $C = [2 \ 1]$, and $D = [11]$

   and $H(s) = \{$pole: -2, zero: -0.1, and gain: $1\}$. Locate only the poles of $Y(s)/R(s)$ for $K = 15$.

7. In figure 2.6(a) the control system functions are the following: $G_1(s) = 20$, $G_2(s) = \dfrac{0.3}{s^2 + 5s + 12}$, $G_3(s) = \dfrac{s + 0.01}{(s + 0.1)(s + 0.02)}$, $G_4(s) = \dfrac{s}{s + 1}$, and $G_5(s) = \{A, B, C, D\}$ where $A = 1$, $B = 1.1$, $C = 10^{-5}$, and $D = K$. Obtain only pole plot of the control system for $Y(s)/R(s)$ on $\Delta K = 0.5$ over $1 \le K \le 10$.

8. In figure 7.1(b) or 7.1(e) consider $G(s) = \dfrac{s^2 + 1}{8s^3 + 6s + 2}$ and $H(s) = \dfrac{1}{2s + 3}$ and exercise the embedded MATLAB function to obtain:

(a) the standard root locus plot of $Y(s)/R(s)$,
(b) the root locus on $\Delta K = 0.05$ over $0 \leq K \leq 12$,
(c) the root locus for single gain $K = 5.5$, and
(d) poles on $\Delta K = 0.25$ over $0 \leq K \leq 0.75$.

9. In figure 7.1(b) or 7.1(e) consider (a) $G(s) = \dfrac{s^2 + 1}{8s^3 + 0.06s + 2}$ (b) $G(s) = \dfrac{s^2 + 1}{8s^3 + 6s + 2}$ (c) $G(s) = \dfrac{s^2 + 1}{8s^3 + 600s + 2}$ and $H(s) = \dfrac{1}{2s + 3}$ and graph the root locus for the three systems in a single plot.

10. Inspect the time domain output waveshape (i.e. $y(t)$ versus t) of each following control system and mention the trajectory trend i.e. stable or unstable: (a) $H(s) = \dfrac{2s}{s^3 - s - 1}$ and $u(t) = 20$ in figure 4.3(a) (b) in part (a) now $H(s) = \dfrac{s^2 - 3s + 3}{s^3 + 5s^2 + 9s + 6}$.

11. Inspect the pole-zero map of each following control system and mention the apparent stability of the system: (a) $H(s) = \dfrac{2s}{s^4 - s - 1}$ in figure 4.3(a) (b) in part (a) now $H(s) = \dfrac{7s^2 + 3s + 4}{s^4 + 5.4s^3 + 19.05s^2 + 20s + 5.85}$. Determine the poles of each system to verify your decision.

12. Determine $Y(s)/R(s)$ of the control system in figure E.2(7) subject to
$G_1(s) = \dfrac{s^2 + 1}{8s^3 + 6s + 2}$, $G_2(s) = \dfrac{s + 1}{4s + 2}$, $G_3(s) = \dfrac{8}{5s^2 + 7s + 2}$, and $H(s) = \dfrac{2s}{s + 1}$.
Inspect the pole-zero map of $Y(s)/R(s)$ and mention the apparent stability of the system. Is common pole-zero present apparently? Determine the poles and zeroes of the system. Eliminate the common pole-zero if there is any and obtain the reduced transfer function. What tolerance did you apply for the reduced system in case reduced transfer function exists? Check the removal of common pole and zero by inspecting the pole-zero map of the reduced transfer function.

13. Test stability on characteristic equations of problem 4(a), 4(b), and 4(e) by applying the Routh table.

14. The characteristic equation of a control system is given as $s^3 + 3s^2 + (K+1)s + 4 = 0$. We wish to check stability of the system on $\Delta K = 0.01$ over $0 \leq K \leq 0.5$.

15. Test the stability of a standard negative feedback control system (figure 2.4(a)) by Nyquist plot for two gains $K = 8$ and $K = 64$ whose GH function is given as $G(s) = \dfrac{K}{(s+2)^3}$ and $H(s) = \dfrac{1}{s+1}$.

16. Approximate the e^{-3s} based on the 4th degree polynomial and find the approximated polynomial expression.

Answers:

(1) (a) C.E.= $2s^4 + 1.01s^3 + 2002.105s^2 + 20.05s + 2100$
 (b) C.E.= $s^4 + 8.5s^3 + 982.5s^2 - 1034s - 4000$
 (c) C.E.= $s^3 + 0.8064s^2 - 5.8101s - 7.3688$
 hint: section 7.1

(2) (a) $-0.2477 \pm j\, 31.6218$ and $-0.0048 \pm j\, 1.0247$
 (b) $-4.7499 \pm j\, 31.2002$, 2.5653, and -1.5655
 (c) 2.57, $-1.6882 \pm j\, 0.1312$
 hint: section 7.1

(3) (a) C.E.= $s^3 - 0.5s^2 - 6.5s - 3$
 (b) C.E.= $s^3 + 13s - 116$ (c) C.E.= $s^5 + s^3$
 hint: section 7.1

(4) (a)
$$\begin{array}{c|ccc} s^4 & 2 & 22.15 & 20 \\ s^3 & 1.01 & 20.05 & \\ s^2 & -17.553 & 20 & \\ s^1 & 21.2008 & & \\ s^0 & 20 & & \end{array}$$

(b)
$$\begin{array}{c|cccc} s^7 & 3 & -3 & 1 & 3.5 \\ s^6 & 4 & 0 & 0 & 7 \\ s^5 & -3 & 1 & -1.75 & \\ s^4 & 1.3333 & -2.3333 & 7 & \\ s^3 & -4.25 & 14 & & \\ s^2 & 2.0588 & 7 & & \\ s^1 & 28.45 & & & \\ s^0 & 7 & & & \end{array}$$

(c)
$$\begin{array}{c|ccc} s^4 & 2 & K & 2 \\ s^3 & 3 & 0 & \\ s^2 & K & 2 & \\ s^1 & -6/K & & \\ s^0 & 2 & & \end{array}$$

(d)
$$\begin{array}{c|ccc} s^4 & 2 & K_1 & K_2 \\ s^3 & 1 & 0 & 0 \\ s^2 & K_1 & K_2 & \\ s^1 & -K_2/K_1 & & \\ s^0 & K_2 & & \end{array}$$

(e) Some element in the first column is zero, to see the sign change you may execute routh(p+eps).
Hint: section 7.2

(5) Figure E.7(a). Hint: section 7.3

(6) Figure E.7(b). Hint: section 7.3

(7) Figure E.7(c) and figure E.6(d) for the codes. Hint: section 7.3

(8) (a) Figure E.7(e) (b) Command: K=0:0.05:12; rlocus(S,K) where S is the series equivalent of $G(s)$ and $H(s)$ (c) Command: rlocus(S,5.5) (d) Table I for the poles. Some graph is not presented for space reason. Hint: section 7.4

(9) Figure E.7(f). Hint: section 7.4

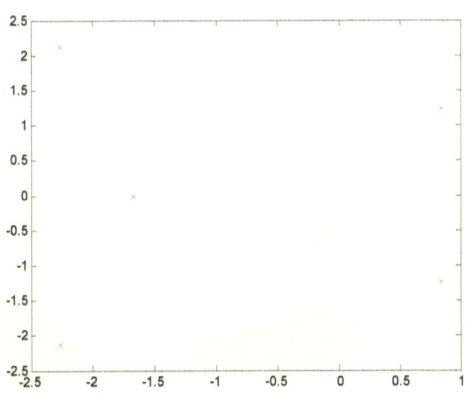

Figure E.7(a) Poles of a control system for a single gain

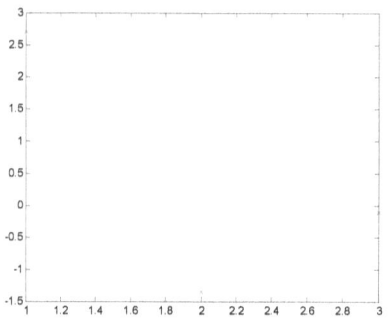

Figure E.7(b) Poles of a control system for a single gain

Figure E.7(c) Poles of a control system over a range of gain

```
            r=[ ];
for K=1:0.5:10
        G1=20;
        G2=tf(0.3,[1 5 12]);
        G3=zpk(-0.01,[-0.1 -0.02],1);
        G4=tf([1 0],[1 1]);
        G5=ss(1,1.1,1e-5,K);
        E=feedback(series(G1,series(feedback(G2,G4,-1),G3)),G5,-1);
        [n,d]=tfdata(E,'v');
        r=[r roots(d)];
end
        plot(r,'x')
```

Figure E.7(d) Codes for figure E.7(c) generation

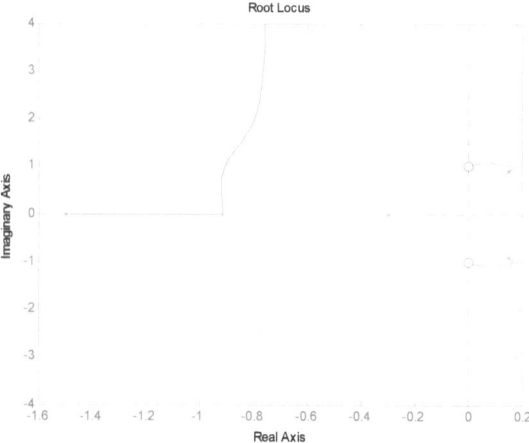

Figure E.7(e) Standard root locus of a control system

Table I Poles of a control system for few gains

$K=0$	$K=0.25$	$K=0.5$	$K=0.75$
-1.5	-1.4879	-1.4756	-1.463
0.149+ j 0.9037	0.15+ j 0.9049	0.1509+ j 0.9061	0.1519+ j 0.9073
0.149- j 0.9037	0.15- j 0.9049	0.1509- j 0.9061	0.1519- j 0.9073
-0.2980	-0.3121	-0.3263	-0.3408

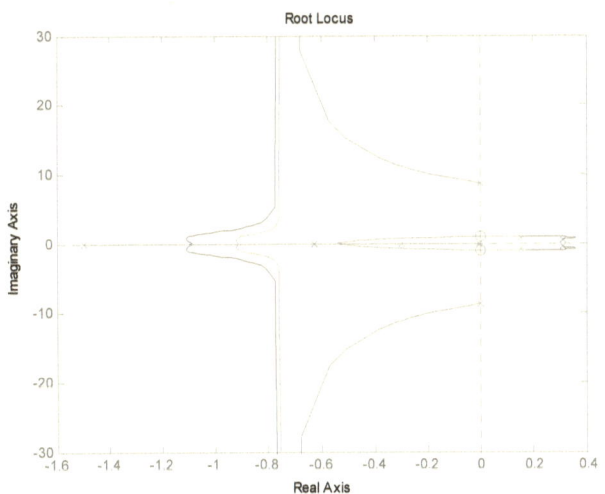

Figure E.7(f) Multiple root locus of a control system

(10) (a) unstable   (b) stable   Hint: section 7.5
(11) (a) unstable and poles: 1.2207, −0.2481+ j 1.0340, −0.2481− j 1.0340, and
     −0.7245   (b) stable and poles: −2+ j 3, −2− j 3, −0.9, and −0.5
     Hint: section 7.6
(12) $Y(s)/R(s) = \dfrac{64s^5 + 160s^4 + 160s^3 + 160s^2 + 128s + 32}{160s^7 + 464s^6 + 728s^5 + 788s^4 + 636s^3 + 468s^2 + 140s + 8}$

Apparently unstable. Yes. Poles: −1.3041, −0.7301+ j 0.9918, −0.7301−
j 0.9918, 0.1645+ j 0.9225, 0.1645− j 0.9225, −0.3912, and −0.0736
Zeroes: 0.1325+ j 0.9555, 0.1325− j 0.9555, −1.3740, −1, and −0.3911

$Y(s)/R(s) \approx \dfrac{0.4s^4 + 0.843s^3 + 0.6701s^2 + 0.7379s + 0.5114}{s^6 + 2.509s^5 + 3.569s^4 + 3.529s^3 + 2.595s^2 + 1.91s + 0.1278}$

on tolerance 10^{-4} Hint: section 7.6
(13) 4(a): Unstable   4(b): Unstable   4(e): The 0 appears in the first column
     of the table. Use of epsilon indicates unstable.   Hint: section 7.7
(14) Unstable: $0 \le K \le 0.33$ Stable: $0.34 \le K \le 0.5$ Hint: section 7.7
(15) Stable for 8 and unstable for 64.   Hint: section 7.8

(16) $e^{-3s} \approx \dfrac{s^4 - 6.667s^3 + 20s^2 - 31.11s + 20.74}{s^4 + 6.667s^3 + 20s^2 + 31.11s + 20.74}$.

Hint: section 7.9

Chapter 8

Control System Projects

In this chapter basic control system design problems trained in undergraduate course are randomly accumulated. Instead of organizing on chapter specific term, we treat every design problem as a mini project because each one needs MATLAB computing or SIMULINK modeling. Frequently practical parameters or concerns are involved in these projects. The project may require gain, transfer function, time/frequency response, or other relevant quantity finding. In every mini project finished solution or answer along with structured explanation would provide the reader some level of confidence on control system analysis through the wing of MATLAB or SIMULINK. Although the mini projects seem plain, they pave the way for insight of much more complicated design problems. However the mini project collection is in the sequel.

◆ **Project 1: Effect of external gain on a second order system**

A prototype second order system is shown in figure 8.1(a). There maybe substantial signal level difference between control input $u(t)$ and output $y(t)$ for various reasons for example due to feedback function $H(s)$, set point $u(t)$ requirement, etc. Introduction of gain K as shown in the figure stabilizes the output $y(t)$. Consider $\zeta=0.6$, $\omega_n=1.1 rad/\sec$, $H(s)=0.9$, and $u(t)=1$ over $0 \leq t \leq 2\sec$. Our objective is to determine the K such that $y(t) = u(t)$.

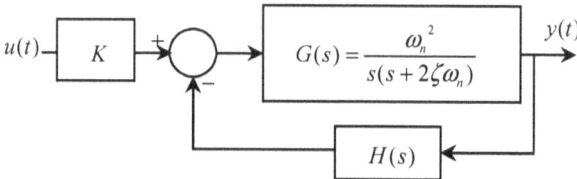

Figure 8.1(a) Prototype second order system with unknown gain

Solution:
Without K we can view the output response (sections 2.4 and 4.7) and have some idea about the output characteristic and do so by:
```
>>z=0.6; w=1.1; G=tf(w^2,[1 2*z*w 0]); H=0.9; ↵
>>S=feedback(G,H,-1); t=0:2/100:2; u=ones(1,length(t)); ↵
```
The symbols have above mentioned section meanings. Also did we decide 100 sample points for the t interval. The **S** holds the equivalent system. If we exercise lsim(S,u,t), we see the plot of output $y(t)$ versus t (graph not shown for space reason). It is evident from the graph that the $y(t)$ has not reached steady state y_{ss} over $0 \le t \le 2$ sec so we increase the last time bound from 2 to 20 (user-chosen):
```
>>t=0:20/100:20; u=ones(1,length(t)); ↵
```
Now lsim(S,u,t) indicates steady state behavior so we need the steady state value and get it by:
```
>>y=lsim(S,u,t); y_ss=y(end) ↵
```

y_ss =
   1.1111

Last return says that steady state y_{ss} is 1.1111. The K is in series with the second order system. If K is chosen as $1/y_{ss}$, overall gain of the system will be 1 thereby making $y(t) = u(t) = 1$. The stipulated K is then computed by:
```
>>K=1/y_ss ↵
```

Figure 8.1(b) A variant of prototype second order system

K =
   0.9000

If the reader says, how to verify that $y(t)$ reaches to $u(t)$ with the above K? The answer is exercise **OS=K*S; lsim(OS,u,t)** to view the overall system behavior where **OS** holds the overall equivalent system and is a user-chosen variable. Note that **series(K,S)** is equivalent to **K*S**.

✦ Project 2: Effect of internal gain on performance parameters

Figure 8.1(b) presents a variant of prototype second order control system with unity negative feedback. Our objective is to find a compromised value of the internal gain K on step response such that

peak time T_p is close to 1.2 secs and

percent overshoot is close to 5.3%.

Solution:

The first question is how to choose K? The answer is by trial and error. We may choose fractional K (e.g. 0.1), single digit K (e.g. 2), double digit K (e.g. 10), etc. Each time we have to view the response where underdamp type behavior of the output is manifested. Again there is no mention about the time interval nor is the step size. That is also selected on trial and error basis. We chose $0 \leq t \leq 10\sec$ with 101 points.

```
Tp=[ ]; PO=[ ];
t=[0:10/100:10]';
u=ones(length(t),1);
H=1;
K1=2:0.05:10;
for K=K1
  G=tf(K,[1 2 0]);
  S=feedback(G,H,-1);
  y=lsim(S,u,t);
  [y_max,l]=max(y);
  Tp=[Tp;t(l)];
  PO=[PO;(y_max-y(end))/y(end)*100];
end
plotyy(K1,PO,K1,Tp)
grid
```

Figure 8.1(c) Script file for peak time and P.O. finding

One solution can be we determine the T_p versus K and plot them. Again for the same range of K we plot P.O. versus K. Intersection of two curves is the compromised value of K.

The T_p computing by programming means is addressed in section 4.7. After several attempts we found K from 2 to 10 will do hence vary the K from 2 to 10 with increment 0.05 and generate it as a vector by K1=2:0.05:10; where K1 is a user-chosen variable. For every value of K in K1 we form the transfer function by tf and get output samples by lsim (section 4.3). Data accumulation technique of appendix D.3 helps us store the peak time and P.O. for every K. If the output $y(t)$ samples are in y, y(end) indicates approximate y_{ss}. Based on this y_{ss}, the P.O. is defined as ($y(t)|_{\max} - y_{ss})/y_{ss} \times 100$ whose code is (y_max-y(end))/y(end)*100. However figure 8.1(c) depicts the complete codes on the problem.

In the figure Tp, PO, and y_max refer to T_p, P.O., and $y(t)|_{\max}$ respectively. For loop counter K selects K sequentially from K1. We have here two dissimilar quantities – P.O. and T_p; this sort of graph is best plotted by plotyy of appendix E. The workspace variables PO, Tp, and K1 retain the data for P.O., T_p, and K upon execution of codes in figure 8.1(c) respectively. At the end of the codes plotyy(K1,PO,K1,Tp) is exercised in order to bring the figure 8.1(d) before us along with grid lines (done by the command grid). The left and right vertical axes of figure 8.1(d) refer to P.O. and T_p respectively whereas the horizontal axis is the K variation.

Next how do we get the compromised value of K? In the figure window menu bar you find an icon by name Data Cursor, click the icon (figure 1.2(c)), bring the mouse pointer at the intersection of two curves in figure 8.1(d), and find the value as K =4.25.

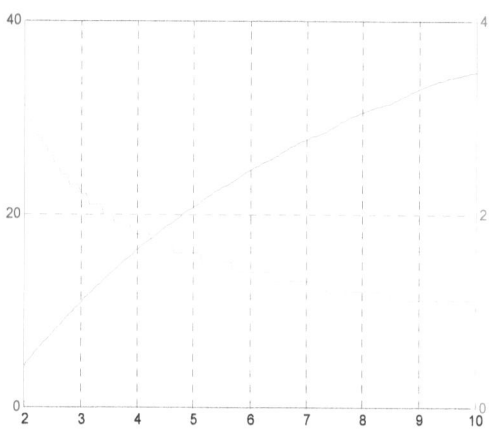

Figure 8.1(d) Plots of T_p and P.O. versus K

✦ Project 3: Identification of damping from time domain control output

Waveshape matters because it speaks of control system input/output trajectory. Think about the second order prototype control system subject to ω_n =1rad/sec, $H(s)$=1, and $u(t)$=1. We intend to choose four damping ratio; ζ =0 (undamped), ζ =0.3 (underdamped, $0<\zeta<1$), ζ =1 (critically damped), and ζ =2 (overdamped, $\zeta >1$). Our aim is to inspect the step response outputs i.e. $y(t)$ versus t for these four cases over $0 \leq t \leq 20\sec$.

After that we intend to see what happens to the output when a third pole at $s = -0.1$ is added to the system for ζ =0.5 i.e. $G(s) = \dfrac{1}{s(s+1)(s+0.1)}$.

Prerequisite: sections 2.1, 2.4, 4.5, 4.7, and 7.5.

Solution:

For each damping the $G(s)$ becomes different but $H(s)$ remains same however $Y(s)/R(s)$ equivalent is different. Feedback equivalent finding is required for each damping. The $G(s)$ is best entered by pole-zero-gain form. If we use step for $y(t)$ versus t, there is no need to be concerned for $R(s)$ or $u(t)$. The step also keeps provision for multiple systems handling. Following table shows all related commands:

Control System Analysis & Design in MATLAB and SIMULINK

Line #	Comment	MATLAB code
1	assigning $H(s)$	H=1;
2	undamped	G1=zpk([ ],[0 0],1); S1=feedback(G1,H,-1);
3	underdamped	G2=zpk([ ],[0 -0.6],1); S2=feedback(G2,H,-1);
4	critically damped	G3=zpk([ ],[0 -2],1); S3=feedback(G3,H,-1);
5	overdamped	G4=zpk([ ],[0 -4],1); S4=feedback(G4,H,-1);
6	for step response	step(S1,S2,S3,S4,20)
7	put a mark of distinction	legend('Un','Under','Critical','Over')
8	assigning p	p=-0.1;
9	$G(s)$ with p	G=zpk([ ],[0 -1 p],1); S=feedback(G,H,-1);
10	for step response	figure,step(S,20)

Execution of above MATLAB codes (lines 1 through 7) results in the figure 8.1(e). Sorry the text is black-white, you find the figure in MATLAB as color discerning. The text (line 7) inside **legend** (e.g. **Un**) is our-chosen.

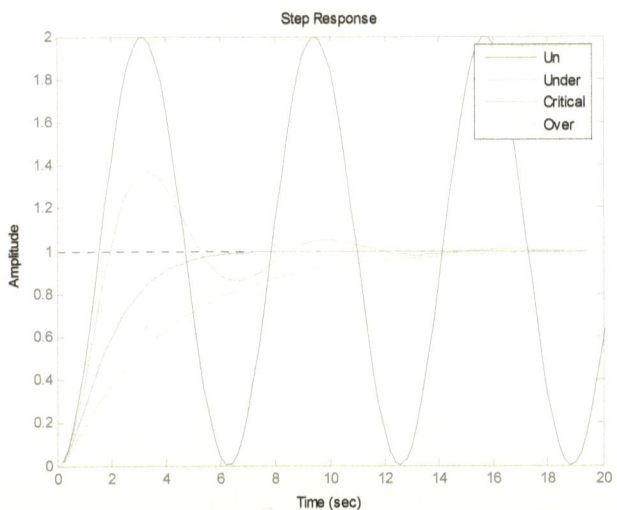

Figure 8.1(e) Plots of outputs ($y(t)$ versus t) for various damping

Inspection briefing is as follows:
    Undamped: The $y(t)$ is a perfect sine wave.
    Underdamped: The $y(t)$ oscillatorily settles to steady state.
    Critically damped: This is the boundary between under and over damped responses. Infinitesimal oscillation is associated with this.
    Over damped: No oscillation is perceived. Slowly $y(t)$ goes to steady state.

Realistic control systems are mostly underdamped or overdamped. The un and critically damped are basically hypothetical ones and used for design parameter specifications. All four outputs are stable.

Extra pole addition related commands are in lines 9 and 10 of previous table. The command **figure** merely opens a new window for the next graph otherwise you lose the last graph. However figure 8.1(f) shows $y(t)$ response due to

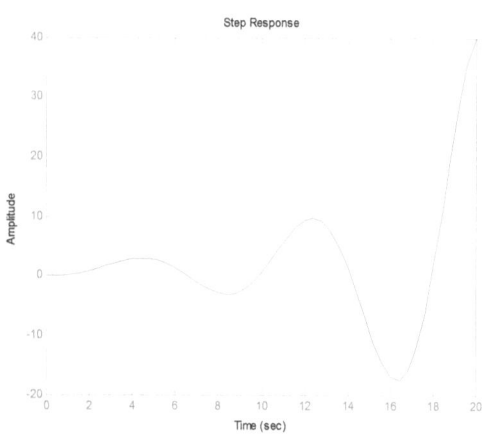

Figure 8.1(f) Plot of output ($y(t)$ versus t) for extra pole

the pole addition whose amplitude is oscillatorily increasing that is case of instability so damping is meaningless here.

Whatever waveshape response you get is a qualitative measure not a quantitative one.

✦ Project 4: Effect of a third order pole on performance parameters

Figure 8.2(a) shows the prototype second order system with a third order pole p. We wish to compute the following performance parameters; peak time T_p, steady state value y_{ss}, 10%-90% rise time T_r, percent overshoot P.O., and settling time T_s on 2% criterion for different p subject to $\zeta = 0.5$, $\omega_n = 1 rad/sec$, $H(s)=1$, and $u(t)=1$.

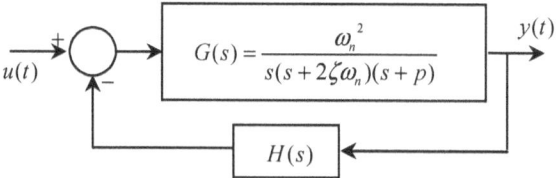

Figure 8.2(a) Prototype second order system with a third order pole

Prerequisite: sections 4.3 and 4.7 and project 3.

Solution:

There is one problem in computing on multiple quantities; the ranges are different. Each quantity had better be taken care of one at a time. Let us choose different poles at; $p = 0$, 0.1, 1, 10, and 100. The selection includes small fractional, integers, hundreds, or may be thousands as a start. The time interval also needs to be selected say $0 \le t \le 20$ sec with 1000 points.

Stability test from time domain waveshape:
Following table shows $y(t)$ versus t graphing for $p=0$:

Line #	Comment	MATLAB code
1	assigning p	p=0;
2	t and $u(t)$ samples	t=[0:20/1000:20]'; u=ones(length(t),1);
3	assign $H(s)$ and ω_n	H=1; w=1;
4	form $G(s)$	G=zpk([ ],[0 -2*1*w -p],w^2);
5	finding equivalent S	S=feedback(G,H,-1);
6	calling the grapher	lsim(S,u,t)

Concerning above table, execution of command line 1 through 6 results an output similar to the one in figure 8.1(f) indicating unstable output. In line 1 now consider **p=0.1**; for $p=0.1$ and execute again lines 2 through 6; the result is similar to figure 8.1(f) again meaning instability. We do not ponder unstable output. Similar execution with **p=1**; causes stable output to appear (like underdamped of figure 8.1(e)). With **p=10** or **100**, output manifests slow increase. In order to perceive wide range time domain behavior we have to increase the last time bound from 20 secs to higher say 100 secs. In line 2 of above table change **t** to **t=[0:100/1000:100]'**; and execute the rest commands, you should be viewing overdamped waveshape similar to figure 8.1(e). Let us summarize our findings:

p	0	0.1	1	10	100
Wave shape pattern	unstable	unstable	stable, under-damped	stable, over-damped	stable, over-damped
Chosen time bound	20 secs	20 secs	100 secs	200 secs	1000 secs

Steady state value y_{ss}:
Steady state value is pertinent for under or over damped case. But the time span of the output should be wide (say 100 secs then t=[0:100/1000:100]';) in order to get the closest steady state value. Now we need the samples of output $y(t)$. Referring to above stability discussion, the line 6 of the command table should be modified to **y=lsim(S,u,t)**; from which the last sample is the y_{ss} and we get it by **y(end)**. Found results are the following.

p	0	0.1	1	10	100
Steady state value	irrelevant	irrelevant	1	1	1

The 10%-90% rise time T_r :

Rise time is relevant to under and over damped cases meaning $p=1$ onwards. In rise time we need the steady state value so get it first by y_ss=y(end). Marking the changes the commands for $p=1$ are the following:

```
p=1;
t=[0:100/1000:100]'; u=ones(length(t),1);
H=1; w=1;
G=zpk([ ],[0 -2*1*w -p],w^2);
S=feedback(G,H,-1);
y=lsim(S,u,t);
y_ss=y(end); t1=c_cross(t,y,0.1*y_ss); t2=c_cross(t,y,0.9*y_ss); Tr=t2-t1
```

Similarly for $p=10$:

```
p=10;
t=[0:200/1000:200]'; u=ones(length(t),1);
H=1; w=1;
G=zpk([ ],[0 -2*1*w -p],w^2);
S=feedback(G,H,-1);
y=lsim(S,u,t);
y_ss=y(end); t1=c_cross(t,y,0.1*y_ss); t2=c_cross(t,y,0.9*y_ss); Tr=t2-t1
```

and for $p=100$:

```
p=100;
t=[0:1000/1000:1000]'; u=ones(length(t),1);
H=1; w=1;
G=zpk([ ],[0 -2*1*w -p],w^2);
S=feedback(G,H,-1);
y=lsim(S,u,t);
y_ss=y(end); t1=c_cross(t,y,0.1*y_ss); t2=c_cross(t,y,0.9*y_ss); Tr=t2-t1
```

In a summary the rise times are:

p	0	0.1	1	10	100
Rise time in secs	irrelevant	irrelevant	2.7	42.6	426.9

Percent overshoot P.O.:

Percent overshoot is relevant only to the under damped case so necessary codes are the following:

```
p=1;
t=[0:100/1000:100]'; u=ones(length(t),1);
H=1; w=1;
G=zpk([ ],[0 -2*1*w -p],w^2);
S=feedback(G,H,-1);
y=lsim(S,u,t);
[y_max,I]=max(y);
y_ss=y(end);
(y_max-y_ss)/y_ss*100
```

In a summary the P.O.s are:

p	0	0.1	1	10	100
P.O.	irrelevant	irrelevant	14.4604	irrelevant	irrelevant

Settling time T_s:

Settling time is relevant to under and over damped cases. Settling time commands for $p=1$ are the following:

```
p=1;
t=[0:100/1000:100]'; u=ones(length(t),1);
H=1; w=1;
G=zpk([ ],[0 -2*1*w -p],w^2);
S=feedback(G,H,-1);
y=lsim(S,u,t);
y_ss=y(end);
ts1=c_cross(t,y,0.98*y_ss);
ts2=c_cross(t,y,1.02*y_ss);
```

Now exercise plot(t,y,t,1.02*y_ss,t,0.98*y_ss) in order to see the settling phenomenon from which T_s is from the lower or $0.98\, y_{ss}$ bound so the settling time is

Ts=ts1(end)

Ts =
        12.3500

Figure (similar to figure 4.6(c)) is not shown from the **plot** for space reason. If it is necessary, use zooming facility of figure window to perceive the settling distinctively. For the overdamped case, delete the line **ts2= c_cross(t,y,1.02*y_ss);** because $1.02\, y_{ss}$ does not exist. Pick the value after viewing the settling phenomenon. However in a summary the settling times are:

p	0	0.1	1	10	100
Settling time in secs	irrelevant	irrelevant	12.35	76.5	724.5

✦ Project 5: Characteristics of a phase lag/lead network

When we design controller in frequency domain, phase lag/lead controller receives importance. Phase lag is called lag because output voltage lags the input. Similar interpretation goes for the other. Let us see how we view their different characteristics.

Prerequisite: sections 2.1, 6.3, and 6.4.

Figure 8.2(b) A phase lag network

Phase-lag network:

Figure 8.2(b) presents a phase lag network where $\alpha = \dfrac{R_1 + R_2}{R_2}$, $\tau = R_2 C$, $z = \dfrac{1}{\tau}$,

-215-

$p = \frac{1}{\alpha\tau}$, and $\alpha > 1$. Clearly the α, p, and z are called attenuation, zero, and pole respectively. With these parameters the controller has the transfer function $G_c = \frac{V_o(s)}{V_i(s)} = \frac{s+z}{\alpha(s+p)}$.

Suppose $R_1 = 1\ K\Omega$, $R_2 = 2\ K\Omega$, and $C = 1\ \mu F$ and we intend to see the pole-zero and frequency responses of the controller. Just carry out the following at the command prompt:

```
>>R1=1e3; R2=2e3; C=1e-6; a=(R1+R2)/R2; t=R2*C;
>>z=1/t; p=1/a/t; Gc=zpk(-z,-p,1/a);
```

In above lines we exercised like name symbology e.g. R1$\Leftrightarrow R_1$, t$\Leftrightarrow \tau$, a$\Leftrightarrow \alpha$, etc and just coded the expression (appendix A). Following two lines yield the frequency and pole-zero responses respectively.

```
>>bode(Gc)
>>figure, pzmap(Gc)    ← figure for viewing the output in another window
```

Graph is not shown for the space reason. The angular frequency corresponding to minimum phase is $\omega_m = \sqrt{pz}$ that you get by sqrt(z*p) and the answer is $408.2483\ rad/\sec$.

Phase-lead network:

Figure 8.2(c) shows a phase lead network where the related parameters or expressions are $\alpha = \frac{R_1 + R_2}{R_2}$, $\tau = \frac{R_1 R_2}{R_1 + R_2} C$, $z = \frac{1}{\alpha\tau}$, $p = \frac{1}{\tau}$, and $G_c = \frac{V_o(s)}{V_i(s)} = \frac{s+z}{s+p}$.

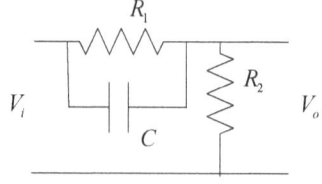

Figure 8.2(c) A phase lead network

With the parameters of lag counterpart you need the following code for the lead network frequency and pole-zero responses:

```
>>R1=1e3; R2=2e3; C=1e-6; a=(R1+R2)/R2; t=R1*R2*C/(R1+R2);
>>z=1/t/a; p=1/t; Gc=tf([1 z],[1 p]);
>>bode(Gc)
>>figure, pzmap(Gc)          ← Graph is not shown for space reason
```

The $\omega_m = \sqrt{pz}$ is also applicable here (sqrt(z*p) yields $1.2247\times 10^3\ rad/\sec$) but for maximum in phase spectrum. Note that phase spectrum is negative and positive for the lag and lead networks respectively.

✦ Project 6: Control system model order reduction

Given a control system, we reduce its order by embedded function **modred** but the function works only on state-space model. The order reduction happens in two steps; first view Gramian on the state-space model and then decide what states you will discard. Usually smaller value Gramians are discarded to get approximately reduced model. The Gramian values you get by another embedded function **balreal**. The **balreal** has the syntax [user-supplied variable for the state space model of given system, user-supplied variable for the Gramians]=**balreal**(system representing variable).

Prerequisite: section 2.1.

Let us consider the control system $H(s) = \dfrac{s^3 + 2s^2 - 3s + 8}{2s^5 + 3s^4 + 4s^3 + s + 1}$ and enter the system as follows:

N=[1 2 -3 8]; D=[2 3 4 0 1 1]; H=tf(N,D);

The last H holds $H(s)$, based on the H call the Gramian finder as:

[S,g]=balreal(H)

a =

	x1	x2	x3	x4	x5
x1	0.3426	0.7309	0	0	0
x2	-0.421	0.3426	0	0	0
x3	0	0	-0.4143	-0.5363	-0.1569
x4	0	0	0.5363	-1.264	-1.198
x5	0	0	-0.1569	1.198	-0.5072

b =

	u1
x1	-0.6464
x2	2.691
x3	1.585
x4	-0.9138
x5	0.3094

c =

	x1	x2	x3	x4	x5
y1	0.294	-0.5877	1.585	0.9138	0.3094

d =

	u1
y1	0

Continuous-time model.

g =

Inf
Inf
3.0303
0.3304
0.0944

The $H(s)$ in state-space form is returned to S and Gramians as a column matrix are returned to g where S as well as g is user-chosen. The lowest Gramian is 0.0944 which can be neglected. The one above the lowest is 0.3304 that can be discarded too. There are 5 states, so are the Gramians. The last two Gramians you select by **4:5** (appendix D.12). The **modred** has three input arguments; the state-space system, the Gramians need to be discarded, and the reserve word **Truncate** under quote respectively. The output of **modred** is the reduced system which you can assign to some variable so execute the following:

    Hn=modred(S,4:5,'Truncate');

The **Hn** is a user-chosen variable which retains the reduced system in state-space form. If we exercise the **tf** on the **Hn**, we view the transfer function:

    tf(Hn)

```
Transfer function:
-0.4721 s^3 + 1.719 s^2 - 2.349 s + 1.927
-----------------------------------------
     s^3 - 0.1185 s^2 + 0.03677 s + 0.2409
```

You could have assigned the **tf(Hn)** to another variable for the rational form. Anyhow above is the reduced form of $H(s)$ we started with. Depending on machine algorithm you may get another transfer function.

✦ Project 7: Liquid level control system design

Figure 8.2(d) shows the schematic diagram of a liquid level control system in which the principal elements are valve, hydraulic actuator, float, and tank. Control engineers performed some tests and determined the following: tank transfer function $G(s) = \dfrac{2.98}{29.8s+1}$, hydraulic actuator transfer function $G_a(s) = \dfrac{10.1}{s+1.01}$, and float transfer function $H(s) = \dfrac{1}{s^2/9.61 + s/3.1 + 1}$. The flow rate toward the valve is $v = 3\ m^3/\sec$. The cross sectional area of the pipe connected to valve and the pipe length are $A = 2\ m^2$ and $d = 1.5\ m$ respectively.

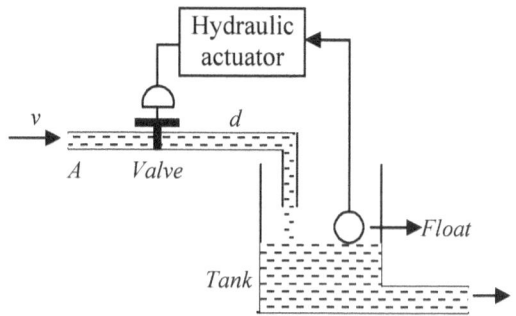

Figure 8.2(d) Liquid level control system

Determine the suitable gain for this liquid level control system which needs to be stable in the long run with phase margin 35^0.

Prerequisite: sections 2.1, 2.3, 2.4, 6.6, and 7.9.

Solution:

From the descriptions of various elements we infer that the float acts as the feedback. The hydraulic actuator works instantaneously because of electrical element involvement but the liquid needs some time to reach to the tank hence there is a time delay towards the tank. Actuator opens the valve such that the input and output flow rates remain the same.

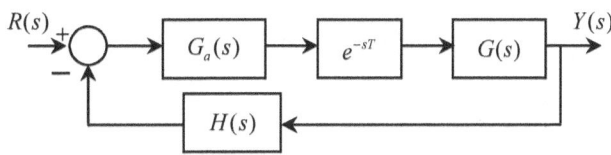

Figure 8.2(e) Liquid level control system

Considering all these, the liquid level control system block diagram is presented in figure 8.2(e). The time delay we obtain from the flow rate that is $T = \frac{v}{Ad} = 1$ sec. The input and output flow rates are here $R(s)$ and $Y(s)$ respectively what we are familiar with from control system theory.

We can test the stability in various ways, out of which let us choose the Nyquist plot. The reader might ask why did we choose Nyquist plot? The answer is all you need is $GH(s)$ function for the stability test without the $Y(s)/R(s)$ equivalent. Before testing the stability we need to approximate the delay element e^{-sT} by certain order say 2 so execute the following:

[N,D]=pade(1,2); D1=tf(N,D);

Above D1 holds the approximate e^{-sT} where D1 is a user-chosen variable. Let us enter the other transfer functions:

G=tf(2.98,[29.8 1]); Ga=tf(10.1,[1 1.01]); H=tf(1,[1/9.61 1/3.1 1]);

Above user-chosen variables G, Ga, and H retain the $G(s)$, $G_a(s)$, and $H(s)$ respectively. The $GH(s)$ is actually the series combination of $G_a(s)$, e^{-sT}, $G(s)$, and $H(s)$ and obtain the combination (series(G,H)⇔G*H) by:

GH=Ga*D1*G*H;
nyquist(GH)

In the Nyquist plot you find the (−1,0) point inside the loop hence the system is unstable (similar to figure 7.4(c), graph is not shown for space reason). One solution to the problem is insert some gain in the forward path of figure 8.2(e) say K. Choose different K e.g. fractional (0.1), integer (2), large integer (100), etc and check the stability each time. Well the K is again in series with $GH(s)$. Suppose $K=0.1$, then the command we need is K=0.1;T=K*GH; nyquist(T) where T is user-chosen that retains the series combination of K and $GH(s)$. The $K=1$ is basically the $GH(s)$ we started with. Found results are stable, unstable, and unstable for $K=0.1$, 1, and 2 respectively. Clearly K can be chosen from 0 to 1 for intended gain.

We can change the K with an increment 0.01 by a for-loop. The **margin** of section 6.6 returns both the gain and phase margins, out of which pick only the phase one because that is what we require. Each time we get the phase margin and store the margin as a column matrix by the data accumulation technique of appendix D.3. The whole programming tactic is the following:

```
P=[ ];
for K=0:0.01:1
    [Gm,Pm,wg,wp]=margin(K*GH);
    P=[P;Pm];
end
```

Ignore the warning if you receive any. The user-chosen variable P holds all phase margins as a column matrix for different K. There is no K vector generated before so generate it as a column matrix by:
K=[0:0.01:1]';

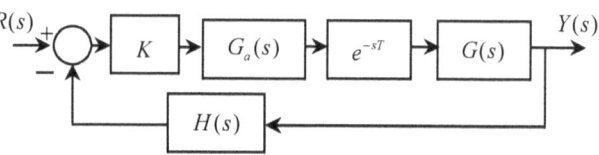

Figure 8.2(f) Liquid level control system with a proportional gain for stability

Finally we call gain and its related phase margin side by side by:
[K P]

```
ans =
         0      Inf
    0.0100    Inf
      ⋮
    0.4800   37.0169
    0.4900   35.9368    ← This is our phase margin
    0.5000   34.8684
      ⋮
    0.9900   -7.3037
    1.0000   -8.0107
```

As you see in above displayed data, the closest phase margin is 35.9368^0 for $K=0.49$. The stability you can verify by nyquist(0.49*GH). Anyhow figure 8.2(f) presents the block diagram considering added series gain K.

✦ Project 8: DC motor transfer function and its response

DC motor is an indispensable part of a control system which works on speed or rotary components. In order to determine the transfer function of a DC motor, control engineers have to perform some test for motor parameters which is beyond the scope of the text. We will start from some example parameters and address two motors – armature and field controlled DC motors.

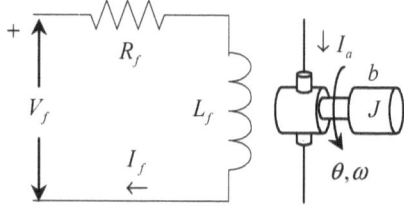

Figure 8.2(g) Elements of a field controlled DC motor

Field Controlled DC motor:

Figure 8.2(g) presents the necessary elements of a field controlled DC motor. Motor angular speed $\omega = \dfrac{d\theta}{dt}$ is controlled by the field circuit applied voltage V_f without hampering the armature circuit. The output to input transfer function of the field controlled DC motor in figure 8.2(g) is given by:

$$\dfrac{\theta(s)}{V_f(s)} = \dfrac{K_m}{s(sJ+b)(sL_f+R_f)} \quad \text{where}$$

K_m =motor torque constant in Newton-meter/Ampere,

J =rotational inertia on the motor in Newton-meter-square second/radian,

b =rotational friction coefficient in Newton-meter-second/radian,

R_f =field circuit resistance in Ohm, and

L_f =field circuit inductance in Henry.

Sometimes it is preferable to express the transfer function in time constant form which is

$$\dfrac{\theta(s)}{V_f(s)} = \dfrac{K_m}{bR_f s(s\tau_f+1)(s\tau_L+1)} \quad \text{where } \tau_f = \dfrac{L_f}{R_f} \text{ and } \tau_L = \dfrac{J}{b} \text{ and } \tau_f \text{ and } \tau_L$$

are called field circuit and rotational load or rotor time constants respectively.

Armature controlled DC motor:

An armature controlled DC motor you see in figure 8.2(h) in which the field circuit is not altered. Motor speed is now controlled by armature applied voltage V_a. The output to input transfer function of the armature controlled DC motor is given by:

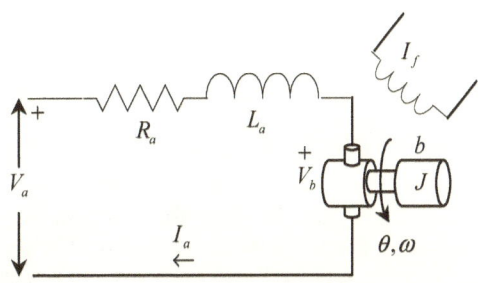

Figure 8.2(h) Elements of an armature controlled DC motor

$$\dfrac{\theta(s)}{V_a(s)} = \dfrac{K_m}{s[(sJ+b)(sL_a+R_a)+K_bK_m]} \quad \text{where } K_m, J, \theta, \text{ and } b \text{ have the}$$

earlier mentioned meanings and

R_a =armature resistance in Ohm,

L_a =armature inductance in Henry, and

K_b =motor back electromotive force constant in Volt/radian/sec.

Also during steady state we get $K_m = K_b$. In terms of time constant we get $\dfrac{\theta(s)}{V_a(s)} = \dfrac{K_m}{R_a bs\left[(s\tau_L + 1)(s\tau_a + 1) + \dfrac{K_b K_m}{bR_a}\right]}$ where $\tau_a = \dfrac{L_a}{R_a}$ and τ_L with previous meaning.

If we know the motor and load parameters, we can easily implement the transfer function by the **tf** or **zpk** of section 2.1.

Table 8.A: Typical constants of a fractional HP DC motor

Motor constant	$51 \times 10^{-3}\ N \cdot m / A$
Rotor inertia	$12 \times 10^{-3}\ N \cdot m \cdot s^2 / rad$
Field time constant	1.25 msecs
Field circuit resistance	2.5 Ω
Rotor time constant	110 msecs
Maximum output power	½ HP, 110 V

Example on a field controlled DC motor:

Suppose a field controlled fractional HP DC motor has the table 8.A displayed parameters. Obtain θ and ω responses of the motor over $0 \le t \le 2\,\text{secs}$.

Let us enter the table 8.A parameters in standard unit as follows:

>>t_L=110e-3; t_f=1.25e-3; J=12e-3; Km=51e-3; Rf=2.5; ↵

The symbology in above is the following: J⇔ J, t_f⇔ τ_f, t_L⇔ τ_L, Km⇔ K_m, and Rf⇔ R_f. The b needs to be calculated and do so by:

>>b=J/t_L; ↵

The $\dfrac{\theta(s)}{V_f(s)}$ can be viewed as pole-zero-gain form in which we see zero: none, pole: 0, $-1/\tau_L$, and $-1/\tau_f$, and gain: $\dfrac{K_m}{bR_f\tau_f\tau_L}$ therefore enter the transfer function by:

>>T=zpk([ ],[0 -1/t_f -1/t_L],Km/b/Rf/t_f/t_L); ↵

The **T** is a user-chosen variable which holds the $\dfrac{\theta(s)}{V_f(s)}$. Concerning the example 1 of section 4.3, **T** is basically $H(s)$ of figure 4.3(a). From the table the input voltage $v_f(t) = 110\,V$ is over $0 \le t \le 2\,\text{secs}$ and output $\theta(t)$ response we obtain by:

>>t=0:0.01:2; u=110*ones(1,length(t)); lsim(T,u,t) ↵

Figure 8.3(a) depicts the $\theta(t)$ response in which the **Amplitude** refers to $\theta(t)$. The **lsim** by default graphs both the input and output. In order to view only the output, rightclick the mouse on the plot

area and uncheck **Show Input** in the popup. For the ω response the transfer function is $\dfrac{K_m}{(sJ+b)(sL_f+R_f)}$ because $\omega = \dfrac{d\theta}{dt}$ hence the modified transfer function (exclude the 0 pole) is:
>>T1=zpk([ ],[-1/t_f -1/t_L],Km/b/Rf/t_f/t_L); ↵

The **T1** is a user-chosen variable which holds the transfer function $\dfrac{\Omega(s)}{V_f(s)}$ (i.e. in time domain $\dfrac{\omega(t)}{v_f(t)}$). In a similar fashion the ω response we view by:
>>figure,lsim(T1,u,t) ↵   ← figure is for another window keeping the previous one

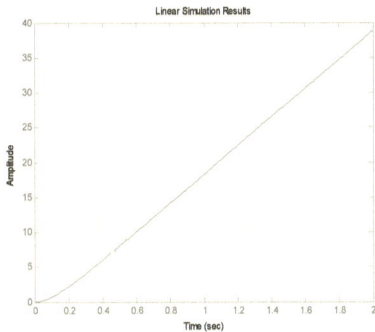
Figure 8.3(a) θ versus t response

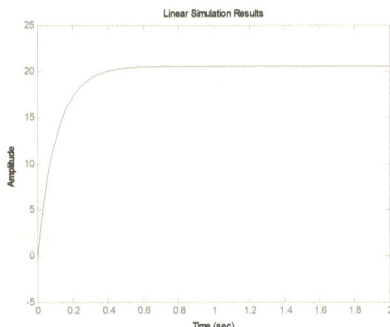
Figure 8.3(b) ω versus t response

Figure 8.3(b) indicates the result in which the **Amplitude** now corresponds to ω. From the response one might be interested in steady state value for which we need to have the samples:
>>y=lsim(T1,u,t); y(end) ↵

ans =
     20.5700

As the return says $\omega_{ss} = 20.57 \, rad/\sec$. Sometimes RPM ($N$) is necessary in motor operation which is linked to ω by $\omega = \dfrac{2\pi N}{60}$. In case RPM is required for the above motor, exercise y(end)*60/2/pi which provides 196.4290.

Example on an armature controlled DC motor with disturbance:
In ongoing example we assumed that there is no disturbance on the motor shaft. In the presence of a disturbance the transfer function becomes different. The disturbance is basically a sudden torque $T_d(s)$ or load on the shaft. Figure 8.3(c) depicts the armature controlled DC

motor block diagram with a disturbance torque where $G_a(s) = \dfrac{K_m}{sL_a + R_a}$, $G_m(s) = \dfrac{1}{sJ + b}$, and $H(s) = K_b$ with earlier symbol meanings. Typical parameters of the motor you find in table 8.B. Determine the angular position and speed responses of the motor shaft in the presence of a disturbance at the 5th sec over $0 \le t \le 12 \sec s$.

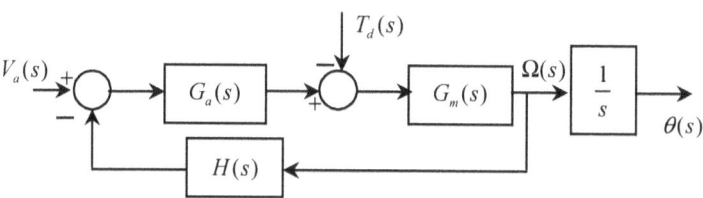

Figure 8.3(c) Armature controlled DC motor with a disturbance torque

Table 8.B: Typical constants of an armature controlled DC motor

Motor constant	$51 \times 10^{-3}\ N \cdot m / A$
Rotor inertia	$12 \times 10^{-3}\ N \cdot m \cdot s^2 / rad$
Armature time constant	1.25 msecs
Armature circuit resistance	$1.5\ \Omega$
Rotor time constant	110 msecs
Maximum output power	½ HP, 110 V

The diagram of figure 8.3(c) is best modeled in SIMULINK (chapters 3 and 5). The $G_a(s)$, $G_m(s)$, and $H(s)$ of figure 8.3(c) are modeled by **Transfer Fcn, Transfer Fcn,** and **Gain** blocks respectively. Both the armature voltage and torque disturbance are modeled by a **Step** block. The **Integrator** simulates $\dfrac{1}{s}$ which provides $\theta(t)$ from $\omega(t)$. Four **Scopes** are necessary for $V_a(s)$, $T_d(s)$, $\theta(s)$, and $\Omega(s)$. A **Display** block shows steady state value of $\omega(t)$.

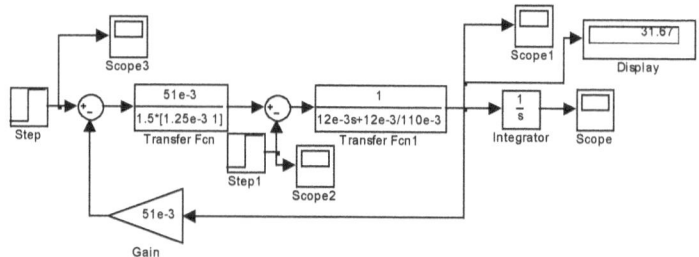

Figure 8.3(d) SIMULINK model of the armature controlled DC motor in figure 8.3(c)

How do we select the disturbance? First let us compute the rated torque. The motor power factor changes during running say from 0.3 to 0.8. Since 1 HP=746 Watts, with the least power factor the rated motor current is $0.5 \times 746/(110 \times 0.3) = 11.303\, A$. From the motor constant the rated torque is $51 \times 10^{-3} \times 11.303 = 0.5765\, N \cdot m$. Figure 8.3(d) presents SIMULINK model we need for the simulation. Settings in various blocks are the following:

For $G_a(s)$; **Transfer Fcn**: Numerator: **51e-3** and Denominator: **1.5*[1.25e-3 1]**

For $G_m(s)$; **Transfer Fcn1**: Numerator: **1** and Denominator: **[12e-3 12e-3/110e-3]**

For $H(s)$; **Gain: 51e-3**

For $V_a(s)$; **Step**; Step time: 0 and Final value: 110

For $T_d(s)$; **Step1**; Step time: 5 and Final value: **0.4*0.5765**

Disturbance torque should never be beyond the rated torque usually that is simulated up to certain percentage of the full load torque say 40%. Since the $T_d(s)$ appears at the 5th sec, the torque in time domain then becomes $0.4 \times 0.5765\, u(t-5)$ that is why the **Step1** setting is so. If you wish the machine to perform calculation, exercise

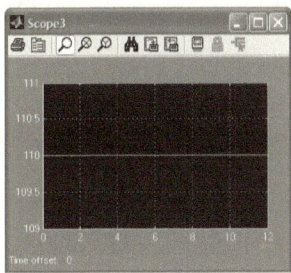

Figure 8.3(e) Armature input voltage

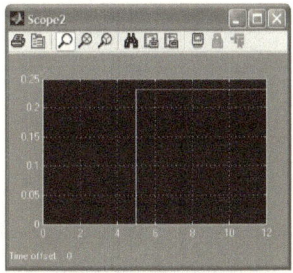

Figure 8.3(f) Motor disturbance torque

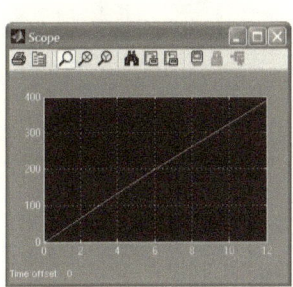

Figure 8.3(g) Motor θ versus t response

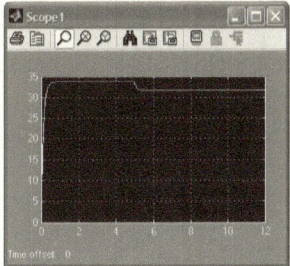

Figure 8.3(h) Motor ω versus t response

0.5*746/110/0.3*51e-3*0.4 as the Final value of Step1. In order to simulate the problem, open a new SIMULINK model file and bring all blocks of figure 8.3(d) in the file. Doubleclick each block and enter its related setting. In addition you need two Sum blocks, each one needs default sign change from ++ to +-. The Gain block needs flipping. You may need to enlarge some block to see its contents. Connect the blocks as in figure 8.3(d), enter the Stop time of the solver as 12, run the model, and finally doubleclick the Scopes to view waveshapes like figures 8.3(e), 8.3(f), 8.3(g), and 8.3(h) which show armature applied $v_a(t)$, torque disturbance $t_d(t)$, shaft position $\theta(t)$, and shaft angular speed $\omega(t)$ in Scope3, Scope2, Scope, and Scope1 respectively. Each Scope needs autoscale setting.

Each waveshape is over $0 \le t \le 12 \text{ secs}$. The disturbance appears at the 5th second, so there is a sudden drop in $\omega(t)$ in figure 8.3(h) which in turn reduces the steady state value. The Display shows steady state value of the $\omega(t)$ which is $\omega_{ss} = 31.67 \, rad/\sec$. Without the $t_d(t)$ or $T_d(s)$, the ω_{ss} would have been larger than 31.67.

◆ Project 9: Root sensitivity to a control system parameter

Given a control system, roots of the system may change slowly or sharply subject to the change of a system parameter. The parameter can be pole, zero, gain, or other. Root sensitivity S of a control system is given by $S = \dfrac{\partial r}{\partial \beta}$ where β is the parameter meant for change and r is the root in s plane. The root is a complex quantity, so is the S. Without s plane root, sensitivity problem can not be enumerated. The solving technique of root sensitivity is interactive.

Figure 2.4(a) shows a feedback system with $G(s) = \dfrac{21(s+3.2)}{s(s+2.5)(s+\beta)}$ and $H(s) = \dfrac{1}{s}$. We wish to obtain root sensitivity of the characteristic equation on every single root if operating β is 8 and if it increases by 5%.

Prerequisite: sections 2.1, 2.3, 7.1, and appendices D.9 and D.11.

Solution:

Let us enter $G(s)$ and $H(s)$ to G and H respectively:
G=zpk(-3.2,[-2.5 -8 0],21); H=tf(1,[1 0]);
Assign characteristic polynomial $1+G(s)H(s)$ to C by:
C=1+G*H;

Note that **series(G,H)**⇔**G*H** and **parallel(G,H)**⇔**G+H**. Extract the numerator and denominator to **n** and **d** respectively by:
    [n,d]=tfdata(C,'v');
Roots of the characteristic equation are found by:
    r=roots(n)

    r =
            -8.2698
            -2.7428
            0.2563 + 1.7021i
            0.2563 - 1.7021i

The parameter β increases by 5% meaning new value of β must be 1.05×8. With the new value of β, reexecute the above for later computing:
    G1=zpk(-3.2,[-2.5 -8*1.05 0],21); H1=tf(1,[1 0]);
    C1=1+G1*H1;
    [n1,d1]=tfdata(C1,'v');
    r1=roots(n1)

    r1 =
            -8.6488
            -2.7322
            0.2405 + 1.6691i
            0.2405 - 1.6691i

Every symbol in above is followed by 1 to indicate the later for example G by G1. Roots of the characteristic equation are stored in r and r1 before and after the change of β respectively. For example the first root is −8.2698 and −8.6488 respectively. Knowing so and reiterating $\frac{\partial \beta}{\beta}$=0.05, sensitivity on the first root is:
    (r1(1)-r(1))/0.05

    ans =
            -7.5794

The third and fourth roots are complex conjugate. Usually for complex roots we consider one say the third one. Similar computing for the third root provides:
    B=(r1(3)-r(3))/0.05

    B =
            -0.3161 - 0.6585i

In above we assigned the result to some user-chosen variable B. In order to find the magnitude-phase angle, exercise the following:
    [abs(B) rad2deg(angle(B))]

    ans =
            0.7304   -115.6456

The sensitivity S for the third root is read off as $0.7304 \angle -115.6456°$. In a similar fashion you may compute other sensitivities.

♦ Project 10: Space telescope pointing control system design

Figure 8.4(a) shows the block diagram of a space telescope pointing control system. The $r(t)$, $y(t)$, and $d(t)$ are called position input, attitude, and attitude error respectively. The problem statement is to determine the value of K such that the effect of disturbance $d(t)$ is minimized or in other words the integral square error (ISE=$\int y^2(t)dt$) is minimized for a unit step disturbance.

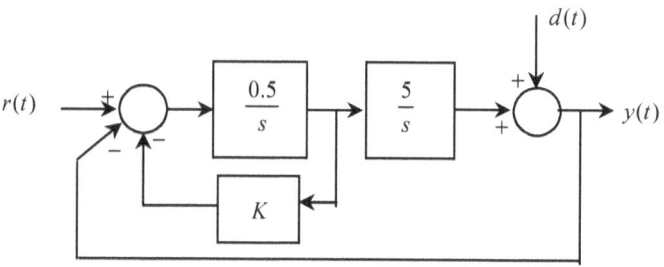

Figure 8.4(a) Space telescope pointing control system

Prerequisite: sections 3.1, 5.1, 5.2, 5.8, and 5.10 and appendix D.4.

Solution:

In order to see the effect of $d(t)$ on $y(t)$ we set the position input $r(t)$ as 0. There is no mention about the K range. The $K=0$ is not chosen because the feedback loop is broken. Let us choose start K as 0.1 with increment 0.1 and last value 10. One way to solve this problem is to model the figure 8.4(a) in SIMULINK and pass the K from MATLAB besides hold the ISE index return from the model for every single K.

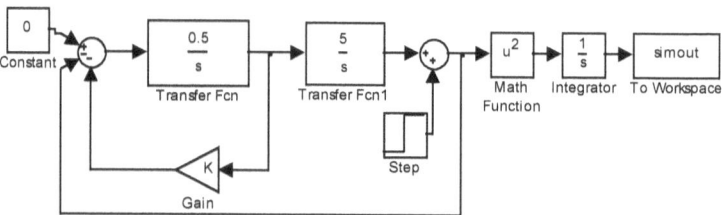

Figure 8.4(b) SIMULINK model of the space telescope pointing control system in figure 8.4(a)

Figure 8.4(b) presents the SIMULINK model of the control system in figure 8.4(a). Open a new SIMULINK model file. The $r(t)$ we model by a **Constant** block with value setting 0. The **Transfer Fcn** models $\frac{0.5}{s}$ and needs setting **Numerator:** 0.5 and **Denominator:** [1 0] similarly **Transfer Fcn1** does for $\frac{5}{s}$ with **Numerator:** 5 and **Denominator:** [1 0]. A **Gain** block simulates K. Doubleclick the **Gain**, enter its setting as **K**, and flip the block. Two **Sum** blocks are needed, one with default **List of Signs** and the other with **+--** for the three inputs. The $y(t)$ is squared by **Math Function** (doubleclick the block and change its setting from default to square). The **To Workspace** transports SIMULINK data i.e. ISE index to MATLAB for every K. The **To Workspace** needs **Save Format** setting as **Array**. The **Step** simulates $d(t)$ with **Step time** setting as 0. Connect the blocks like figure 8.4(b) but do not run the model because **K** is unknown to SIMULINK. The integration $\int y^2(t)dt$ limit should be chosen say 0 to 100. For the limit enter **Stop time** of SIMULINK as 100. Save the file by name **Project10** (user-chosen) at your working path of MATLAB.

Now move on to MATLAB. Data accumulation of appendix D.3 will be applied to hold the ISE index for different K. We can generate different K as a row vector by **K1=0.1:0.1:10**; where **K1** is user-chosen. SIMULINK solver adaptively selects step size by default, from which the last time bound refers to the ISE index. The **simout** variable of **To Workspace** keeps the index for adaptive steps but **simout(end)** provides $\int_0^{100} y^2(t)dt$. We select every **K** of **K1** by a for-loop. Staying in MATLAB you can run the **Project10** model by the command **sim**. Anyhow the complete command line is the following:

    I=[ ]; K1=0.1:0.1:10; for K=K1,sim('Project10'),I=[I simout(end)];end

Ignore the SIMULINK warning. The variable **I** is the one that keeps all ISE indices, graph of which you see by **plot** of appendix E:
    plot(K1,I)

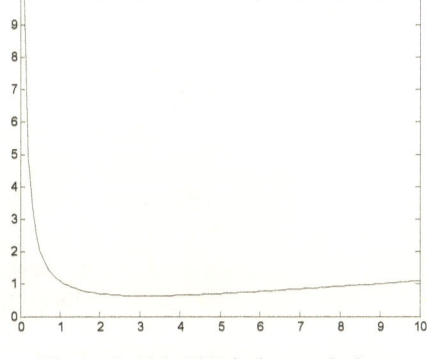

Figure 8.4(c) is the result from above execution. The vertical and horizontal axes refer to ISE index and K respectively. The gain corresponding to minimum ISE we obtain by:
    [M,p]=min(I);
    K1(p)

Figure 8.4(c) ISE index variation against gain

    ans =
        3.1000

As the return indicates the intended gain is $K=3.1$.

♦ Project 11: Elements of a root locus

Although in chapter 7 we presented root locus related computing or graphing, some elements of root locus might be useful for design purpose which we address in this project.

Prerequisite: sections 2.1, 6.4, 7.1, 7.4, and appendices D.9, D.11, and D.14.

Angle of departure from a pole:

Suppose loop transfer function of the control system in figure 7.1(b) or 7.1(e) is $GH(s) = \dfrac{1}{s^4 + 12s^3 + 64s^2 + 128s}$. We wish to determine the angle of departure at every pole of $GH(s)$.

Enter the $GH(s)$ to user-chosen **GH** and find its poles by:
GH=tf(1,[1 12 64 128 0]); p=pole(GH)

p =
    0
   -4.0000 + 4.0000i
   -4.0000 - 4.0000i
   -4.0000

Above user-chosen **p** holds the four poles of $GH(s)$. For example angle of departure of the root locus (which is $-135°$) from the second pole (which is $-4 + j\,4$) we obtain by:
180-sum(rad2deg(angle(p(2)-p)))

ans =
  -135.0000

The second pole we get by **p(2)**. The second pole minus all other poles is computed by **p(2)-p**. The **angle(p(2)-p)** determines phase angle of the complex difference vector **p(2)-p** in radian from which the degree counterpart is found by **rad2deg**. As we know the angle of departure from any pole is $180°$ −sum of angles of all other difference vectors, the **180-sum(rad2deg(angle(p(2)-p)))** is exercised. Again if the angle of departure on the fourth pole were required, we would have exercised **180-sum(rad2deg(angle(p(4)-p)))**. What if a loop function has both the zero and pole for instance $GH(s) = \dfrac{5s^2 + 6s + 7}{9s^3 - 6}$? Carry out similar execution by:
GH=tf([5 6 7],[9 0 0 -6]); p=pole(GH)

p =
   -0.4368 + 0.7565i
   -0.4368 - 0.7565i
    0.8736

z=zero(GH)

z =
 -0.6000 + 1.0198i
 -0.6000 - 1.0198i

If pole and zero both are present, the angle of departure we obtain by $\sum$(distance vector angle for zero)–$\sum$(distance vector angle for pole)= $180°$. At any pole the angle of departure becomes $\sum$(distance vector angle from the concerned pole to zero) $-180°$ $-\sum$(distance vector angle from the concerned pole to other poles). Angle of departure in degree (which is $33.4526°$) from the second pole (which is $-0.4368- j\,0.7565$) is computed by:

rad2deg(sum(angle(p(2)-z)))-180-rad2deg(sum(angle(p(2)-p)))

ans =
 33.4526

Angle of arrival towards a zero:

You can apply codes similar to angle of departure but maintaining the phase criterion $\sum$(distance vector angle for zero)–$\sum$(distance vector angle for pole)=$-180°$.

Break away point:

In order to find break away point on the loop function $GH(s)$, we take derivative of the $1/GH(s)$ and set the derivative to 0. For example we have $GH(s)$ $=\dfrac{5s+6}{9s^4+2s^3-6s+8}$ and the $\dfrac{d[1/GH(s)]}{ds}=0$ results s $=-1.7053$, $-0.3147+j\,0.6810$, $-0.3147-j\,0.6810$, and 0.5866 which we wish to obtain.

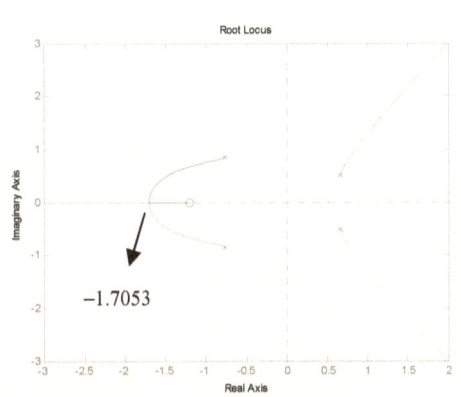

Figure 8.4(d) Breakaway point on a root locus

First enter the $GH(s)$ function by:
GH=tf([5 6],[9 2 0 -6 8]);
Then extract the polynomial coefficients of $GH(s)$ by:
[n,d]=tfdata(GH,'v');
The $GH(s)$ numerator and denominator coefficients are stored in user-chosen n and d respectively. Polynomial derivative of a division

function we find by **polyder** that applies the syntax [user-supplied variable for resulting numerator, user-supplied variable for resulting denominator]=**polyder**(given numerator coefficients as a row matrix, given denominator coefficients as a row matrix). Since we are after $1/GH(s)$, denominator and numerator positions are interchanged that is:

[N,D]=polyder(d,n);

Hence user-chosen **N** and **D** hold the numerator and denominator coefficients of $\dfrac{d[1/GH(s)]}{ds}$ respectively from which the break away points we determine by:

roots(N)

ans =

-1.7053
-0.3147 + 0.6810i
-0.3147 - 0.6810i
0.5866

If we exercise rlocus(GH), we see the root locus like figure 8.4(d). Not all points so computed are break away. Out of all roots only $s = -1.7053$ is the break away one which is arrow indicated in figure 8.4(d).

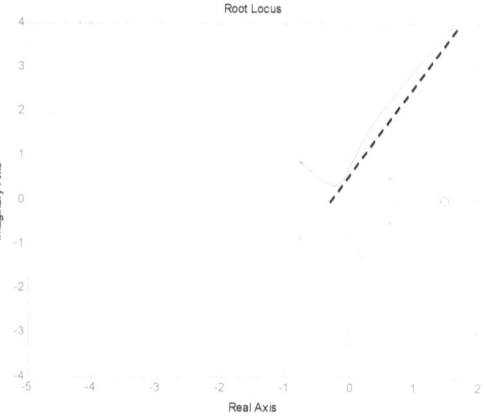

Figure 8.4(e) Asymptote on a root locus

Real axis intercept and angle of the asymptote:

If a root locus asymptote intersects the real axis at $s = \sigma_A$, we determine that by $\sigma_A = \dfrac{\sum pole - \sum zero}{n_p - n_z}$ where n_p and n_z are numbers of poles and zeroes respectively. For example we have $\sigma_A = -0.5741$ for $GH(s) = \dfrac{6s - 9}{9s^4 + 2s^3 - 6s + 8}$ and intend to implement.

We enter first given system to user-chosen **GH** by:
GH=tf([6 -9],[9 2 0 -6 8]);

Determine then the poles and zeroes and assign them to user-supplied **p** and **z** respectively by:

```
p=pole(GH); z=zero(GH);
```
Command **length** finds the number of elements in **p** or **z**:
```
np=length(p); nz=length(z);
```
Above user-supplied **np** and **nz** hold the n_p and n_z respectively on account of that the σ_A (assigned to some user-chosen variable **sig**) computing is:
```
sig=(sum(p)-sum(z))/(np-nz)
```

sig =
        -0.5741

Now angle of the asymptote on a root locus is given by $\varphi_A = \dfrac{2q+1}{n_p - n_z} 180°$ where $q = 0, 1, 2, \ldots, n_p - n_z - 1$. All the values are at the workspace hence obtain the q as a row matrix by:
```
q=0:np-nz-1;
pa=(2*q+1)/(np-nz)*180
```

pa =
        60   180   300

Above **pa** is a user-chosen variable which holds computed φ_A. The loop function $GH(s)$ is kept in variable **GH** so execute **rlocus(GH)** for the standard root locus. Figure 8.4(e) is the root locus in which we included a dotted line indicating the asymptote at $\sigma_A = -0.5741$ and $\varphi_A = 60°$.

Imaginary axis intercept:

Suppose the loop function of a standard control system (figure 7.1(b) or 7.1(e)) is $GH(s) = \dfrac{5s+6}{9s^3 + 2s^2 - 8}$. The standard root locus you may view by **GH=tf([5 6],[9 2 0 -8]);rlocus(GH)** which shows locus of the roots of the characteristic equation $1 + K\, GH(s) = 0$ over $0 \le K \le \infty$. The graph is shown in figure 8.4(f).

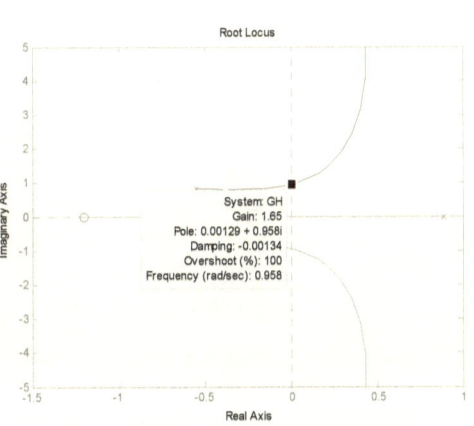

Figure 8.4(f) Root locus showing imaginary axis crossing data

Bring your mouse pointer on the imaginary axis crossing point and click the point in order to see the figure attached box properties. In the box you find $K=1.65$ which corresponds to the root locus crossing of the imaginary axis. Frequency of sustained oscillation is also displayed which is $0.958\ rad/\sec$.

◆ Project 12: Designing a phase lag network for compensation

Phase lag compensator is designed in frequency domain to fulfill phase margin and velocity error coefficients. There is a process in the design that is stated as follows:

(a) phase margin and error constant of uncompensated system are evaluated,
(b) a $5°$ phase lag is allowed in the phase margin for determining the new crossover frequency i.e. $\angle GH(j\omega_c) = -180° + 5° +$ wanted phase margin,
(c) zero of the compensator is placed one decade below the new crossover frequency ω_c,
(d) parameter α of the compensator is chosen as $|G(j\omega_c)|$, and
(e) pole of the compensator is found from the zero and α.

Prerequisite: sections 2.1, 2.4, 4.5, 6.2, 6.3, and 6.6, project 5, and appendices D.8 and D.11.

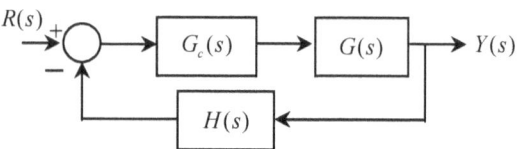

Figure 8.4(g) Feedback system with a series compensator

As an example the standard feedback system of figure 2.4(a) has $G = \dfrac{K}{s(s+2)}$ and $H = 1$. Design a series lag compensator like figure 8.4(g) such that the velocity error coefficient is 20 and phase margin is greater than or equal to $45°$. Verify your design and plot the output response before and after the design.

Solution:

Velocity error coefficient is defined as $\underset{s \to 0}{Lt}\ sGH(s)$ which provides $K = 40$. Let us enter the system to workspace G by:
G=zpk([ ],[0 -2],40);

You can exercise the command **margin** to find the phase margin of the uncompensated system (with the symbol meaning of section 6.6):
[Gm,Pm,wg,wp]=margin(G);

Pm

ans =

17.9642

Since the phase margin is not $45°$, we have to add compensator so the phase angle considering allowance should be $-180° + 5° + 45° = -130°$. The problem with discrete computing is we may not have the phase sample available for $-130°$ in samples of the phase spectrum that is why we get the samples of phase spectrum and find $-130°$ crossing by the **c_cross** function. The bode plot (i.e. magnitude spectrum in dB and phase spectrum in deg) you can view by **bode(G)**, graph is not shown for space reason. From the graph we see the ω variation as 10^{-1} to $10^2 \, rad/sec$. With a user-chosen step 0.1 for ω we generate the ω vector by:

    w=0.1:0.1:100;

The complex $G(j\omega)$ samples are obtained by (assign to some user-chosen **R**):

    R=freqresp(G,w); R=squeeze(R);

The phase spectrum $\angle G(j\omega)$ samples in degree we get to user-chosen **P** by:

    P=rad2deg(angle(R));

Obtain the $-130°$ crossing of the phase spectrum $\angle G(j\omega)$ by:

    wcp=c_cross(w,P,-130)   ← wcp⇔ ω_c

wcp =
    1.6500   i.e. at $\omega_c = 1.65 \, rad/sec$, $\angle G = -130°$

The $G(j\omega)$ at $\omega_c = 1.65 \, rad/sec$ we get by:

    Mc=freqresp(G,wcp)   ← Mc⇔ $G(j\omega_c)$

Mc =
    -5.9502 - 7.2123i   ← i.e. $G(j\omega)$ at $\omega_c = 1.65 \, rad/sec$

The α of series compensator is just $|G(j\omega_c)|$ so get it:

    a=abs(Mc)   ← a⇔ α

a =
    9.3500

The zero z is one decade below of ω_c and the pole is $p = z/\alpha$ hence likewise computing is:

    z=wcp/10; p=z/a;

The compensator transfer function G_c (put to **Gc**) is then formed by:

    Gc=zpk(-z,-p,1/a)
    Zero/pole/gain:
    0.10695 (s+0.165)

    (s+0.01765)

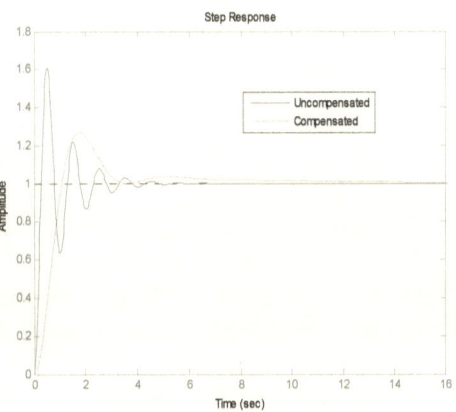

Figure 8.5(a) Step response with and without the lag compensator

As displayed the compensator has the system function $G_c(s) = \dfrac{0.10695(s+0.165)}{s+0.01765}$. The compensated system will have the function $G_c(s)\,G(s)$ and that we get by:

```
Gc*G
Zero/pole/gain:
   4.2781 (s+0.165)
---------------------------
   s (s+0.01765) (s+2)
```

Should you need the phase margin with the compensator, carry out the following:
```
[Gmc,Pmc,wgc,wpc]=margin(Gc*G);Pmc
```

Pmc =
   45.2994

Should you need the step response of without and with the compensator, obtain the feedback equivalent transfer function respectively by:
```
E=feedback(G,1,-1);
Ec=feedback(G*Gc,1,-1);
```

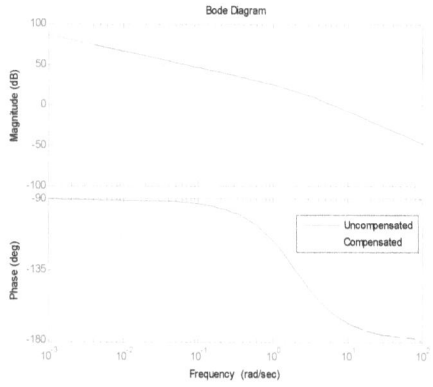

Figure 8.5(b) Bode plot with and without the lag compensator

Call the step response plotter as:
```
step(E,Ec)
```
Include the distinction label by:
```
legend('Uncompensated','Compensated')
```
Figure 8.5(a) depicts the two responses together. Again the bode plot of the uncompensated and compensated systems you see by:
```
figure,bode(G,Gc*G)
legend('Uncompensated','Compensated')
```
Figure 8.5(b) is the outcome from the last two lines. Whatever variable is used is user-selected for example E, Ec, etc.

✦ Project 13: Linking gain and phase margins of a control system

In chapter 6 we explained the computing of gain and phase margins by using the function **margin**. For design reason we may have to link the two margins. In this project our objective is to graph (a) gain margin versus gain (b) phase margin versus gain (c) gain margin versus phase margin for a given control system. As an example plot the three graphs for control system $G(s) = \dfrac{K}{s(s+4)^2}$.

Prerequisite: sections 2.1 and 6.6 and appendices A, D.3, D.4, D.13, and E.

Solution:

The gain range needs to be selected say over $1 \le K \le 1000$. The problem with specific step size is the necessity of many points. That type of selection puts computing burden on machine not to mention for every single K computing also requires for $0 \le \omega \le \infty$. Specific number of samples is useful in reducing the computing time. Instead of linear frequency logarithmic scale is used in order to cover wide range of ω. Let us choose 500 samples for the K. Since **logspace** needs power of 10, we enter beginning and ending of K as $\log_{10} 1$ and $\log_{10} 1000$ or 0 and 3 respectively.

Generate the K samples as a logarithmically spaced row vector (available in user-chosen **k**):

k=logspace(0,3,500);

Our strategy is we change the K by a for-loop counter and form the variable transfer function by **zpk**. For each K **margin** returns the gain and phase margins. To gather the margins we apply the data accumulation technique to user-chosen variables **Gmm** and **Pmm** respectively. Variable gain and phase margins are returned to user-chosen **Gm** and **Pm** respectively. The whole programming tactic is presented as follows:

```
Gmm=[ ]; Pmm=[ ];
for K=k
   G=zpk([ ],[0 -4 -4],K);
   [Gm,Pm,wg,wp]=margin(G);
   Gmm=[Gmm Gm];
   Pmm=[Pmm Pm];
end
```

Note that above programming may take a while so be patient. Ignore the warning because for some gain the margins can be irrelevant. Our intended gains, gain margins, and phase margins are available in **k**, **Gmm**, and **Pmm** respectively, each of which is a row matrix.

Conventionally gain margin versus gain is plotted in logarithmic scale for which we exercise the command **loglog** with syntax **loglog**(x data as a row matrix, y data as a row matrix) and do so by:

loglog(k,Gmm)

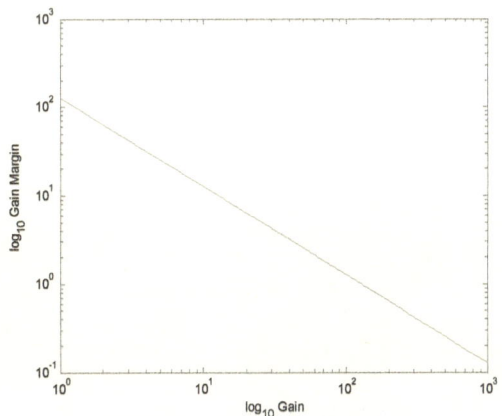

Figure 8.5(c) Gain margin versus gain both in log scale

Figure 8.5(c) presents the variation which is a linear one. Horizontal and vertical axes of figure 8.5(c) indicate $\log_{10}$Gain and $\log_{10}$Gain Margin respectively. If you intend to add the axis labels, exercise the following:
> xlabel('log_1_0 Gain')
> ylabel('log_1_0 Gain Margin')

If gain margin in dB is required, you may exercise **plot** (log10(k),20*log10(Gmm)).

Phase margin versus gain we view by the command **semilogx** with the syntax semilogx(x data as a row matrix, y data as a row matrix) where the logarithmic scale is applied only in the x data while keeping the y data in the same scale. In a new window see the graph by:
> figure,semilogx(k,Pmm)

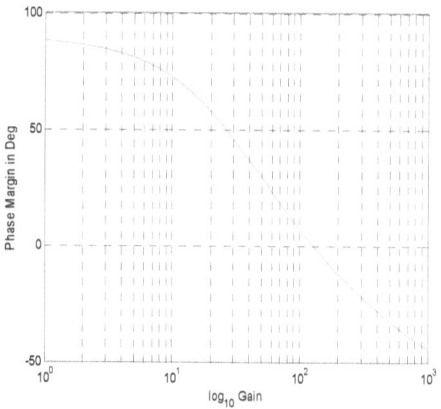

Figure 8.5(d) Phase margin versus gain

Axis labeling and grid line inclusion are conducted by:
> xlabel('log_1_0 Gain')
> ylabel('Phase Margin in Deg')
> grid

Figure 8.5(d) is the outcome from above commands. Gain margin versus phase margin we graph by the **plot** as follows:
> figure,plot(Pmm,20*log10(Gmm))

The gain margin is in the vertical axis and in dB. Add the axis label and grid by:
> xlabel('Phase Margin in Deg')
> ylabel('Gain Margin in dB')
> grid

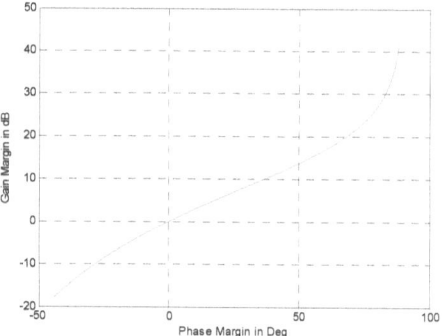

Figure 8.5(e) Gain margin versus phase margin

Figure 8.5(e) depicts the variation.

◆ Project 14: Linking damping ratio and phase margin

The concept of damping ratio is established primarily on a second order prototype control system. If a higher order control system is to be dealt with, its response is compared with the second order counterpart. This is not appropriate but for the design reason it is okay. Controller design in frequency space sometimes needs the relation between damping ratio and phase margin which we intend to implement here.

Prerequisite: appendices A, D.8, and E.

Solution:

The second order prototype system of figure 4.6(a) has the phase margin $\varphi_{pm} = \tan^{-1} \dfrac{2\zeta}{\sqrt{\sqrt{4\zeta^4 + 1} - 2\zeta^2}}$ on $H(s) = 1$. If you are only interested to view the graph of φ_{pm} in degree versus ζ, exercise the following:

ezplot('atand(2*z/sqrt(sqrt(4*z^4+1)-2*z^2))',[0 1])

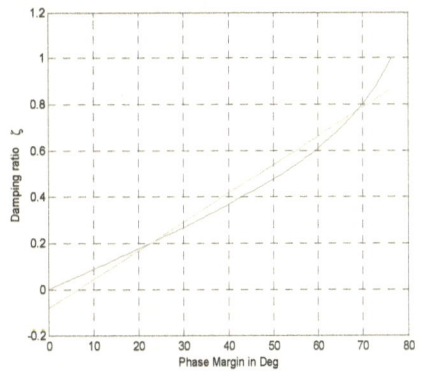

Figure 8.5(f) Damping ratio versus phase margin

In above we wrote the vector code concerning the appendix A without finding the samples. Obviously z stands for the ζ and damping variation $0 \le \zeta \le 1$ is considered. Figure is not shown for the space reason.

If ζ is given, φ_{pm} is easily computed from above expression. Conversely how do we get ζ from φ_{pm}? First make the samples available for ζ and φ_{pm} by using $\Delta\zeta$ and scalar code. Then call the c_cross to determine the specific ζ.

As an example subject to $\Delta\zeta = 0.01$, what is the ζ when $\varphi_{pm} = 35.78°$? The answer is $\zeta = 0.325$. Its implementation is in the following:

```
z=0:0.01:1;
pm=atand(2*z./sqrt(sqrt(4*z.^4+1)-2*z.^2));
c_cross(z,pm,35.78)
```

ans =
    0.3250

Now we wish to find a linear relationship between ζ and φ_{pm}. We need first order approximation of φ_{pm} for that. Given the x and y data, a best fit line or curve one finds by the function **polyfit** that applies the syntax polyfit(x samples as row matrix, y samples as a row matrix, user-chosen degree) and the return from **polyfit** is the approximate polynomial coefficients. Samples of ζ and φ_{pm} are available in z and pm respectively, on degree 1 (for linearity) we call the function as:

p=polyfit(pm,z,1)

p =
    0.0124   -0.0811

The p is a user-chosen variable which holds the first degree polynomial coefficients as a row matrix. The last return indicates that the first degree approximation is linked by $\zeta = 0.0124 \varphi_{pm} - 0.0811$. Note that we treated ζ and φ_{pm} as y and x respectively. The converse is also possible. For this reason many authors apply $\zeta \approx 0.01 \varphi_{pm}$.

What if we intend to see the two graphs together i.e. ζ versus φ_{pm} for actual and approximated one? Samples for approximated ζ you get by:
        za=p(1)*pm+p(2);
The za is a user-chosen variable which holds approximate ζ samples as a row matrix. The p(1) and p(2) are the two polynomial coefficients. Call the plotter for the two graphs:
        plot(pm,z,pm,za)
Include the labels and grid by:
        xlabel('Phase Margin in Deg')
        ylabel('Damping ratio \zeta')
        grid
Figure 8.5(f) depicts the combined graph.

✦ Project 15: Controller of an inverted pendulum

Figure 8.6(a) presents an inverted pendulum on a moving base. The pendulum system is described by $G(s) =$ $\dfrac{-1}{ML\left[s^2 - \dfrac{g(m+M)}{ML}\right]}$ where M, L, g, and m are base mass, base displacement, acceleration due to gravity, and pendulum mass respectively. Figure 8.6(b) depicts block diagram of the pendulum

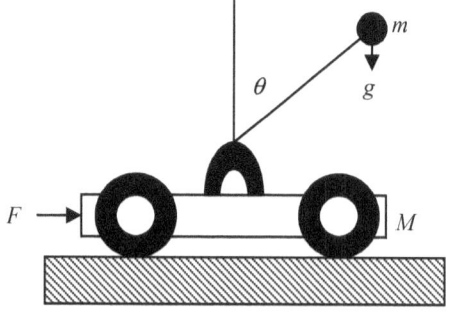

Figure 8.6(a) An inverted pendulum

system in conjunction with a controller which stabilizes the system in case force F creates a disturbance causing a displacement L. The controller $G_c(s)$ is given by $G_c(s) = \dfrac{-K(s+p)}{s+q}$. Typical values of the system parameters are M =99 Kg, L=1 m, g =9.81 m/s^2, p=5.1, q =10.2, and m = 9.5 Kg. Objective of the design is to balance the pendulum i.e.

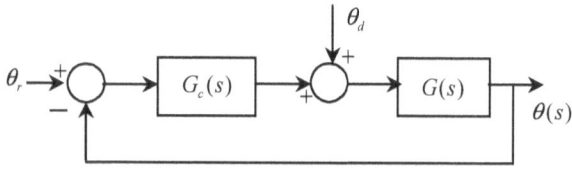

Figure 8.6(b) Block diagram of an inverted pendulum

turning $\theta(t)$ close to 0 in the presence of a disturbance $\theta_d(t)$. Of coarse source of the $\theta_d(t)$ is the F. Clearly desired $\theta_r(t)$ is 0. Subject to a unit step disturbance (i.e. $\theta_d(t)=1°$) the controller design should meet the following criteria:
  (a) steady state error less than $0.1°$,
  (b) settling time (2% criterion) less than 10 sec, and
  (c) percent overshoot less than 40%.

Prerequisite: sections 3.1, 4.7, 5.1, 5.2, 5.8, and 5.9.

Solution:
What is the adjustable parameter in the design? Certainly the gain parameter K of the controller. What is the challenge? The answer is satisfying three criteria by varying a single quantity. The best approach should be first satisfy the part (a) then check the other two.
We intend to exercise MATLAB-SIMULINK combined implementation which makes the design easier. Figure 8.6(c) depicts SIMULINK model for the inverted pendulum system. Open a new SIMULINK model file and bring two **Step**, one **Zero-Pole**, one **Transfer Fcn**, one **To Workspace**, two **Sum**, and one **Scope** blocks in the model file. The **Step, Step1, Zero-Pole, Transfer Fcn**, and **Scope** are intended for $\theta_r(t)$, $\theta_d(t)$, $G_c(s)$, $G(s)$, and error $e(t)=\theta_r(t)-\theta(t)$ respectively. The **To Workspace** is to transport the data of $e(t)$ to MATLAB workspace.

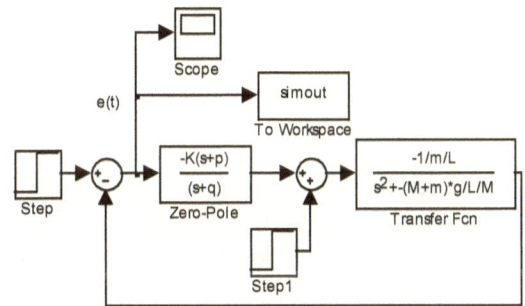

Figure 8.6(c) SIMULINK model of the inverted pendulum system

Settings of various blocks are to be entered which are the following; **Step**: Step time as 0 and Final Value as 0, **Step1**: Step time as 0 and Final Value as 1, **Zero-Pole**: zero as -p, pole as -q, and gain as -K, **Transfer Fcn**: Numerator as -1/m/L and Denominator as [1 0 -g*(M+m)/L/M], **To Workspace**: Save Format as **Array**, and one **Sum**: List of signs as +-. Save the model file by some name **Project15**. Since settling time requirement is less than 10 secs, analyzing time interval 20 secs will be okay so enter the **Stop time** as 20 in SIMULINK. Let us enter the given parameters to like name variables (e.g. M to **M**):
M=99;m=9.5;L=1;g=9.81;p=5.1;q=10.2;

The $K=0$ is meaningless say some small value 0.1. At the beginning we should not take fine increment because machine may take longer. Say $K=0.1$ with increment 50 and continue until steady state error less than $0.1°$ is reached. This sort of finding may be conducted by **while-end** programming which applies the following syntax:

    while logical statement
        executable statements
    end

Since inside the **while-end** the parameter K has to change that is why some initial value has to be assigned to **K** as well as steady state error and let it be:
    e_ss=1; K=0.1;

You could have chosen any other value for the steady error (**e_ss**) which is more than 0.1, we did 1. The logical statement of **while** works on inversion of the requirement. We need steady state error less than 0.1 so the statement for the **while** should be greater than or equal to 0.1. The complete code is in the following:
    while abs(e_ss)>=0.1 K=K+50; sim('Project15'); e_ss=simout(end); end

For every single **K** model **Project15** is run by **sim('Project15')**. Ignore the warnings. The **simout** holds the samples of error signal $e(t)$ from which the last sample **simout(end)** indicates the approximate steady state value. If you call **K** and **e_ss**, you see the K and its corresponding steady state error as follows:

    K   e_ss

    K =   e_ss =
        250.1000   0.0436

We got coarse estimate of K and steady state $e(t)$ now we can go for the fine K say increment 0.5:
    e_ss=1; K=0.1;
    while abs(e_ss)>=0.1 K=K+0.5; sim('Project15'); e_ss=simout(end); end

Call like before:
    K   e_ss

    K =   e_ss =
        224.6000   0.0984

Above result indicates that part (a) condition is satisfied if we choose $K=224.6$. Following the symbology of section 4.7 now we find P.O. by:
    y=simout; t=tout; y_max=max(y); y_ss=simout(end);
    (y_max-y_ss)/y_ss*100

    ans =
        9.9154

Hence the percent overshoot with $K=224.6$ is 9.9154%. The next is the settling time:
    ts1=c_cross(t,y,0.98*y_ss);
    ts2=c_cross(t,y,1.02*y_ss);
    ts2(end)

ans =
        5.1892
From the last return we have $T_s=5.1892$ secs. Fortunately parts (b) and (c) are satisfied with the condition of part (a). Should you intend to see the $e(t)$ signal, exercise following at the prompt:
    plot(t,y,t,1.02*y_ss,t,0.98*y_ss)
Graph is not shown for the space reason.

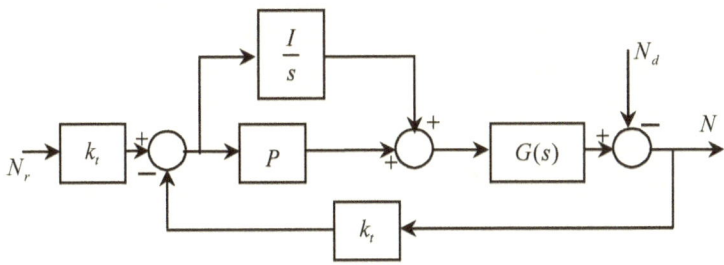

Figure 8.6(d) Block diagram of a speed control system with PI controller

✦ Project 16: Speed control by using a PI controller

A first order motor is characterized by $G(s)=\dfrac{K_m}{\tau s+1}$ where motor parameters are time constant $\tau=1.3$ secs and torque constant $K_m=0.25$ rpm/mV. The tachometer used for RPM detection has constant $k_t=3.4\ mV/rpm$. The desired RPM is $N_r=800\ rpm$. A subtractive disturbance is expected to occur at $t=4$ secs which is tantamount to an RPM $N_d=400$. A PI controller is used as shown in figure 8.6(d). Determine the following:
  (a) steady state RPM without the PI controller and N_d,
  (b) waveshape for N and its steady state value without the PI controller and with N_d,
  (c) value of P if only proportional controller is applied to attain the 2% criterion RPM subject to the disturbance, and
  (d) value of proportional and integral gains to attain the 2% criterion RPM subject to the disturbance.

Prerequisite: sections 3.1, 5.8, and 5.9 and project 15.

Solution:
We intend to adopt MATLAB-SIMULINK strategy to solve the problem. Figure 8.6(e) shows the SIMULINK model in which **Constant, Gain, Gain1, Gain2, Transfer Fcn, Transfer Fcn1,** and **Step** refer to N_r, P, k_t, k_t, $G(s)$, $\dfrac{I}{s}$, and N_d respectively.

Open a new SIMULINK model file and bring all quoted blocks along with three **Sums** in the file. Enter the settings of various blocks as follows; **Constant:** Value as **800**, **Gain:** Gain as **P**, **Gain2:** Gain as **kt**, **Gain1:** Gain as **kt**, **Transfer Fcn:** Numerator as **Km** and Denominator as **[T 1]**, **Transfer Fcn1:** Numerator as **I** and Denominator as **[1 0]**, one **Sum:** List of signs as **+-**, and another **Sum:** List of signs as **+-**. Save the model file by some name **Project16**. The **Scope** and **Display** in the model show the N waveshape and steady state N respectively. In addition **To Workspace** is to transport N data through **simout** for manipulation, doubleclick the block and set the Save Format as **Array**. The **Gain1** block needs to be flipped. Enter the given K_m, τ, and k_t to user-chosen **Km, T,** and **kt** respectively as follows (obviously at the MATLAB command prompt):
    Km=0.25; T=1.3; kt=3.4;

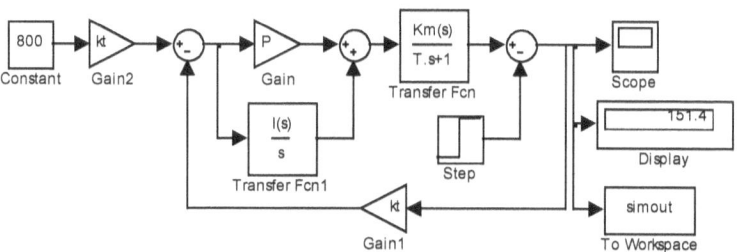

Figure 8.6(e) SIMULINK model of speed control system with PI controller

SIMULINK model time interval needs to be selected say over $0 \le t \le 15 \sec s$ hence enter the SIMULINK **Stop time** as 15.

If we select proportional and integral controller gains differently, those render different implementations. For instance $P=1$ and $I=0$ indicate speed control without the PI controller so enter those at the prompt:
    P=1; I=0;

Enter the **Step** setting as Step time: 0 and Final value: 0 for deactivating N_d. Run the SIMULINK model and find the steady state N in the **Display** as 367.67 *rpm* that is the answer of part (a).

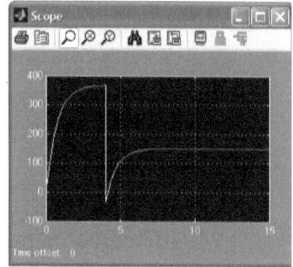

Figure 8.6(f) **Scope** output with disturbance

For the part (b) enter the **Step** setting as Step time: 4 and Final value: 400 for activating N_d. Run the model and find the steady state N in the **Display** as 151.4 *rpm*. The **Scope** shows the N waveshape like figure 8.6(f) with autoscale setting.

For the part (c) we can apply the **while-end** programming like project 15. Consider the starting value of P and steady state N as:
P=0.1; N_ss=1;

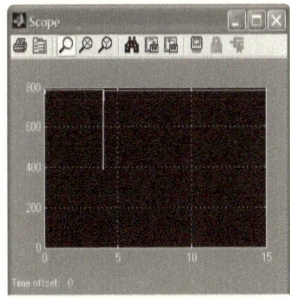

Figure 8.6(g) **Scope** output
with only P controller

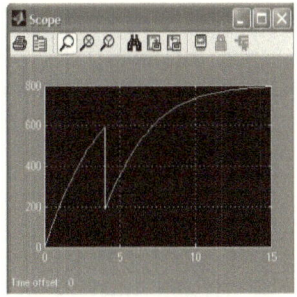

Figure 8.6(h) **Scope** output
with P and I controllers

Our objective is to obtain $N_r = 800\ rpm$ so logical statement of the **while** should be $N_r < 2\%$ criterion on RPM 800. Since the system is first order most likely the waveshape will have similarity with the overdamped counterpart of figure 8.1(e). The 2% of 800 is 16 so we have to use $N_r < 800-16$ or 784. Again increment of the P is required say 5 hence the complete code like project 15 is:
while N_ss<784 P=P+5; sim('Project16'); N_ss=simout(end); end

Ignore warnings and call the **P** and **N_ss**:
P   N_ss

P =   N_ss =
    90.1000   784.2471

That is for the part (c) only proportional controller reach $N = 784.2471\ rpm$ with $P = 90.1$. Run the model and view the waveshape like figure 8.6(g).

For the part (d) both controllers are present so we have to consider **P** as well as **I**. Let us choose increments as 0.1 and initials as 0.1 for both controllers. The **while-end** calling with the initials and increments are the following:
P=0.1; I=0.1; N_ss=1;
while N_ss<784 P=P+0.1; I=I+0.1; sim('Project16'); N_ss=simout(end); end

Ignore warnings and call the **P, I, and N_ss**:
P   I   N_ss

P =   I =   N_ss =
    0.4000   0.4000   792.0080

Run the model and view the waveshape like figure 8.6(h). All workspace variables are user-chosen for example **N_ss**.

How do you perceive the four cases? Without any controller the steady state RPM is 367.67 *rpm* which is reduced to 151.4 *rpm* in the presence of disturbance. When only proportional controller is applied, a large gain $P=90.1$ is required to stabilize the RPM on 2% criterion with the disturbance. When both the proportional and integral controllers are applied, the gain requirement is substantially reduced to $P=0.4$ and $I=0.4$. Application of high gain saturates the amplifier which is not desired. Application of controller helps the system return to the rated RPM even with the disturbance.

MATLAB and SIMULINK jointly keep powerful programming resources for study and analysis of control system problems which we demonstrated sampling a few in this chapter. We do not wish to overwhelm the reader with project materials given our involvement in mobile-internet-computer trio. We hope our comprehensive introduction in the subject would inspire the reader to simulate his/her control system design and analysis problems in MATLAB /SIMULINK platform. We intend to terminate the chapter with this.

Appendices

Appendix A

Coding in MATLAB

MATLAB executes the code of an expression in terms of string which is the set of keyboard characters placed consecutively. One distinguishing feature of MATLAB is that the workspace variable itself is a matrix. The strings adopted for computation are divided into two classes – scalar and vector. The scalar computation results the order of the output matrix same as that of the variable matrix. On the contrary, the order for the vector computation is determined in accordance with the matrix algebra rules. Some symbolic functions and their MATLAB counterparts are presented in table A.1. The operators for arithmetic computations are as follows:

addition	+
subtraction	−
multiplication	*
division	/
power	^

The operation sequence of different operators in a scalar or vector string observes the following order:

enclosing braces	( )	first,
power operator	^	then,
division operator	/	next,
multiplication operator	*	after that,
addition operator	+	then, and
subtraction operator	−	finally.

The syntax of the scalar computation urges us to use .*, ./, and .^ in lieu of *, /, and ^ respectively. The operators *, /, and ^ are never preceded by . for the vector computation. The vector string is the MATLAB code of any symbolic expression or function often found in mathematics. In the sequel we present some examples on writing an expression in MATLAB.

◆ Write MATLAB codes both in scalar and vector forms on following functions

A. $\sin^3 x \cos^5 x$

B. $2 + \ln x$

C. $x^4 + 3x - 5$

D. $\dfrac{x^3 - 5}{x^2 - 7x - 7}$

E. $\sqrt{|x^3| + \sec^{-1} x}$

F. $(1 + e^{\sin x})^{x^2 + 3}$

G. $\dfrac{\cosh x + 3}{\sqrt{\dfrac{x+4}{\log_{10}(x^3 - 6)}}}$

H. $\dfrac{1}{(x-3)(x+4)(x-2)}$

I. $\dfrac{1}{1 + \dfrac{1}{1 + \dfrac{1}{x}}}$

J. $\dfrac{a}{x+a} + \dfrac{b}{y+b} + \dfrac{c}{z+c}$

K. $\dfrac{u^2 v^3 w^9}{x^4 y^7 z^6}$

In tabular form, they are coded as follows:

Example	String for scalar computation	String for vector computation
A	sin(x).^3.*cos(x).^5	sin(x)^3*cos(x)^5
B	2+log(x)	2+log(x)
C	x.^4+3*x-5	x^4+3*x-5
D	(x.^3-5)./(x.^2-7*x-7)	(x^3-5)/(x^2-7*x-7)
E	sqrt(abs(x.^3)+asec(x))	sqrt(abs(x^3)+asec(x))
F	(1+exp(sin(x))).^(x.^2+3)	(1+exp(sin(x)))^(x^2+3)
G	(cosh(x)+3)./sqrt((x+4)./log10(x.^3-6))	(cosh(x)+3)/sqrt((x+4)/log10(x^3-6))
H	1./(x-3)./(x+4)./(x-2)	1/(x-3)/(x+4)/(x-2)
I	1./(1+1./(1+1./x))	1/(1+1/(1+1/x))
J	a./(x+a)+b./(y+b)+c./(z+c)	a/(x+a)+b/(y+b)+c/(z+c)
K	u.^2.*v.^3.*w.^9./x.^4./y.^7./z.^6	u^2*v^3*w^9/x^4/y^7/z^6

Control system programming circumstance dictates the type of code – whether scalar or vector should be employed.

Table A.1 Some mathematical functions and their MATLAB counterparts

Mathematical notation	MATLAB notation	Mathematical notation	MATLAB notation	Mathematical notation	MATLAB notation
$\sin x$	sin(x)	$\sin^{-1} x$	asin(x)	π	pi
$\cos x$	cos(x)	$\cos^{-1} x$	acos(x)	A+B	A+B
$\tan x$	tan(x)	$\tan^{-1} x$	atan(x)	A−B	A−B
$\cot x$	cot(x)	$\cot^{-1} x$	acot(x)	A×B	A*B
$\cosec x$	csc(x)	$\sec^{-1} x$	asec(x)	e^x	exp(x)
$\sec x$	sec(x)	$\cosec^{-1} x$	acsc(x)	A^B	A^B
$\sinh x$	sinh(x)	$\sinh^{-1} x$	asinh(x)	$\ln x$	log(x)
$\cosh x$	cosh(x)	$\cosh^{-1} x$	acosh(x)	$\log_{10} x$	log10(x)
$\sec hx$	sech(x)	$\sec h^{-1} x$	asech(x)	$\log_2 x$	log2(x)
$\cosechx$	csch(x)	$\cosech^{-1} x$	acsch(x)	Σ	sum
$\tanh x$	tanh(x)	$\tanh^{-1} x$	atanh(x)	Π	prod
$\coth x$	coth(x)	$\coth^{-1} x$	acoth(x)	$|x|$	abs(x)
10^A	1e A e.g. 1e3	10^{-A}	1e- A e.g. 1e-3	$\sqrt{x}$	sqrt(x)

* In the six trigonometric functions for example sin(x), the x is in radian. If the x is in degree, we use sind(x). The other five functions also have the syntax cosd(x), tand(x), cotd(x), cscd(x), and secd(x) when the x is in degree. The default return from asin(x) is in radian, if you need the return to be in degree, use the command asind(x). Similar degree return is also possible from acosd(x), atand(x), acotd(x), asecd(x), and acscd(x).

Numerical examples to point out the difference between scalar and vector computations are in the following.

We have the matrices $A = \begin{bmatrix} 3 & 5 \\ 7 & 8 \end{bmatrix}$, $B = \begin{bmatrix} 5 & 2 & 1 \\ 0 & 1 & 7 \end{bmatrix}$, and $C = \begin{bmatrix} 3 & 2 & 9 \\ 4 & 0 & 2 \end{bmatrix}$. The scalar computation is not possible between the matrices A and B because of their unequal order, nor is between the matrices A and C for the same reason. On the contrary the scalar multiplication can be conducted between the B and C for having the same order, which is $B.*C = \begin{bmatrix} 15 & 4 & 9 \\ 0 & 0 & 14 \end{bmatrix}$ (element by element multiplication).

Matrix algebra rule says that any matrix A of order $M \times N$ can only be multiplied with another matrix B of order $N \times P$ so that the resulting matrix has the order $M \times P$. For the last paragraph cited A and B, we have $M=2$, $N=2$, and $P=3$ and obtain the vector-multiplied matrix as $A \times B = \begin{bmatrix} 3\times5+5\times0 & 3\times2+5\times1 & 3\times1+5\times7 \\ 7\times5+8\times0 & 7\times2+1\times8 & 7\times1+8\times7 \end{bmatrix} = \begin{bmatrix} 15 & 11 & 38 \\ 35 & 22 & 63 \end{bmatrix}$, which has the MATLAB code A*B not A.*B. Similar interpretation follows for the operators * and /.

Whenever writing the scalar codes A.*B, A./B, and A.^B, we make it certain that both the A and B are identical in matrix size. The 3*A means all elements of matrix A are multiplied by 3 and we do not use 3.*A. Also do we not use A./3 but do A/3. The signs + and - are never preceded by the operator . in scalar codes. The command 4./A means 4 is divided by all elements in A. The A.^4 means power on all elements of A is raised by 4 and so on.

✦ Scale factors or units

In science and engineering physical quantity measurement always requires the understanding of units. Measured unit of a physical quantity can be far apart from the standard unit very often in the power of 10 that is why scale factors are important. Table A.2 presents the engineering scale factor units and their MATLAB equivalences.

Table A.2 Engineering unit scale factors and their MATLAB counterparts

Scale factor	Symbol	As power of 10	MATLAB code
giga	G	10^9	e9
mega	M	10^6	e6
kilo	K	10^3	e3
milli	m	10^{-3}	e-3
micro	μ	10^{-6}	e-6
nano	n	10^{-9}	e-9
pico	p	10^{-12}	e-12

For example the time 10.7 *msec* is coded as **10.7e-3**. Again a distance of 4.7 *km* is entered by writing **4.7e3** in standard unit.

Appendix B
MATLAB functions exercised in the text

Function name	Purpose	Page
abs	extracts magnitude values from $H(j\omega)$ samples	157
angle	extracts phase angles in radians from $H(j\omega)$ samples	157
any	returns 1 and 0 for any nonzero element present and absent in a matrix respectively	196
append	forms unconnectedly stacked control system from component ones	45
balreal	provides Gramian values of a control system	217
bandwidth	computes the cutoff frequency of a lowpass control system	172
bode	computes and graphs the bode plot of $H(s)$	159
c_cross	author written script file which finds functional value crossing from x and y samples	270
connect	forms an interconnected control system	46
conv	multiplies two polynomials in terms of coefficients	31
damp	computes damping ratio and natural angular frequency at every pole of $H(s)$	164
dcgain	computes the DC gain of a control system	171
end	terminates the execution of a for-loop or if-else checking	262
eps	lowest numerical quantity in MATLAB treated as epsilon	197
eval	evaluates a function	197
ezplot	draws y versus x type graph from y expression and x interval	283
feedback	determines equivalent transfer function of elementary negative/positive feedback control system	39
figure	opens a new window for graphics	194
find	determines the element position in a matrix subject to condition	268
for-end	beginning and ending statements of a for-loop	265
frd	defines transfer function from frequency response data	30

Continuation of the last table:

Function name	Purpose	Page
freqresp	computes and graphs the frequency response of $H(s)$	156
grid	adds horizontal and vertical grid lines to a drawn graph	162, 240
imag	extracts imaginary values from $H(j\omega)$ samples	157
impulse	computes and graphs impulse response of a control system	100
legend	includes distinctive words for multiple traces on a drawn graphics	111
length	determines the number of elements in a row or column matrix	208, 233
linspace	generates row or column matrix of linearly spaced elements	279
loglog	graphs both the x and y data in logarithmic scale	237
logspace	generates a row or column matrix of logarithmically spaced elements	279
lsim	generates samples of output from a control system and graphs too	95
margin	computes gain and phase margins on bode plot of $H(s)$	167
max	finds the maximum from numerical data supplied as a row or column matrix	267
min	finds the minimum from numerical data supplied as a row or column matrix	267
minreal	cancels identical pole-zero subject to some user-defined tolerance	195
modred	reduces a control system order defined by state space	217
nichols	computes and graphs the Nichol's plot of $H(s)$	169
nyuist	computes and graphs the Nyquist plot of $H(s)$	168
ones	generates matrix of ones	269
ord2	defines second transfer function from damping ratio and natural frequency	32
pade	approximates e^{-sT} by a rational transfer function on user-supplied degree	199

Continuation of the last table:

Function name	Purpose	Page
parallel	determines parallel equivalent of two transfer functions	38
plot	graphs y versus x data in continuous sense	281
plotyy	graphs two dissimilar y traces over a common x variation	209
pole	extracts poles from $H(s)$	163
poly	forms a polynomial from its roots	185
poly2str	displays a polynomial as expression from its coefficient form	184
polyder	computes derivative of a polynomial in coefficient form	232
polyfit	finds best fit line or curve on user-definition	239
polyval	evaluates a polynomial	199
pretty	displays a mathematical readable form from coded expression	188
pzmap	computes and locates poles and zeroes of $H(s)$ in the s plane	162
rad2deg	converts radians to degrees	157
real	extracts real values from $H(j\omega)$ samples	157
rlocus	graphs root locus only for standard control system	191
rmodel	generates transfer function of random order	32
roots	computes the roots of a polynomial	185
routh	author written script file which finds the Routh table up to the 9th degree	186
sawtooth	generates samples of triangular waves of varying characteristics	93
semilogx	graphs only x data in logarithmic scale	238
series	determines series equivalent of two transfer functions	37
sgrid	includes damping ratio and natural angular frequency grid in s plane on pole-zero plot	165
sim	simulates a SIMULINK model from MATLAB	229
square	generates samples of rectangular waves of varying characteristics	91
squeeze	compresses three dimensional array if empty dimension exists	157
ss	defines transfer function from state space representation	29

Continuation of the last table:

Function name	Purpose	Page
ss2tf	converts a transfer function from state space to polynomial form	42
ss2zp	converts a transfer function from state space to pole-zero-gain form	43
step	computes and graphs step response of a control system	100
sum	sums all elements in a row or column matrix	280
syms	reserve word for declaring variables	187
tf	defines transfer function from numerator-denominator polynomial coefficients	28
tf2ss	converts a transfer function from polynomial to state space form	42
tf2zp	converts a transfer function from polynomial to pole-zero-gain form	41
tfdata	extracts transfer function data – numerator and denominator polynomial coefficients	272
trapz	computes numerical integration from x and y samples employing basic trapezoidal rule	273
while-end	continues looping until some logical statement is satisfied	242
xlabel	includes x or horizontal label in a drawn graph	237
ylabel	includes y or vertical label in a drawn graph	237
zero	extracts zeroes from $H(s)$	163
zp2ss	converts a transfer function from pole-zero-gain to state space form	43
zp2tf	converts a transfer function from pole-zero-gain to polynomial form	41
zpk	defines transfer function from pole-zero-gain	28

Appendix C

SIMULINK block links for modeling control systems

When you open a new or work in a previously saved SIMULINK model file, the very next step is to know the exact location or link of a block which will be employed in modeling for control system problems. Table C accumulates icon appearance, brief function, and location or link of SIMULINK blocks exercised in the text.

Table C. SIMULINK blocks and their links for modeling control systems

Block name	Icon Outlook	Function of the block or operation	Link or location
Abs	|u| Abs	It returns the absolute value of the function to its input port	SIMULINK → Math Operations → Abs
Constant	1 Constant	It generates user-defined constant values	SIMULINK → Sources → Constant
Demux		It demultiplexes signals entering into input port	SIMULINK → Commonly Used Blocks → Demux
Derivative	du/dt Derivative	It performs the numerical differentiation of the signal to its input port in continuous sense	SIMULINK → Continuous → Derivative
Display	0 Display	It shows the instantaneous value of the signal at the end of the simulation which it is connected to	SIMULINK → Sinks → Display
Divide	× ÷ Divide	It performs user-defined multiplication or division of input signals	SIMULINK → Math Operations → Divide
Fcn	f(u) Fcn	It performs user-defined mathematical operations on the signal to its input port assuming that the input signal name is u	SIMULINK → User-Defined Functions → Fcn

Continuation of the last table:

Block name	Icon Outlook	Function of the block or operation	Link or location
From Workspace	simin (From Workspace)	It imports data from MATLAB to SIMULINK	SIMULINK → Sources → Constant
Gain	Gain	It multiplies the signal to its input port according to user-supplied gain	SIMULINK → Math Operations → Gain
Integrator	$\frac{1}{s}$ (Integrator)	It performs the numerical integration of the signal to its input port in continuous sense	SIMULINK → Continuous → Integrator
Math Function	e^u (Math Function)	It conducts various mathematical operations on user-selected functional popup	SIMULINK → Math Operations → Math Function
MinMax Running Resettable	min(u,y) y, R (MinMax Running Resettable)	It finds the minimum or maximum value of the input signal after simulation	SIMULINK → Math Operations → MinMax Running Resettable
Mux	(Mux icon)	It multiplexes two or more signals from single vector to group vector form	SIMULINK → Commonly Used Blocks → Mux
Product	× (Product)	It multiplies two or more signal values connected to its input ports	SIMULINK → Math Operations → Product
Pulse Generator	(Pulse Generator)	It generates rectangular pulse waves of various characteristics	SIMULINK → Sources → Pulse Generator
Ramp	(Ramp)	It generates straight line functions of various characteristics	SIMULINK → Sources → Ramp
Saturation	(Saturation)	It clips positive and negative portions of a wave on user-definition	SIMULINK → Discontinuities → Saturation

Continuation of the last table:

Block name	Icon Outlook	Function of the block or operation	Link or location
Scope	Scope	It shows the functional variation (s) of some signal (s) which it is connected to	SIMULINK → Sinks → Scope
Signal Builder	Signal 1 / Signal Builder	It provides window interface for designing signal of user's choice	SIMULINK → Sources → Signal Builder
Sine Wave	Sine Wave	It generates sine waves of various characteristics	SIMULINK → Sources → Sine Wave
State-Space	x' = Ax+Bu / y = Cx+Du / State-Space	It implements continuous system which follows the dynamic equations $\dot{x} = Ax + Bu$ and $y = Cx + Du$	SIMULINK → Continuous → State-Space
Step	Step	It generates unit step function and its derived signals	SIMULINK → Sources → Step
Subtract	Subtract	It subtracts two or more signal values connected to its input ports	SIMULINK → Math Operations → Subtract
Sum		It adds two or more signal values connected to its input ports	SIMULINK → Math Operations → Sum
To Workspace	simout / To Workspace	It exports signal data from SIMULINK to MATLAB workspace which it is connected to	SIMULINK → Sinks → To Workspace
Transfer Fcn	$\frac{1}{s+1}$ / Transfer Fcn	It implements Laplace transform system function in rational form when numerator and denominator in polynomial form	SIMULINK → Continuous → Transfer Fcn

Continuation of the last table:

Block name	Icon Outlook	Function of the block or operation	Link or location
Transport Delay	Transport Delay	It translates a function $f(t)$ to t_0 that is the input and output of the block are $f(t)$ and $f(t-t_0)$ respectively where t_0 is user-defined	SIMULINK → Continuous → Transport Delay
Zero-Pole	$\frac{(s-1)}{s(s+1)}$ Zero-Pole	It implements Laplace transform system function in rational form when numerator and denominator in factored form	SIMULINK → Continuous → Zero-Pole

Appendix D

MATLAB functions/statements for control system study

While working on control system problems in MATLAB, we come across lots of built-in MATLAB functions or programming statements. In order to employ these elements for control system analysis, we need to understand their input and output argument types and purpose of the elements. Functions or program elements exercised in the text with brief descriptions are in the following.

D.1 Comparative and logical operators

Comparative operators are used for comparison on two scalar elements, one scalar and one matrix elements, or two identical size matrix elements. There are six comparative operators as presented in table D.1.

Table D.1 Equivalence of comparative operators

Comparative operation	Mathematical notation	MATLAB notation
equal to	=	==
not equal to	≠	~=
greater than	>	>
greater than or equal to	≥	>=
less than	<	<
less than or equal to	≤	<=

The output of expression pertaining to the comparative operators is logical – either true (indicated by 1) or false (indicated by 0). For example when A=3 and B=4, the comparisons A=B, A≠B, A>B, A≥B, A<B, and A≤B should be false(0), true(1), false(0), false(0), true(1), and true(1) respectively. We implement these comparative operations as presented in table D.2.

Table D.2 Scalar comparative operation

>>A=3; B=4; ↵ >>A==B ↵ ans = 0 >>A~=B ↵ ans = 1	>>A>B ↵ ans = 0 >>A>=B ↵ ans = 0	>>A<B ↵ ans = 1 >>A<=B ↵ ans = 1

There are two operands A and B in table D.2, each of which is a single scalar. Each of the operands can be a matrix in general. In that case the logical decision takes place element by element on all elements in the matrix. For instance if $A = \begin{bmatrix} 5 & 8 \\ 5 & 7 \end{bmatrix}$ and $B = \begin{bmatrix} 2 & 1 \\ -2 & 9 \end{bmatrix}$, A>B should be $\begin{bmatrix} 5>2 & 8>1 \\ 5>-2 & 7>9 \end{bmatrix} = \begin{bmatrix} 1 & 1 \\ 1 & 0 \end{bmatrix}$. Again if the A happens to be a scalar (say A=4), the single scalar is compared to all elements in the B therefore A≤B should be $\begin{bmatrix} 4 \leq 2 & 4 \leq 1 \\ 4 \leq -2 & 4 \leq 9 \end{bmatrix} =$

$\begin{bmatrix} 0 & 0 \\ 0 & 1 \end{bmatrix}$. In a similar fashion the B also operates on A however the scalar and matrix related comparative implementation is presented in the table D.3.

Some basic logical operations are NOT, OR, and AND. The characters ~, |, and & of the keyboard are adopted for the logical NOT, OR, and AND respectively. In all logical outputs the 1 and 0 stand for true and false respectively. All logical operators apply to the matrices in general. For the matrix $A = \begin{bmatrix} 0 & 0 \\ 0 & 1 \end{bmatrix}$, NOT(A) operation should provide $\begin{bmatrix} 1 & 1 \\ 1 & 0 \end{bmatrix}$ (see table D.4). The logical OR and AND operations on the like positional elements of the two matrices $A = \begin{bmatrix} 1 & 1 \\ 0 & 1 \end{bmatrix}$ and $B = \begin{bmatrix} 0 & 1 \\ 1 & 1 \end{bmatrix}$ must return $\begin{bmatrix} 1 & 1 \\ 1 & 1 \end{bmatrix}$ and $\begin{bmatrix} 0 & 1 \\ 0 & 1 \end{bmatrix}$ respectively. Table D.4 shows both implementations.

Table D.3 Scalar and matrix comparative operations

when A and B are matrices, >>A=[5 8;5 7]; ↵ >>B=[2 1;-2 9]; ↵ >>A>B ↵ ans = 1 1 1 0	when A is scalar and B is matrix, >>A=4; ↵ >>B=[2 1;-2 9]; ↵ >>A<=B ↵ ans = 0 0 0 1

Table D.4 Basic logical operations on matrix elements

for NOT(A) operation, >>A=[0 0;0 1]; ↵ >>~A ↵ ans = 1 1 1 0	for A OR B, >>A=[1 1;0 1]; ↵ >>B=[0 1;1 1]; ↵ >>A|B ↵ ans = 1 1 1 1	for A AND B, >>A&B ↵ ans = 0 1 0 1	for A XOR B, >>xor(A,B) ↵ ans = 1 0 1 0

If the A or the B is a single 1 or 0, it operates on all elements of the other.

Sometimes we need to check the interval of the independent variable of mathematical functions for instance $-6 \leq x \leq 8$. The interval is split in two parts $-6 \leq x$ and $x \leq 8$. In terms of the logical statement one expresses the $-6 \leq x \leq 8$ as (-6<=x)&(x<=8).

There is no operator for the XOR logical operation instead the MATLAB function xor syntaxed by xor(A,B) implements the operation as presented in the table D.4.

D.2 Simple if/if-else/nested if syntax

Conditional commands are exercised by the **if-else** statements (reserve words). Also comparisons and checkings may need **if-else** statements. We can have different **if-else** structures namely simple-if, if-else, or nested-if depending on programming circumstances, some of which we discuss in the following.

Simple if

The program syntax of simple-if is as follows:

> if *logical expression*
>   *Executable MATLAB command(s)*
> end

Logical expression usually requires the use of comparative operators which are explained in appendix D.1. If the logical expression beside the **if** is true, the command between the **if** and **end** is executed otherwise not. In tabular form a simple-if implementation is as follows:

Example: If $x \geq 1$, we compute $y = \sin x$. When $x = 2$, we should see $y = \sin 2 = 0.9093$.	Executable M-file: x=2; if x>=1 y=sin(x); end	Steps: Save the statements in a new M-file (section 1.1) by the name test and execute the following: >>test ↵	Check from the command window after running the M-file: >>y ↵ y = 0.9093

If-else

General program syntax for the **if-else** structure is as follows:

> if *logical expression*
>   *Executable MATLAB command(s)*
> else
>   *Executable MATLAB command(s)*
> end

If the logical expression beside the **if** is true, the command between if and else is executed else the command between else and end is executed. In tabular form, an **if-else-end** implementation is the following:

Example: When $x = 1$, we compute $y = \sin\dfrac{x\pi}{2} = 1$ otherwise $y = \cos\dfrac{x\pi}{2} = 0$.	Executable M-file: x=1; if x==1 y=sin(x*pi/2); else y=cos(x*pi/2); end	Steps: Save the statements in a new M-file by the name test and execute the following: >>test ↵	Check from the command window after running the M-file: >>y ↵ y = 1

If we had x=2; in the first line of M-file in last exercise, we would see y= cosπ =–1.

⊟ Nested-if

The third type of if structure is the nested-if whose general program syntax is attached in the right side text box. Clearly the syntax takes care of multiple logical expressions which we demonstrate by one example as shown in the following table.

```
if logical expression
    Executable MATLAB command(s)
elseif logical expression
    Executable MATLAB command(s)
    ⋮
elseif logical expression
    Executable MATLAB command(s)
else
    Executable MATLAB command(s)
end
```

Example: The best example can be taking the decision of grades out of 100 based on the achieved number of a student. The grading policy is stated as if the achieved number of a student is greater than or equal 90, greater than or equal to 80 but less than 90, greater than or equal to 70 but less than 80, greater than or equal to 60 but less than 70, greater than or equal to 50 but less than 60, and less than 50, then the grade is decided as A, B, C, D, E, and F respectively.	Executable M-file:
	
	N=77;
	if N>=90
	  g='A';
	elseif (N<90)&(N>=80)
	  g='B';
	elseif (N<80)&(N>=70)
	  g='C';
	elseif (N<70)&(N>=60)
	  g='D';
	elseif (N<60)&(N>=50)
	  g='E';
	else
	  g='F';
	end
		In the executable M-file, the N and g refer to the number achieved and the grade respectively. If the number N is 77, the grade g should be C. Any character is argumented under the single inverted comma.
		
		Steps: Save the left statements in a new M-file by the name test and execute the following:
		>>test ↵
			Check from the command window after running the M-file:
			
			>>g ↵
			
			g =
			
			C

D.3 Data accumulation

Sometimes it is necessary that we perform appending operation on an existing matrix at MATLAB workspace.

✦ Appending rows

Assume that the $A = \begin{bmatrix} 1 & 3 & 5 \\ 2 & 6 & 8 \\ 9 & 5 & 0 \\ 4 & 7 & 8 \end{bmatrix}$ is formed by appending two row matrices [9 5 0] and [4 7 8] with the matrix $B = \begin{bmatrix} 1 & 3 & 5 \\ 2 & 6 & 8 \end{bmatrix}$.

We first enter the matrix B (section 1.1) into MATLAB and append one row after another by using the command as presented below:

for entering B,	for appending the first row,	for appending the second row,
>>B=[1 3 5;2 6 8] ⏎	>>B=[B;[9 5 0]] ⏎	>>A=[B;[4 7 8]] ⏎
B =	B =	A =
1 3 5	1 3 5	1 3 5
2 6 8	2 6 8	2 6 8
	9 5 0	9 5 0
		4 7 8

The command **B=[B;[9 5 0]]** in above execution says that the row [9 5 0] is to be appended with the existing **B** (inside the third bracket) and that the result is again assigned to **B**. You can append as many rows as you want. The important point is the number of elements in each row that is to be appended must be equal to the number of columns in the matrix **B**.

◆ Appending columns

Suppose $C = \begin{bmatrix} 1 & 3 & 5 & 9 & 3 \\ 2 & 6 & 8 & 0 & 1 \\ 9 & 5 & 0 & 1 & 9 \end{bmatrix}$ is formed by appending two column matrices $\begin{bmatrix} 9 \\ 0 \\ 1 \end{bmatrix}$ and $\begin{bmatrix} 3 \\ 1 \\ 9 \end{bmatrix}$ with matrix $D = \begin{bmatrix} 1 & 3 & 5 \\ 2 & 6 & 8 \\ 9 & 5 & 0 \end{bmatrix}$. We get the matrix D into MATLAB and append one column after another as follows:

for entering D,	for appending the first column,	for appending the second column,
>>D=[1 3 5;2 6 8;9 5 0] ⏎	>>D=[D [9 0 1]'] ⏎	>>C=[D [3 1 9]'] ⏎
D =	D =	C =
1 3 5	1 3 5 9	1 3 5 9 3
2 6 8	2 6 8 0	2 6 8 0 1
9 5 0	9 5 0 1	9 5 0 1 9

The column matrix [9 0 1]' and D in above execution has one space gap within the third bracket. In the second of above implementation, the resultant matrix is again assigned to D. Append as many columns as you want just remember that the number of elements in each column that is to be appended must be equal to the number of rows in the matrix D.

◆ Data accumulation by using the two appending techniques

Suppose initially there is nothing in the f matrix, which in MATLAB we write by the statement f=[ ]; (an empty matrix is

assigned to f). An empty matrix does not have any size and completely empty, it follows the null symbol Ø of matrix algebra. Let us say k=2 and perform the assignment as follows:
>>f=[ ]; k=2; ↵

Now if we execute f=[f k] time and again first f=[f k] returns 2, second f=[f k] returns [2 2], third f=[f k] returns [2 2 2], and so on. This is called row directed data accumulation. Column directed data accumulation occurs by executing f=[f;k] each time.

The demonstrated k is just a scalar but it can be a return from some function, scalar, row matrix, column matrix, or rectangular matrix.

D.4 For-loop syntax

A for-loop performs similar operations for a specific number of times and must be started with the **for** and terminated by an **end** statements. Following the **for** there must be a counter. The counter of the for-loop can be any variable that counts integer or fractional values depending on the increment or decrement. If the MATLAB command statements between the **for** and **end** of a for-loop are few words lengthy, one can even write the whole for-loop in one line. The programming syntax and some examples on the for-loop are as follows:

◆ **Program syntax**

for *counter*=starting value:increment or decrement of the
counter value:final value
Executable MATLAB command(s)
end

◆ **Example 1**

Our problem statement is to compute $y = \cos x$ for $x = 10^0$ to 70^0 with the increment 10^0. Let us assign the computed values to some variable y where y should be $[\cos 10^0 \quad \cos 20^0 \quad \cos 30^0 \quad \cos 40^0 \quad \cos 50^0 \quad \cos 60^0 \quad \cos 70^0] = [0.9848 \quad 0.9397 \quad 0.866 \quad 0.766 \quad 0.6428 \quad 0.5 \quad 0.342]$.

In the programming context, y(1) means the first element in the row matrix y, y(2) means the second element in the row matrix y, and so on. MATLAB code for the $\cos x$ is **cosd(x)** where x is in degree. The for-loop counter expression should be k=1:1:7 or k=1:7 to have the control on the position index in the row matrix y (because there are 7 elements or indexes in y). Since the computation needs 10 to 70, one generates that by writing k*10. Following is the implementation:

Executable M-file:	*Or, as a one line:*
for k=1:1:7 　　y(k)=cosd(k*10); end	for k=1:1:7 y(k)=cosd(k*10); end

Steps we need:
Open a new M-file (section 1.1), type the executable M-file statements in the M-file editor, save the editor contents by the name **test** in your working path, and call the **test** as shown below.
>>test ↵
>>y ↵

y =
    0.9848 0.9397 0.8660 0.7660 0.6428 0.5000 0.3420

✦ Example 2

A for-loop helps us accumulate data (appendix D.3) controlled by the consecutive loop index. In this example we accumulate some data row directionally according to the for-loop counter index.

For $k = 1, 2$, and 3, we intend to accumulate the k^2 side by side. At the end we should be having [1 4 9] assigned to some variable f – this is our problem statement.

for the right shifting,	for the left shifting,
>>f=[ ]; for k=1:3 f=[f k^2]; end ↵	>>f=[ ]; for k=1:3 f=[k^2 f]; end ↵
>>f ↵	>>f ↵
f =	f =
1 4 9	9 4 1

The for-loop for the accumulation is presented above. The accumulation may occur as right or left shifting. Corresponding to the right shifting, the vector code (appendix A) for k^2 is k^2. The statement f=[ ]; means that an empty matrix is assigned to f outside the for-loop but at the beginning. The k variation in our problem is put as the for-loop counter. How the for-loop accumulates is shown below:

When k=1, f=[f k^2]; returns f=[[ ] 1^2]; ⇒ f=1;
When k=2, f=[f k^2]; returns f=[1 2^2];   ⇒ f=[1 4];
When k=3, f=[f k^2]; returns f=[1 4 3^2]; ⇒ f=[1 4 9];

The accumulation is happening from the left to the right. A single change provides the shifting from the right to the left which is f=[k^2 f];. The complete code and its execution result are also shown above by the heading 'for the left shifting'.

✦ Example 3

Another accumulation can be column directed that is we wish to see the output like $\begin{bmatrix} 1 \\ 4 \\ 9 \end{bmatrix}$ in example 2.

We just insert the row separator of a rectangular matrix (done by the operator ;) in the command f=[f k^2];. Again the shifting can happen either from the up to down or from the down to up. Both implementations are shown below:

for the down shifting,	for the up shifting,
>>f=[ ]; for k=1:3 f=[f;k^2]; end ↵	>>f=[ ]; for k=1:3 f=[k^2;f]; end ↵
>>f ↵	>>f ↵
f =	f =
1	9
4	4
9	1

✦ Example 4

Many control problems need writing multiple for-loops. Usually one loop is for one dimensional function, two loops are for two dimensional function, and so on. One dimensional function data takes the form of a row or column matrix.

Suppose we have the one dimensional data as $y = [9\ 6\ 7\ 4\ 6]$. We wish to access to every data in y. A single for-loop helps us conduct that as shown below:

>>y=[9 6 7 4 6]; for k=1:length(y) v=y(k); end ↵

First we assign the data to workspace y as a row matrix. The command **length** finds the number of elements in the row matrix y. The y(k) means the k-th element in the y which we assign to workspace v (any user-chosen variable). Every single data of the y is available sequentially in the v. The contents of y can be a column matrix too.

D.5 Finding the maximum/minimum numerically

Given a matrix, one finds the maximum element from the matrix by using the command **max** (**min** for the minimum). Let us say we have three matrices $R = [1\ -2\ 3\ 9]$, $C = \begin{bmatrix} 23 \\ -20 \\ 30 \\ 8 \end{bmatrix}$, and $A = \begin{bmatrix} 2 & 4 & 7 \\ -2 & 7 & 9 \\ 3 & 8 & -8 \end{bmatrix}$ whose maxima are 9, 30, and 9 (from all elements in the matrix) and minima are -2, -20, and -8 respectively. We find the maxima first entering (section 1.1) the respective matrices as follows:

for the row matrix,	for the column matrix,	for the rectangular matrix,
>>R=[1 -2 3 9]; ↵	>>C=[23;-20;30;8]; ↵	>>A=[2 4 7;-2 7 9;3 8 -8]; ↵
>>max(R) ↵	>>max(C) ↵	>>max(max(A)) ↵
ans =	ans =	ans =
9	30	9
>>min(R) ↵	>>min(C) ↵	>>min(min(A)) ↵
ans =	ans =	ans =
-2	-20	-8

Font equivalence is maintained by using the same letter for example A⇔ A in last implementation. If the matrix is a row or column one, we apply one **max** or **min**. For a rectangular matrix, the **max** or **min** separately operates on each column that is why two **max** or **min** functions are required. The functions are equally applicable on decimal number elements.

In the row matrix R, the maximum 9 is occurring as the fourth element in the matrix. Suppose we intend to find the position index (that is 4) of the maximum element in the R. Now we need two output arguments – one for the maximum and the other for its index. Its implementation is shown in the right side attached text box of this paragraph in which the two output arguments M and I correspond to the maximum and its integer index respectively.

```
for index finding in R,
>>[M,I]=max(R) ↵
M =
      9
I =
      4
```

The function **min** keeps this type of integer index returning option in a similar fashion.

D.6 Position indexes of matrix elements with conditions

MATLAB command **find** looks for the position indexes of matrix elements subject to some logical condition whose general format is [R C]= find(condition) where the indexes returned to the R and C are meant to be for the row and column directions respectively. The R and C are user-chosen workspace variables. Let us consider $A = \begin{bmatrix} 11 & 10 & 11 & 10 \\ 12 & 10 & -2 & 0 \\ -7 & 17 & 1 & -1 \end{bmatrix}$ which we enter by the following:

>>A=[11 10 11 10;12 10 -2 0;-7 17 1 -1]; ↵ ← A is assigned to A

We would like to know what the position indexes of A where the elements are greater than 10 are. In matrix A the left-upper most element has the position index (1,1). The elements of A being greater than 10 have the position indexes (1,1), (2,1), (3,2), and (1,3). MATLAB finds the required index in accordance with columns. Placing the row and column indexes vertically, we have $\begin{bmatrix} 1 \\ 2 \\ 3 \\ 1 \end{bmatrix}$ and $\begin{bmatrix} 1 \\ 1 \\ 2 \\ 3 \end{bmatrix}$ respectively. The output arguments R and C of the **find** receive these two column matrices respectively. The input argument of the **find** must be a logical statement, any element in A greater than 10 is written as $A>10$ (appendix D.1). The position indexes are found as shown in the right side attached text box.

```
where elements of A are
greater than 10,
>>[R C]=find(A>10) ↵
R =
      1
      2
      3
      1
C =
      1
      1
      2
      3
```

where elements of	where elements of	for the row matrix D:
A =10:	A ≤0:	>>D=[-10 34 1 2 8 4]; ↵
>>[R C]=find(A==10) ↵	>>[R C]=find(A<=0) ↵	>>R=find(D>=8) ↵
R =	R =	R =
1	3	2 5
2	2	for the column matrix E:
1	2	>>E=[-2 8 -2 7]'; ↵
C =	3	>>C=find(E~=-2) ↵
2	C =	
2	1	C =
4	3	2
	4	4
	4	

To exercise more conditions, what are the position indexes in the matrix A where the elements are equal to 10? The answer is (1,2), (2,2), and (1,4). Again the position indexes where the elements are less than or equal to zero are (3,1), (2,3), (2,4), and (3,4).

The comparative operators >, <, ≥, ≤, and ≠ have the MATLAB counterparts >, <, >=, <=, and ~= respectively.

So far we considered a rectangular matrix for demonstration on position index finding. Let us see how the find works for a row or column matrix. Let us take $D=[-10\ \ 34\ \ 1\ \ 2\ \ 8\ \ 4]$ from which we find the position indexes of the elements where they are greater than or equal to 8. Obviously they are the 2nd and 5th elements. Here we do not need to place two output arguments to the find.

Again let us find the position indexes of the elements in column matrix $E=\begin{bmatrix}-2\\8\\-2\\7\end{bmatrix}$ where the elements are not equal to −2. The 2nd and 4th elements are not equal to −2.

The output of find is a row one for the row matrix input and a column one for the column matrix input.

Presented above are the executions on all these conditional findings.

D.7 Matrix of ones, zeroes, and constants

MATLAB built-in commands ones and zeros implement user-defined matrix of ones and zeroes respectively. Each function conceives two input arguments, the first and second of which are the required numbers of rows and columns respectively. Let us say we intend to form the matrices $A = \begin{bmatrix}1&1&1\\1&1&1\\1&1&1\\1&1&1\end{bmatrix}$, $B = \begin{bmatrix}1&1&1\\1&1&1\\1&1&1\end{bmatrix}$, and $C = \begin{bmatrix}1&1&1&1\\1&1&1&1\end{bmatrix}$. Their orders are 4×3, 3×3, and 2×4 respectively and the implementations are as follows:

for A,
\>\>A=ones(4,3) ⏎

A =
```
     1   1   1
     1   1   1
     1   1   1
     1   1   1
```

for B,
\>\>B=ones(3) ⏎

B =
```
     1   1   1
     1   1   1
     1   1   1
```

for C,
\>\>C=ones(2,4) ⏎

C =
```
     1   1   1   1
     1   1   1   1
```

Either the number of rows or columns will do if the matrix is a square. For the row and column matrices of ones for example of length 6, the commands would be ones(1,6) and ones(6,1) respectively.

Formation of the matrix of zeroes is quite similar to that of the matrix of ones. Replacing the function ones by zeros does the formation. Matrix of zeroes like $A=\begin{bmatrix}0&0&0\\0&0&0\\0&0&0\\0&0&0\end{bmatrix}$, $B=\begin{bmatrix}0&0&0\\0&0&0\\0&0&0\end{bmatrix}$, and $C=\begin{bmatrix}0&0&0&0\\0&0&0&0\end{bmatrix}$ (whose orders are 4×3, 3×3, and 2×4) we form by the commands A=zeros(4,3), B=zeros(3), and C=zeros(2,4) respectively. A row and a column matrices of 6 zeroes are formed by the commands zeros(1,6) and zeros(6,1) respectively.

A matrix of constants is obtained by first creating a matrix of ones of the required size and then multiplying by the constant number. For example the matrix $\begin{bmatrix}0.2&0.2&0.2\\0.2&0.2&0.2\\0.2&0.2&0.2\\0.2&0.2&0.2\end{bmatrix}$ is generated by the command 0.2*ones(4,3).

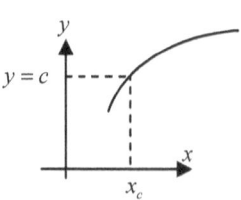

Figure D.1 Crossing of y through a particular value c

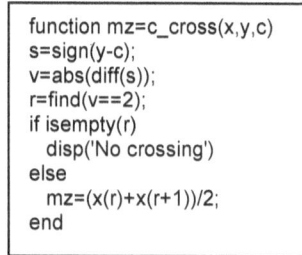

Figure D.2 Author-written function file for finding the crossing point of a function

D.8 Functional value crossing

Suppose we have a graph y versus x. The y may cross a particular value c. Figure D.1 shows the value crossing of the graph when $x = x_c$. We wish to determine x_c.

What data should be available? – undoubtedly the samples of y and x along with c.

Author-written function file **c_cross** conducts this sort of finding. Figure D.2 presents the complete code of the file. Type the codes in a script file (section 1.1.2) and save the file by the name **c_cross** in your working path. The **c_cross** has three input arguments; x samples as a row matrix, y samples as a row matrix, and c which are indicated by (x,y,c).

Consider the function $y = xe^{-x}$ over $0 \le x \le 3$. Its graph you may view by exercising the command **ezplot('x*exp(-x)',[0 3])** (appendix E). The graph is not shown for space reason. We wish to determine the value of x when $y = 0.1$. In order to make the samples available we have to choose some step size say $\Delta x = 0.01$, smaller is better. Generate the x samples by:
```
>>x=0:0.01:3; ↵
```
Generate the y samples by scalar code (appendix A):
```
>>y=x.*exp(-x); ↵
```
Just call the **c_cross** as:
```
>>c_cross(x,y,0.1) ↵

ans =
        0.1150
```
Above return says that $x_c = 0.115$. There is no y value as $c = 0.4$ and the response is:
```
>>c_cross(x,y,0.4) ↵
No crossing
```
There are two crossings with $c = 0.2$ over the given bounds which we find by:
```
>>c_cross(x,y,0.2) ↵

ans =
        0.2550    2.5450
```
i.e. the first x_c is 0.255 and the second x_c is 2.545.

D.9 Coefficient extraction from transfer function

Transfer function entering reference (rational, state-space, pole-zero-gain, or other form) you find in section 2.1. Having a control system entered, its polynomial coefficient data we extract by the function **tfdata**. The **tfdata** has a syntax [user-supplied variable for numerator, user-supplied variable for denominator]=**tfdata**(system representing variable, 'v') where the system is defined by **tf**, **ss**, or **zpk**. The second input argument 'v' is a reserve letter to make the function operational.

Suppose $G(s) = \dfrac{s+5}{s^3+9}$ and enter it to **G** by:
```
>>G=tf([1 5],[1 0 0 9]); ↵
```
The polynomial coefficient data then find by:
```
>>[N,D]=tfdata(G,'v') ↵

N =
        0    0    1    5
```

D =
1 0 0 9

As a state-space example say $G(s) = \{A, B, C, D\}$ where $A = \begin{bmatrix} -2 & -1 \\ 2 & 3 \end{bmatrix}$, $B = \begin{bmatrix} -1 \\ -2 \end{bmatrix}$, $C = [2\ 1]$, and $D = [9]$ and enter this $G(s)$ by:

>>A=[-2 -1;2 3]; B=[-1;-2]; C=[2 1]; D=[9]; ↵
>>G=ss(A,B,C,D); ↵

Since D is used for **ss**, let us use another variable **D1** for the denominator and call the extractor as follows:
>>[N,D1]=tfdata(G,'v') ↵

N =
    9.0000 -13.0000 -32.0000
D1 =
    1.0000   -1.0000   -4.0000

Again as a pole-zero-gain example choose $G(s) = \dfrac{5(s+1)}{(2s+3)(s+j+5)(s-j+5)}$ and enter the $G(s)$ by:

>>z=-1; p=[-3/2 -i-5 i-5]; k=5/2; G=zpk(z,p,k); ↵

Determine the polynomial coefficient data by:
>>[N,D]=tfdata(G,'v') ↵

N =
    0   0   2.5000   2.5000
D =
    1.0000   11.5000   41.0000   39.0000

Should the reader need s related expression (example 1 of section 7.1), exercise **poly2str** for instance the last denominator we see as:
>>poly2str(D,'s') ↵

ans =
    s^3 + 11.5 s^2 + 41 s + 39

Note that the returned coefficients are preceded by zeroes up to the higher degree between numerator and denominator e.g. in the last numerator the N has the first two coefficients as zero.

D.10 Numerical integration from samples

Numerically we compute definite integration of the type $\int_{x=a}^{x=b} f(x)\,dx$ employing a function **trapz** (abbreviation for trapezoidal). First one needs to make x and $f(x)$ samples available at the workspace. If $f(x)$ is an expression, use scalar code for the samples (appendix A). The step size of x is selected by the user. If $f(x)$ samples are given as a row or column matrix, there is no need to exercise scalar code. The syntax is **trapz**(x samples as a row/column matrix, $f(x)$ samples as a row/column matrix) and the return is the integration value. We present two examples on **trapz** as follows:

$$A. \int_{x=-\frac{\pi}{3}}^{x=\frac{\pi}{2}} \sin^3 x \cos^3 x \, dx = 0.013 \qquad B. \int_{x=3}^{x=9} \frac{1}{x(x^2+1)^2} \, dx = 0.0026$$

Example A:
   Let us choose step size 0.01 for x and generate the samples of x as a row matrix (section 1.1) as follows:
   >>x=-pi/3:0.01:pi/2; ↵
Compute the $f(x)$ samples for every sample in **x** by the scalar code:
   >>f=sin(x).^3.*cos(x).^3; ↵
Call the integrator as:
   >>trapz(x,f)

   ans =
         0.0130

Example B:
   Similar to example A the codes are:
   >>x=3:0.01:9; f=1./x./(x.^2+1).^2; trapz(x,f) ↵

Example on given data:
   Sometimes samples are given as follows:

x	0	1	1.5	2	3	5
$f(x)$	3	4	5	7	8	8.5

For this circumstance enter each x and $f(x)$ as a row matrix and then call the integrator as follows:
   >>x=[0 1 1.5 2 3 5]; f=[3 4 5 7 8 8.5];trapz(x,f) ↵

   ans =
         32.7500

D.11 Complex number entering into MATLAB

   Symbolically the imaginary unit of a complex number is denoted by i, j, or $\sqrt{-1}$ whose MATLAB representation is **i**, **j**, or **sqrt(-1)**. As an example the complex number $4+j5$ is entered into MATLAB by any of the following expressions **4+5i, 4+5*i, 4+i*5, 4+5*j,** or **4+5*sqrt(-1)**.
   Complex number matrix follows similar entering style to that of the integer or real number with little difference in conjugateness. Let us enter the complex number matrices $R=[3-j \ 4j \ -4]$, $C=\begin{bmatrix}7j \\ -4+5j \\ 8j\end{bmatrix}$, and $A=\begin{bmatrix}2 & 5-j & 9j \\ 7j & 2+j & 11j\end{bmatrix}$ into MATLAB by the following:
   for R,
   >>R=[3-i 4i -4] ↵

-273-

R =
           3.0000 - 1.0000i   0 + 4.0000i   -4.0000
for C,
>>C=[7i   -4+5i 8i].' ↵

C =
           0 + 7.0000i
          -4.0000 + 5.0000i
           0 + 8.0000i
for A,
>>A=[2 5-i 9i;7i 2+i 11i] ↵

A =
           2.0000   5.0000 - 1.0000i   0 + 9.0000i
           0 + 7.0000i   2.0000 + 1.0000i   0 +11.0000i

The operators .' and ' mean transpose without and with conjugate respectively. In the column matrix case if we use the operator ' at the end, we would assign $\begin{bmatrix} -7j \\ -4-5j \\ -8j \end{bmatrix}$ to C.

Modulus or absolute value of a complex number $A+jB$ is given by $\sqrt{A^2 + B^2}$. To take the modulus of a complex number, we call the command **abs** (abbreviation for absolute value) with the syntax **abs**(complex scalar or matrix). For example the modulus of $4+j3$ and elements in $R=[12+j5 \quad -4-j3 \quad -8+j6]$ are 5 and [13 5 10] respectively which we compute by using the right side attached command. In both cases we assigned the return to workspace A which can be any user-supplied variable.

modulus for the single complex number,
>>A=abs(4+3i) ↵

A =
     5
modulus for the complex row matrix elements in R,
>>R=[12+5i -4-3i -8+6i]; ↵
>>A=abs(R) ↵

A =
     13   5   10

Argument of a complex number $A+jB$ is given by $\tan^{-1}\frac{B}{A}$. To find the argument, we call the function **angle** with the syntax **angle**(complex scalar or matrix name). The function returns any value from $-\pi$ to π. For instance the arguments of $4+j3$ and each element in $R=[12+j5 \quad -4-j3 \quad -8+j6]$ are $\tan^{-1}\frac{3}{4} = 0.6435^c$ and $[0.3948^c \quad -2.4981^c \quad 2.4981^c]$ respectively which we implement by right side attached text box commands. In both cases we assigned the return to the workspace P which can be any user-given variable. For degree to radian and radian to degree conversions we call the commands **deg2rad** and **rad2deg** respectively for

argument for the single complex number:
>>P=angle(4+3i) ↵

P =
     0.6435
argument for the complex row matrix elements in R,
>>P=angle(R) ↵

P =
     0.3948 -2.4981   2.4981

example on P as rad2deg(P) for the degree.

The conjugate of a complex number $A+jB$ is given by $A-jB$. To find the conjugate of a complex number, we apply the function conj with the syntax conj(complex scalar or matrix name). As an example the conjugate of $4+j3$ and all elements in $R=[12+j5 \quad -4-j3 \quad -8+j6]$ are $4-j3$ and $[12-j5 \quad -4+j3 \quad -8-j6]$ respectively. Both implementations are shown below and assigned to the workspace C (user-given variable):

conjugate for the single complex number,
```
>>C=conj(4+3i) ↲
```

C =
            4.0000 - 3.0000i

conjugate of the elements in the row matrix R,
```
>>R=[12+5i -4-3i -8+6i]; ↲
>>C=conj(R) ↲
```

C =
            12.0000-5.0000i -4.0000+3.0000i -8.0000- 6.0000i

A complex number $A+jB$ has the real part A and the imaginary part B. To find the real and imaginary parts from complex number(s), we apply the functions real and imag with the syntax real(complex scalar or

```
>>real(R) ↲         ← for the real elements in R
ans =
        12  -4  -8
>>imag(R) ↲         ← for the imaginary elements in R
ans =
        5  -3  6
```

matrix name) and imag(complex scalar or matrix name) respectively. The real and imaginary parts for the elements in $R=[12+j5 \quad -4-j3 \quad -8+j6]$ are [12 −4 −8] and [5 −3 6] respectively whose findings are attached in the upper right side text box of this paragraph. The returns could have been assigned to some user-supplied variables.

D.12 Matrix manipulations

New matrices can be formed from the matrix we have at workspace of MATLAB. We pick a portion of matrix by using colon operator (:).

Let us see some coloning by first assigning the row matrix $A=[2 \quad 4 \quad 3 \quad -10 \quad 0 \quad 9 \quad 73 \quad 29 \quad -31 \quad 50]$ to the workspace variable A:
```
>>A=[2 4 3 -10 0 9 73 29 -31 50]; ↲
```

Suppose we wish to form a matrix B where B will be the second, third, and ninth element of A i.e. $B=[4 \quad 3 \quad -31]$:
```
>>B=A([2 3 9]) ↲    ← The input argument of A is a row matrix indicating
                       position indices
```
B =

$$\begin{matrix} 4 & 3 & -31 \end{matrix} \quad \leftarrow B \text{ holds the required matrix}$$

A matrix C is to be formed from the third through eighth elements of A i.e. $C = [3\ -10\ 0\ 9\ 73\ 29]$. We execute the following:

    >>C=A(3:8) ↵   ← The input argument 3:8 indicates the third through eighth

    C =

        3 -10 0 9 73 29   ← C holds the required matrix

What if we have a column matrix like $D = \begin{bmatrix} 2 \\ 4 \\ 5 \\ -10 \\ 0 \\ 6 \\ 73 \\ 7 \\ -31 \\ 50 \end{bmatrix}$, enter the matrix into MATLAB workspace as follows:

    >>D=[2 4 5 -10 0 6 73 7 -31 50]'; ↵   ← D holds the D

We wish to form a matrix E from the tenth and seventh elements of D i.e. $E = \begin{bmatrix} 50 \\ 73 \end{bmatrix}$ and F from the first five elements of D i.e. $F = \begin{bmatrix} 2 \\ 4 \\ 5 \\ -10 \\ 0 \end{bmatrix}$ and do so by:

    formation of matrix E,   formation of matrix F,
    >>E=D([10 7]) ↵   >>F=D(1:5) ↵

    E =   F =
        50   2
        73   4
                                                                       5
                                                                   -10
                                                                     0

Now we present how the coloning of a rectangular matrix is accomplished.

Let us input the $G = \begin{bmatrix} 8 & 64 & 27 & 56 & 98 & 43 & 4 \\ -64 & 216 & 729 & 40 & 12 & 23 & 568 \\ 678 & -90 & 70 & 61 & 67 & 445 & 3 \\ 1 & 47 & 45 & 72 & 34 & -5 & -7 \\ 3 & 87 & 82 & 29 & 10 & -16 & -59 \end{bmatrix}$ to G by:

    >>G=[8 64 27 56 98 43 4;-64 216 729 40 12 23 568;678 ... ↵
    -90 70 61 67 445 3;1 47 45 72 34 -5 -7;3 87 82 29 10 -16 -59] ↵

    G =   ← G holds the G
          8 64 27 56 98 43 4

$$\begin{bmatrix} -64 & 216 & 729 & 40 & 12 & 23 & 568 \\ 678 & -90 & 70 & 61 & 67 & 445 & 3 \\ 1 & 47 & 45 & 72 & 34 & -5 & -7 \\ 3 & 87 & 82 & 29 & 10 & -16 & -59 \end{bmatrix}$$

In last command the last word of the first line is 678. After typing 678 we leave one space by pressing spacebar and then type three dots from the keyboard. These three dots mean continuation of any MATLAB command. Press Enter key and type the other matrix elements of the row which were interrupted. The three dots (…) are called ellipsis.

Required matrix elements from G are shown by elements inside the dotted box in the following.

Matrix H is to be formed from the second and the fourth columns of G:

$$\begin{bmatrix} 8 & \boxed{64} & 27 & \boxed{56} & 98 & 43 & 4 \\ -64 & \boxed{216} & 729 & \boxed{40} & 12 & 23 & 568 \\ 678 & \boxed{-90} & 70 & \boxed{61} & 67 & 445 & 3 \\ 1 & \boxed{47} & 45 & \boxed{72} & 34 & -5 & -7 \\ 3 & \boxed{87} & 82 & \boxed{29} & 10 & -16 & -59 \end{bmatrix}$$

Matrix K is to be formed from the third and the fifth rows of G:

$$\begin{bmatrix} 8 & 64 & 27 & 56 & 98 & 43 & 4 \\ \boxed{-64 & 216 & 729 & 40 & 12 & 23 & 568} \\ 678 & -90 & 70 & 61 & 67 & 445 & 3 \\ 1 & 47 & 45 & 72 & 34 & -5 & -7 \\ \boxed{3 & 87 & 82 & 29 & 10 & -16 & -59} \end{bmatrix}$$

Matrix L is to be formed from the fourth through seventh columns of G:

$$\begin{bmatrix} 8 & 64 & 27 & \boxed{56} & 98 & 43 & 4 \\ -64 & 216 & 729 & 40 & 12 & 23 & 568 \\ 678 & -90 & 70 & 61 & 67 & 445 & 3 \\ 1 & 47 & 45 & 72 & 34 & -5 & -7 \\ 3 & 87 & 82 & 29 & 10 & -16 & -59 \end{bmatrix}$$

Matrix M is to be formed from the third through fifth rows of G:

$$\begin{bmatrix} 8 & 64 & 27 & 56 & 98 & 43 & 4 \\ -64 & 216 & 729 & 40 & 12 & 23 & 568 \\ 678 & -90 & 70 & 61 & 67 & 445 & 3 \\ 1 & 47 & 45 & 72 & 34 & -5 & -7 \\ 3 & 87 & 82 & 29 & 10 & -16 & -59 \end{bmatrix}$$

Matrix N is to be formed from the intersection of the third through fifth rows and the fourth through seventh columns of G:

$$\begin{bmatrix} 8 & 64 & 27 & 56 & 98 & 43 & 4 \\ -64 & 216 & 729 & 40 & 12 & 23 & 568 \\ 678 & -90 & 70 & 61 & 67 & 445 & 3 \\ 1 & 47 & 45 & 72 & 34 & -5 & -7 \\ 3 & 87 & 82 & 29 & 10 & -16 & -59 \end{bmatrix}$$

Formations of the required H, K, L, M, and N assigned to respective workspace variables are presented in the sequel:

for the formation of H,
>>H=G(:,[2 4]) ↵

H =
```
    64   56
   216   40
   -90   61
    47   72
    87   29
```

for the formation of N,
>>N=G(3:5,4:7) ↵

N =
```
    61   67   445    3
    72   34    -5   -7
    29   10   -16  -59
```

for the formation of M,
>>M=G(3:5,:) ↵

M =
```
   678  -90   70   61   67   445    3
     1   47   45   72   34    -5   -7
     3   87   82   29   10   -16  -59
```

for the formation of K,
>>K=G([3 5],:) ↵

K =
```
   678  -90   70   61   67   445    3
     3   87   82   29   10   -16  -59
```

for the formation of L,
>>L=G(:,4:7) ↵

L =
```
    56   98   43    4
    40   12   23  568
    61   67  445    3
    72   34   -5   -7
    29   10  -16  -59
```

Summarizing all, we exercise the commands matrix name(desired row/rows,:), matrix name(:,desired column/columns), and matrix name(desired row/rows, desired column/columns) for selecting row, column, and submatrix from any existing matrix respectively.

D.13 Linear or logarithmic samples from specific number

From user-defined bounds and sample number one can generate linearly or logarithmically spaced vectors. Both are addressed below.

Linear vector:

Linearly spaced vector elements form an arithmetic progression. If the first element in the vector is a and common difference of the progression is d, the vector becomes $[a \quad a+d \quad a+2d \quad .. \quad a+(N-1)d]$ where N is the number of elements in the vector. Clearly d is equal to $\frac{Last\ element - First\ element}{N-1}$. The function linspace (abbreviation for the linear space) forms a linearly spaced vector for which the syntax is linspace(first element, last element, number of points from first to last) and whose output is a row matrix.

Let us form a row vector R from 3 to 13 with 6 points therefore $d=2$ and R should be [3 5 7 9 11 13] and do so by:
>>R=linspace(3,13,6) ↵   ← R holds the required R and is user-chosen

R =

3 5 7 9 11 13

Again a column vector C is to be formed from -7 to 3 with 5 points so that $d = \frac{5}{2}$ and $C = \begin{bmatrix} -7 \\ -\frac{9}{2} \\ -2 \\ \frac{1}{2} \\ 3 \end{bmatrix}$ which one obtains by:

>>C=sym(linspace(-7,3,5))' ↵   ← C holds the required C and is user-chosen

C =
    -7
    -9/2
    -2
    1/2
    3

The command **sym** turns a number from decimal fractional to rational e.g. 4.5 to 9/2. A row matrix is changed to column by using the transpose operator ' (chapter 1). We have to conduct it for C because the **linspace** return is a row one.

Logarithmic vector:
Logarithmically (base of the logarithm is 10) spaced vector elements form a geometric progression. If first element in the vector is a and common ratio of the progression is r with length N, the vector is given by $[a \ ar \ ar^2 .. \ ar^{N-1}]$. The function **logspace** (abbreviation for logarithmically spaced) generates a logarithmically spaced vector with the syntax **logspace**(power of the first element, power of the last element, number of points from the first to last) and whose output is a row vector.

We wish to form a logarithmically spaced vector where power of the elements will be from 3 to 4 and the number of elements will be 5 therefore $a = 10^3$, $N = 5$, and $r = 10^{\frac{1}{4}}$ which results the vector to be $L = [10^3 \quad 10^3 10^{\frac{1}{4}} \quad 10^3 10^{\frac{2}{4}} \quad 10^3 10^{\frac{3}{4}} \quad 10^3 10^{\frac{4}{4}}] = [1000 \quad 1778 \quad 3162 \quad 5623 \quad 10000]$ (neglecting the fractional parts). Following is the execution:

>>L=logspace(3,4,5)' ↵

L =
    1.0e+004 *
    0.1000
    0.1778
    0.3162
    0.5623
    1.0000

If the power of 10 is higher, the return from the **logspace** will be of higher digits that is why the return is in exponential form. Concerning the execution,

1.0e+004 * means $1.0 \times 10^4 \times$ and each of the return elements is multiplied by 10^4. We assigned the column vector to L (any user-chosen variable).

D.14 Summing matrix elements

MATLAB function sum adds all elements in a row, column, or rectangular matrix when the matrix is its input argument. Example matrices are $R = [1 \ -2 \ 3 \ 9]$, $C = \begin{bmatrix} 23 \\ -20 \\ 30 \\ 8 \end{bmatrix}$, and $A = \begin{bmatrix} 2 & 4 & 7 \\ -2 & 7 & 9 \\ 3 & 8 & -8 \end{bmatrix}$ whose all element sums are 11, 41, and 30 for the R, C, and A respectively. We execute the summations as follows (font equivalence is maintained by using the same letter for example A⇔ A):

Sum for the row matrix,	Sum for the column matrix,	Sum for the rectangular matrix,

>>R=[1 -2 3 9]; ↵
 >>sum(R) ↵
 	
 ans =
    11
	>>C=[23 -20 30 8]'; ↵
	 >>sum(C) ↵
	
	 ans =
	    41
		>>A=[2 4 7;-2 7 9;3 8 -8]; ↵
		 >>sum(sum(A)) ↵
		 ans =
		    30

For a rectangular matrix, two functions are required because the inner sum performs the summing over each column and the result is a row matrix. The outer sum provides the sum over the resulting row matrix.

The function is operational for real, complex, even for symbolic variables like x or y.

D.15 Cell arrays

A cell array is composed of cells where the cells can contain previously discussed ordinary arrays (of real, integer, or complex numbers), structure arrays, multidimensional arrays, character arrays ⋯ etc. The cells of a cell array are indexed like a rectangular matrix but using the second brace {..}. For example A{3,4} indicates that A is a cell array and we are addressing the cell with the coordinate (3,4) – third row and fourth column. If we build a cell array A of order 2×3, the position indexes of different cells are A{1,1}, A{1,2}, A{1,3}, A{2,1}, A{2,2}, and A{2,3} in row direction respectively. In every cell you can have any sort of data. For example A{1,1} can have $\begin{bmatrix} 1+2i & 2-3i \\ 4+6i & 9+3i \end{bmatrix}$ and A{1,2} can have $\begin{bmatrix} \text{Shameem} \\ \text{Shimul} \\ \text{Richard} \end{bmatrix}$. More about cell array you can find in [34].

Appendix E

Some graphing functions of MATLAB

One of MATLAB's nicest features is you can have your graphics drawn while programming control system related problems. There are so many easy accessible built-in graphics functions that one finds it very interesting when the input-output argumentation style of these functions is understood. Some graphing functions which we applied frequently in previous chapters are addressed for syntax details in the sequel.

♦ y versus x data

The command **plot** graphs y versus x data. Let us say we have the attached (on the right side in this paragraph) tabular data. We intend to graph these data as y versus x graph.

Tabular data of y versus x type:							Command to graph the y vs x data:
x	-6	-4	0	4	5	7	>>x=[-6 -4 0 4 5 7]; ↵
y	9	3	-3	-5	2	0	>>y=[9 3 -3 -5 2 0]; ↵ >>plot(x,y) ↵

Commands to graph the data are also presented beside the tabular data on the right side in the last paragraph. We first assign the x and y data to workspace **x** and **y** (some user-chosen variables) respectively and then call the command **plot** to see the figure E.1(a). The **plot** has two input arguments, the first and second of which are the x and y data both as a row or column matrix of identical size respectively.

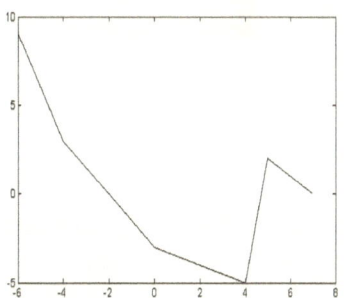

Figure E.1(a) y vs x plot of the tabular data

In order to graph the mathematical expression by using the **plot**, one first needs to calculate the functional values by using the scalar code (appendix A) and then applies the command. During the calculation, computing step selection is mandatory which is completely user-defined.

For instance we wish to graph the function $f(x) = x^2 - x + 2$ over $-2 \le x \le 3$.

Let us choose some x step 0.1. The **x** vector as a row matrix is generated by **x=-2:0.1:3;** (section 1.1). At every element in **x** vector, the functional value is computed and assigned to workspace **f** by **f=x.^2-x+2;**. The **f** is any user-chosen variable. Now we call the grapher as **plot(x,f)** to see the trajectory (not shown for space reason).

The command **plot** just draws the graph, no graphical features such as x axis label or title are added to the graph. It is the user who is supposed to add these graphical features.

◆ Multiple y data versus common x data

The **plot** keeps many options, one of which is just discussed. We graph several y data versus common x data with the help of **plot** but with different number of input arguments. Let us choose the right side attached table for graphing.

Tabular data for multiple y versus common x:

x	-6	-4	0	4	5	7
y_1	9	3	-3	-5	2	0
y_2	0	-2	1	0	5	7.7
y_3	-1	2	8	1	0	-3

We intend to plot the y_1, y_2, and y_3 on common x data. To do so,

>>x=[-6 -4 0 4 5 7]; ↵   ← Assigning the x data as a row matrix to x
>>y1=[9 3 -3 -5 2 0]; ↵   ← Assigning the y_1 data as a row matrix to y1
>>y2=[0 -2 1 0 5 7.7]; ↵   ← Assigning the y_2 data as a row matrix to y2
>>y3=[-1 2 8 1 0 -3]; ↵   ← Assigning the y_3 data as a row matrix to y3
>>plot(x,y1,x,y2,x,y3) ↵   ← Applying the command **plot**

The **plot** now has six input arguments – two for each graph, the first and second of which are the common x and y data to be plotted respectively. If there were four y data, the command would be plot(x,y1,x,y2,x,y3,x,y4). Once the data is plotted for several y, identifying the y traces is obvious which is carried out by the command **legend**. The command **legend('y1', 'y2','y3')** puts identifying marks/colors among various graphs. The input argument of the **legend** is any user-given word but under quote and separated by a comma. The number of y traces must be equal to the number of input arguments of **legend**. We gave the names y1, y2, and y3 for the three y traces respectively. In doing so, we end up with the figure E.1(b). You can even move the legend on the plot area by using mouse.

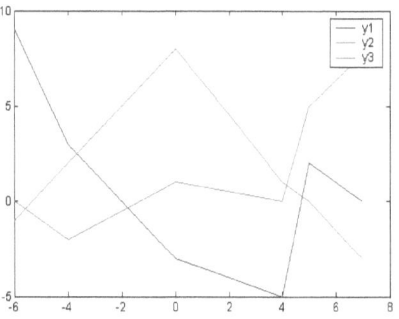

Figure E.1(b) Multiple y vs x for the tabular data

You see all graphics throughout the text as black and white because we did not include color graphics in the text (for expense reason). But MATLAB displays figures in color plots, which you can easily identify.

Another situation can be that we have several functions and intend to plot those on common x variation. For instance we wish to graph $y_1 = x^3 - x^2 + 4$ and $y_2 = x^2 - 7x - 5$ over the common $-1 \le x \le 3$.

Under these circumstances, the step selection of x data is compulsory. Without calculating the functional values of given y curves, we can not graph the functions for which we exercise the scalar code. Let us choose the x step as 0.1. We first generate the common **x** vector as a row matrix by writing **x=-1:0.1:3**; and then calculate the y_1 and y_2 (**y1** $\Leftrightarrow y_1$ and **y2** $\Leftrightarrow y_2$) data by writing **y1=x.^3-x.^2+4**; **y2=x.^2-7*x-5**; and eventually the graph appears by executing **plot(x,y1,x,y2)**, graph is not shown for space reason. Thus you can graph three or more functions.

◆ **Functions of the form** $y = f(x)$

If any function is of the form $y = f(x)$ and the $f(x)$ versus x is to be graphed, the built-in **ezplot** is the best option which uses a syntax **ezplot**(functional vector code under quote according to appendix A, interval bounds as a two element row matrix) where the first and second elements in the row matrix are beginning and ending bounds of the interval respectively. The **ezplot** graphs $y = f(x)$ in the default interval

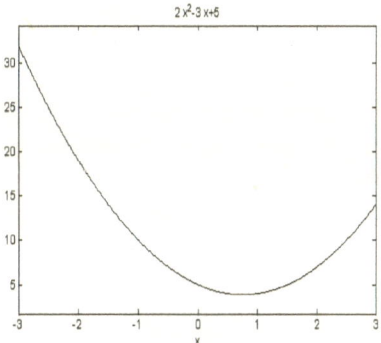

Figure E.1(c) Plot of $y = 2x^2 - 3x + 5$ versus x over $-3 \le x \le 3$

$-2\pi \le x \le 2\pi$ when no interval description is argumented.

We intend to graph the function $y = 2x^2 - 3x + 5$ over the interval $-3 \le x \le 3$. We first give $2x^2 - 3x + 5$ MATLAB vector code and then assign that to **y** as follows:
>>y='2*x^2-3*x+5'; ↵

In above implementation the **y** is any user-chosen variable. The interval $-3 \le x \le 3$ is entered by [-3 3]. To obtain the plot of y in the given interval, we execute the following at the command prompt:
>>ezplot(y,[-3,3]) ↵

Above command results the figure E.1(c).

◆ **Multiple graphs in the same window**

The function **subplot** splits a figure window in subwindows based on the user definition. It accepts three positive integer numbers as the input arguments, the first and second of which indicate the number of subwindows in the horizontal and the number of subwindows in the vertical directions respectively. For example 22 means two subwindows horizontally and two subwindows vertically, 32 means three subwindows horizontally and two subwindows vertically, ... and so on. The third integer in the input argument numbered consecutively offers control on the subwindows so generated. If the first two digits are 32, there should be 6 subwindows and they are numbered and controlled by using 1 through 6. When you plot some graph in a subwindow, as if you are handling an independent figure window.

Commands for the figure E.1(d):
>>subplot(121) ↵   ← It handles the first graph
>>ezplot('x') ↵   ← Plotting $y = x$
>>subplot(122) ↵   ← It handles the second graph
>>ezplot('exp(-x)') ↵ ← Plotting $y = e^{-x}$

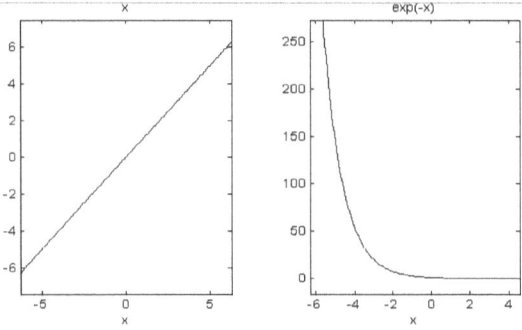

Figure E.1(d) Plots of $y = x$ and $y = e^{-x}$ side by side in the same window

We wish to graph $y = x$ and $y = e^{-x}$ side by side as two different plots by using earlier mentioned **ezplot** but in the same window. If we imagine the subfigures as matrix elements, we have a figure matrix of size 1×2 (one row and two columns). That is why the first two integers of the input argument of **subplot** should be 12. Attached commands in the upper right text box of last paragraph show the figure E.1(d). The third integers 1 and 2 in the **subplot** give the control on the first and second subfigures respectively.

As another example we wish to plot $y = x$ and $y = e^{-x}$ in the upper row and only $y = (1 - e^{-x})$ in the lower row subfigures in the same window

Commands for the figure E.1(e):
>>subplot(221) ↵   ← Subfigure selection for $y = x$
>>ezplot('x') ↵   ← Plotting $y = x$
>>subplot(222) ↵   ← Subfigure selection for $y = e^{-x}$
>>ezplot('exp(-x)') ↵ ← Plotting $y = e^{-x}$
>>subplot(212) ↵   ← Subfigure selection for $y = (1 - e^{-x})$
>>ezplot('1-exp(-x)') ↵ ← Plotting $y = (1 - e^{-x})$

whose implementation needs above attached text box commands and whose final output is the figure E.1(e). We are supposed to have four figures when the integer input argument of **subplot** is 22 (two for rows and two for columns). The arguments 221, 222, 223, and 224 provide handle on the four figures consecutively. The figures could have been plotted on 223 and 224 are absent so we ignore them. The argument 21 creates two subfigures (two rows and one column) handled by 211 and 212, but 211 is absent so we ignore that too.

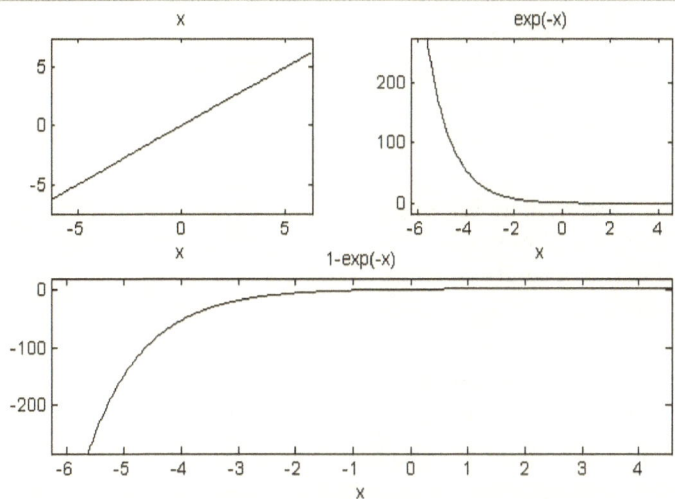

Figure E.1(e) Plots of $y = x$ and $y = e^{-x}$ in the upper row and $y = (1 - e^{-x})$ in the lower row in the same window

Let us see the input arguments of **subplot** for different subfigures (each third brace set [ ] is one subfigure in the following tabular representation) as follows:

Subfigures needed	First two input integers of **subplot**	Third input integer of **subplot**	Commands we need
[ ] [ ] [ ] [ ]	22	[1] [2] [3] [4]	subplot(221) subplot(222) subplot(223) subplot(224)
[ ] [ ] [ ]	22 for upper two (lower two remain empty) 21 for the lower one (upper one remains empty)	[1] [2] [2]	subplot(221) subplot(222) subplot(212)
[ ] [ ] [ ]	21 for the upper one (lower one remains empty) 22 for the lower two (upper two remain empty)	[1] [3] [4]	subplot(211) subplot(223) subplot(224)
[ ] ⎡ ⎤ [ ] ⎣ ⎦	22 for the left two (right two remain empty) 12 for the right one (left one remains empty)	[1] 2 [3]	subplot(221) subplot(223) subplot(122)
⎡ ⎤ [ ] ⎣ ⎦ [ ]	22 for right two (left two remain empty) 12 for the left one (right one remains empty)	[2] 1 [4]	subplot(222) subplot(224) subplot(121)

-285-

◆ **Two dissimilar y - common x i.e.** y_1 **and** y_2 **over** x

Sometimes two dissimilar y data over common x needs to be plotted. For instance the x, y_1, and y_2 data in
$$\begin{Bmatrix} x & 0 & 1 & 2 & 3 \\ y_1 & 5 & 4 & 3 & 5 \\ y_2 & 7 & 2 & 1 & 3 \end{Bmatrix} \text{ are in}$$
second, meter, and kilogram respectively. Obviously the data in meter and kilogram can not be plotted in a single axis or y trace. For comparison reason one must use two different y axes which we implement by **plotyy** with the syntax **plotyy**(x data as a row matrix, y_1 data as a row matrix, x data as a row matrix, y_2 data as a row matrix) and do so by:

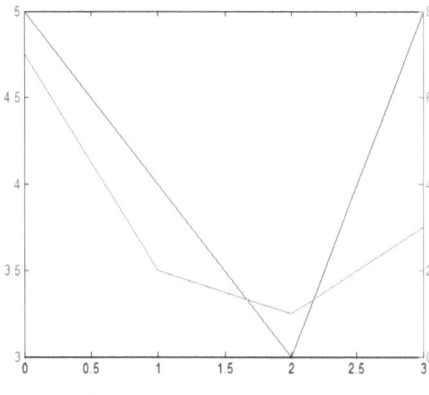

Figure E.1(f) Two dissimilar y data on common x

>>x=[0 1 2 3]; y1=[5 4 3 5]; y2=[7 2 1 3]; plotyy(x,y1,x,y2) ↵

In above we assigned the x, y_1, and y_2 data to the workspace **x**, **y1**, and **y2** respectively, each of the variables is user-chosen. By calling the **plotyy** we view the graph like figure E.1(f). Horizontal axis of the figure refers to x. The left and right vertical axes in the graph correspond to the y_1 and y_2 data respectively.

This particular graph needs some other approach for labeling. Say we intend to include the left and right y axis labels as **Distance in meter** and **Mass in Kg** respectively for which execute the following:
>>h=plotyy(x,y1,x,y2); ↵
>>set(get(h(1),'Ylabel'),'String','Distance in meter') ↵
>>set(get(h(2),'Ylabel'),'String','Mass in Kg') ↵

In above first line, we assigned output of **plotyy** to some user-chosen variable **h**. When a graph is plotted, many object properties become associated with the graphics like axes, background, title, text, etc. The **h** is a graphics handle which holds the axes information. The **h(1)** and **h(2)** provide control on the left and right axes respectively. The command **get(h(1),'Ylabel')** looks for the **Ylabel** (reserve word and put under quote) property in the left y-axis. The command **set** writes the user-supplied string in the y-axis. The **set** has three input arguments, the first, second, and third of which are the y axis object location (found by the **get**), reserve word **String** under quote, and the user-given words put under quote respectively. Figure E.1(g) shows the graph following the execution.

As another example let us say we wish to plot the curves $y_1 = x^3 - x^2 + 4$ and $y_2 = 400(x^2 - 7x - 5)$ on the left and right y axes over the common $-1 \le x \le 3$ with a x step 0.1. First generate a x vector as a row matrix over the given interval and step size by using x=-1:0.1:3; and then calculate each function by using the scalar code (appendix A) to have the samples of y_1 and y_2 that is y1=x.^3-x.^2+4; y2=400*(x.^2-7*x-5);. Eventually the graph appears by the use of the command plotyy(x,y1,x,y2).

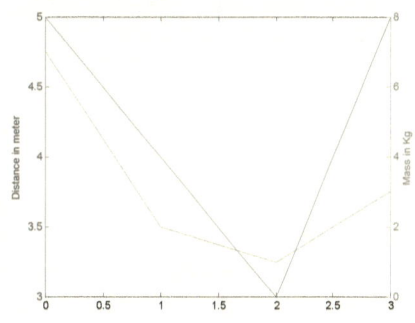

Figure E.1(g) Figure E.1(f) with axis labeling

Appendix F

Creating a function file

A function file is a special type of M-file (section 1.1) which has some user-defined input and output arguments. Both arguments can be single or multiple. The first line in a function file always starts with the reserve word **function**. A function file must be in your working path or its path must be defined in MATLAB. Depending on problem, a function file is written by the user and can be called from the MATLAB command prompt or from another M-file. For convenience, long and clumsy programs are split into smaller modules and these modules are written in a function file. The basic structure of a function file is as follows:

MATLAB Prompt   function file

\>\> g =call f ⟹ $\underbrace{g(y_1, y_2, ...y_m)}_{\text{output arguments}} = \underbrace{f(x_1, x_2, x_3, ...x_n)}_{\text{input arguments}}$

We present following examples for illustration of function files keeping in mind that the arguments' order and type of the caller and function file are identical.

⊟ Example 1

Let us say the computation of $f(x) = x^2 - x - 8$ is to be implemented as a function file. When $x = -3$ and $x = 5$, we should be having 4 and 12 respectively.

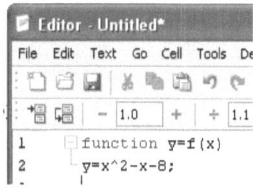

Figure F.1(a) Single input – single output function file

The vector code (appendix A) of the function is **x^2-x-8** assuming scalar **x** and obviously the **x** is for x. We have one input (which is x) and one output (which is $f(x)$). Open a new M-file editor, type the codes of figure F.1(a) exactly as they appear in the M-file, and save the file by the name **f**. The assignee **y** and independent variable **x** can be any variable of your choice, which are the output and input arguments of the function respectively. Again the file and function name **f** can be any user-chosen name only the point is the chosen function or file name should not exist in MATLAB. Let us call the function **f(x)** to verify the programming as shown in the right side text box. You can write dozens of MATLAB executable statements in the file but whatever is assigned to the last **y** returns the function **f(x)** to **g**. Writing the = sign between the **y** and **f(x)** in the function file is compulsory.

```
Calling for example 1:
>>g=f(-3) ↵    ← call f(x) for x = -3
g =
    4
for x = 5,
>>g=f(5) ↵
g =
    12
```

Example 2

Example 1 presents one input-one output function how if we handle multiple inputs and one output? The input argument variables are separated by commas in a function file. A three variable function $f(x_1, x_2, x_3) = x_1^2 - 2x_1 x_2 + x_3^2$ is to be computed by a function file. The input arguments (assuming all scalar) are x_1, x_2, and x_3 and

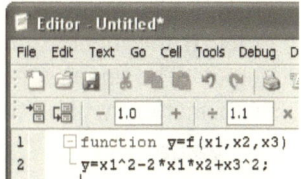

Figure F.1(b) Multiple inputs – single output function file

the output argument is the functional value of the function. The x_1 is written as x1, and so is the others. Follow the M-file procedure of example 1 but the code should be as shown in figure F.1(b). Let us inspect the function (with the specific $x_1 = 3$, $x_2 = 4$, and $x_3 = 5$, the output value of the three variable function must be $f(3,4,5) = 3^2 - 2 \times 3 \times 4 + 5^2 = 10$) as presented in the text box below.

```
Calling for example 2: when input arguments are all scalar:
>>g=f(3,4,5) ↵      ← calling f(x₁, x₂, x₃) for x₁=3, x₂=4, and x₃=5

g =
       10
Calling for the example 2: when input arguments are all column matrix:
>>x1=[2 3 4]'; ↵    ← x₁ values are assigned to x1 as a column matrix
>>x2=[-2 2 5]'; ↵   ← x₂ values are assigned to x2 as a column matrix
>>x3=[1 0 3]'; ↵    ← x₃ values are assigned to x3 as a column matrix
>>f(x1,x2,x3) ↵     ← calling f(x₁, x₂, x₃) using column matrix input arguments

ans =
       13
       -3
       -15
```

The **function** not only works for the scalar inputs but also does for matrices in general for example a set of input argument values are $x_1 = \begin{bmatrix} 2 \\ 3 \\ 4 \end{bmatrix}$, $x_2 = \begin{bmatrix} -2 \\ 2 \\ 5 \end{bmatrix}$, and $x_3 = \begin{bmatrix} 1 \\ 0 \\ 3 \end{bmatrix}$ for which the $f(x_1, x_2, x_3)$ values should be $\begin{bmatrix} 13 \\ -3 \\ -15 \end{bmatrix}$ respectively.

The computing needs scalar code (appendix A) of $f(x_1, x_2, x_3)$ regarding x_1, x_2, and x_3. The modified second line statement of the figure F.1(b) now should be y=x1.^2-2*x1.*x2+x3.^2;. On making the modification and saving the file, let us carry out the commands which are placed in above text box of this page too. If it is necessary, the output can be assigned to user-supplied workspace variable v by writing v=f(x1,x2,x3) at the command prompt. The return from the function file also follows the same input matrix order. If the input arguments of f(x1,x2,x3) are rectangular matrix, so is the output. The input arguments of the function file do not have to be the mathematics symbol. Suppose x_1=ID, x_2=Value, and x_3=Data, one could have written the first and second lines of the function file in the figure F.1(b) as function y=f(ID,Value, Data) and y=ID.^2-2*ID.*Value+ Data.^2; respectively.

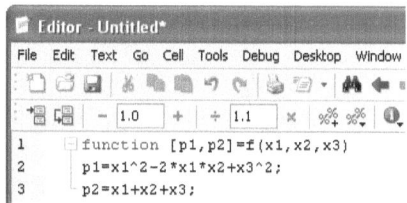

Figure F.1(c) Function file for three input and two output arguments

Example 3

To illustrate a multi-input and multi-output function file, let us consider that p_1 and p_2 are to be found from three variables x_1, x_2, and x_3 (all are scalars) employing the expressions $p_1 = x_1^2 - 2x_1 x_2 + x_3^2$ and $p_2 = x_1 + x_2 + x_3$ whose function file (type the codes in a new M-file editor and save the file by the name f) is presented in figure F.1(c).

Choosing x_1=4, x_2=5, and x_3=6, one should get p_1=12 and p_2=15 for which right side text box commands are conducted at the command prompt. More

Function file calling for the example 3:
>>[p1,p2]=f(4,5,6) ↵ ← calling the function file f for p_1 and p_2 using x_1=4, x_2=5, and x_3=6
p1 =
    12
p2 =
    15

than one output arguments (which are here p_1 are p_2 and represented by p1 and p2 respectively) are separated by commas and placed inside the third bracket following the word function in figure F.1(c).

When we call the function from the command prompt, the output argument writing is similar to that of the function file (that is why we write [p1,p2] as output arguments at the command prompt). The output argument variable names do not have to be p1 and p2 and can be any name of user's choice. If there were three output arguments p_1, p_2, and p_3, the output arguments in the function file would be written as [p1,p2,p3] and their calling would happen in a like manner.

Note: We saved different function files by the same name f just for simplicity and maintaining unifying approach. By this action any previously saved file by the name f disappears. What we suggest is save the function file by other name like f1 and call accordingly for instance the first line of figure F.1(c) would be function [p1,p2]=f1(x1,x2,x3) and calling would take place as [p1,p2]=f1(4,5,6) for the last illustration.

References

>> >> Control System Fundamentals >> >>

[1] Richard C. Dorf and Robert H. Bishop, *"Modern Control Systems"*, 2001, Ninth Edition, Prentice-Hall, Inc., Upper Saddle River, New Jersey.
[2] W. S. Levine, *"The Control Handbook"*, 1996, CRC Press, Boca Raton, FL.
[3] C. A. Canudas De Wit, *"Theory of Robot Control"*, 1996, Springer-Verlag, New York.
[4] F. G. Martin, *"The Art of Robotics"*, 1999, Prentice-Hall, Upper Saddle River, New Jersey.
[5] Christopher T. Kilian, *"Modern Control Technology: Components and Systems"*, 2006, Third Edition, DELMAR CENGAGE Learning, New York.
[6] Enso Ikonen and Kaddour Najim, *"Advanced Process Identification and Control"*, 2002, Marcel Dekker, New York.
[7] Francis J. Hale, *"Introduction to Control System Analysis and Design"*, 1988, Prentice-Hall, Inc.
[8] Chen, Chi-Tsing, *"Linear System Theory and Design"*, 1998, Third Edition, Oxford University Press, New York.
[9] W. J. Palm, *"Modeling, Analysis, Control of Dynamic Systems"*, 2000, Second Edition, John Wiley & Sons, New York.
[10] K. Zhou and J. C. Doyle, *"Essentials of Robust Control"*, 1998, Prentice Hall, Upper Saddle River, New Jersey.

>> >> MATLAB/SIMULINK Basics >> >>

[11] Nuruzzaman, M., *"Tutorials on Mathematics to MATLAB"*, 2003, AuthorHouse, Bloomington, Indiana.
[12] Nuruzzaman, M., *"Modeling and Simulation in SIMULINK for Engineers and Scientists"*, 2005, AuthorHouse, Bloomington, Indiana.
[13] Duffy, Dean G., *"Advanced Engineering Mathematics with MATLAB"*, Second Edition, 2003, Chapman & Hall, CRC, Boca Raton.
[14] Hanselman, Duane C. and Littlefield, Bruce R., *"Mastering MATLAB 5: A Comprehensive Tutorial"*, 1998, Prentice Hall, Upper Saddle River, New Jersey.
[15] Shampine, Lawrence F. and Reichelt, Mark W., *"The MATLAB ODE Suite"*, 1996, The Math-Works, Inc., Natick, MA.
[16] Marcus, Marvin, *"Matrices and MATLAB - A Tutorial"*, 1993, Prentice Hall, Englewood Cliffs, N. J.
[17] Ogata, Katsuhiko, *"Solving Control Engineering Problems with MATLAB"*, 1994, Englewood Cliffs, N. J. Prentice Hall.
[18] Part-Enander, Eva, *"The MATLAB Handbook"*, 1998, Harlow: Addisson Wesley.
[19] Prentice Hall, Inc., *"The Student Edition of MATLAB for MS-DOS Personal Computers"*, 1992, Prentice Hall, Englewood Cliffs, N. J.

[20] Saadat, Hadi., *"Computational Aids in Control Systems Using MATLAB"*, 1993, McGraw-Hill, New York.

[21] Gander, Walter. and Hrebicek, Jiri., *"Solving Problems in Scientific Computing Using MAPLE and MATLAB"*, 1997, Third Edition, Springer Verlag, New York.

[22] Biran, Adrian B and Breiner, Moshe, *"MATLAB for Engineers"*, 1997, Addison Wesley, Harlow, Eng.

[23] D. M. Etter, *"Engineering Problem Solving with MATLAB"*, 1993, Prentice Hall, Englewood Cliffs, N. J.

[24] Shahian, Bahram. and Hassul, Michael., *"Control System Design Using MATLAB"*, 1993, Prentice Hall, Englewood Cliffs, N. J.

[25] Prentice Hall, Inc., *"The Student Edition of MATLAB for Macintosh Computers"*, 1992, Prentice Hall, Englewood Cliffs, N. J.

[26] Ogata, Katshuiko, *"Designing Linear Control Systems with MATLAB"*, 1994, Prentice Hall, Englewood Cliffs, N. J.

[27] Bishop, Robert H., *"Modern Control Systems Analysis and Design Using MATLAB"*, 1993, Addsison Wesley, Reading, MA.

[28] Moscinski, Jerzy and Ogonowski, Zbigniew., *"Advanced Control with MATLAB and Simulink"*, 1995, E. Horwood, Chichester, Eng.

[29] Alberto Cavallo, Roberto Setola, and Francesco Vasca, *"Using MATLAB Simulink and Control Systems Toolbox - A Practical Approach"*, 1996, Prentice Hall, London.

[30] Kuo, Benjamin C. and Hanselman, Duanec., *"MATLAB Tools for Control System Analysis and Design"*, 1994, Prentice Hall, Englewood Cliffs, N. J.

[31] Chipperfield, A. J. and Fleming, P. J., *"MATLAB Toolboxes and Applications for Control"*, 1993, London, New York: Peter Peregrinus on Behalf of the Institute of Electrical Engineers.

[32] Math Works Inc., *"MATLAB Reference Guide"*, Math Works Inc., 1993, Natick, Massachusetts.

[33] Theodore F. Bogart, *"Computer Simulation of Linear Circuits and Systems"*, 1983, John Wiley and Sons, Inc., New York.

[34] Nuruzzaman, M., *"Technical Computation and Visualization in MATLAB for Engineers and Scientists"*, February, 2007, AuthorHouse, Bloomington, Indiana.

[35] Nuruzzaman, M., *"Electric Circuit Fundamentals in MATLAB and SIMULINK"*, October 2007, BookSurge Publishing, Charleston, South Carolina.

[36] Nuruzzaman, M., *"Signal and System Fundamentals in MATLAB and SIMULINK"*, July 2008, BookSurge Publishing, Charleston, South Carolina.

[37] Nuruzzaman, M., *"Modern Approach to Solving Electromagnetics in MATLAB"*, January 2009, BookSurge Publishing, Charleston, South Carolina.

[38] Nuruzzaman, M., *"Finite Difference Fundamentals in MATLAB"*, July, 2013, CreateSpace, South Carolina.

[39] Nuruzzaman, M., *"Digital Audio Fundamentals in MATLAB"*, July, 2010, CreateSpace, California.

Subject Index

Numeric start:
2% criterion   105

A
analysis in frequency domain   155
aperiodic input modeling   138

B
band reject   172
bandpass   172
bandstop   172
bandwidth   171,174
bias modeling   133
bode plot   160

C
C.E.   183
characteristic equation   183
chart Nichol's   169
coding input   89
coding signal   89
combining signal   142
constant damping   165
constant frequency   165
constant input   88
constant signal   88
continuous system   27
control input   87,123
control output   94
control output modeling   141
control system   27
control system bandwidth   171
control system bode plot   160
control system C.E.   183
control system damping   164
control system DC gain   171
control system frequency   164
control system frequency response   156
control system in feedback   39
control system in frequency domain   155
control system in parallel   37
control system in series   37
control system in time   87
control system interconnection   44
control system Nichol's plot   169
control system Nyquist plot   168
control system pole-zero   162
control system root locus   183
control system spectrum   157
control system stability   183
conversion of systems   40
cutoff frequency   172

D
damped sine signal   89
damped sine wave modeling   132
damping   164
damping optimum   106
damping ratio   105
dB magnitude   158
DC gain   171
defining MIMO   29,33
defining SISO   27
Dirac delta function   125
duty cycle   134

E
equivalent on parallel system   37
equivalent on series system   37
error indices   109
error performance   109
error performance modeling   147
exponential signal   89
expression based input modeling   140

F
feedback interconnection   46
feedback system   39
feedback system modeling   72
finite pulse modeling   125
first order performance   103
frequency cutoff   172
frequency domain   155
frequency range response   157
frequency response   156
frequency response normalization   159

frequency response of systems   30, 162
frequency spectrum   157
full rectified sine wave   90
full rectified wave modeling   130
function for root locus   191

G
gain DC   171
gain margin   166
graph impulse response   100
graph input-output   96
graph step response   101
graphing MIMO   99
graphing multiple root locus   192
graphing spectrum   159

H
half rectified sine wave   91
half rectified wave modeling   130
highpass   172

I
IAE   110
imaginary spectrum   157
impulse response for multi output 102
impulse response modeling   143
index comparison   111
index modeling   147
input in time   87
input modeling   123
input periodic   90
input rectified sine wave   90
input response   96
input sample   89
input signal   89
input square wave   91
input triangular wave   93
input-output in time   87
integral absolute error   110
integral square error   110
integral time multiplied absolute error   110
integral time multiplied square error   110
interconnected control system   44
interconnected system modeling   73
interconnection on feedback   46
interconnection on MIMO   48
interconnection on parallel   44
interconnection on series   47
interconnection on SISO   49
ISE   110
ITAE   110
ITSE   110

L
lowpass   172

M
magnitude in dB   158
magnitude spectrum   157
mapping pole-zero   162
margin   166
margin gain   166
margin phase   166
MIMO   33
MIMO graph   100
MIMO interconnection   52
MIMO modeling   66,141
MIMO output   96
mixed system modeling   66
mixed transfer function   31
model to transfer function   76
modeling a damped sine wave   132
modeling a full rectified wave   130
modeling a half rectified wave   130
modeling a MIMO   66
modeling a nonperiodic input   138
modeling a pulse generator   133
modeling a ramp   127
modeling a rectangular wave   133
modeling a sine wave   129
modeling a SISO   63
modeling a square wave   134
modeling a system   63
modeling a transfer function   63
modeling a triangular wave   135
modeling a triggered wave   138
modeling a two frequency wave   131
modeling an aperiodic input   138
modeling an expression based input   140
modeling bias   133
modeling control output   141
modeling Dirac delta   125
modeling error performance   147
modeling feedback system   72
modeling finite pulse   125
modeling impulse response   143

modeling index   147
modeling input   123
modeling interconnected system   73
modeling MIMO   141
modeling mixed system   66
modeling multi input   68
modeling multi output   69
modeling negative feedback   72
modeling P.O.   145
modeling parallel system   67
modeling peak time   145
modeling peak value   144
modeling percent overshoot   145
modeling performance   144
modeling positive feedback   72
modeling rise time   145
modeling series system   70
modeling settling time   146
modeling signal   123
modeling single index   147
modeling SISO   141
modeling state space   64
modeling steady state   144
modeling step input   124
modeling step response   143
modeling zero-pole-gain   64
multi input modeling   68
multi input multi output   33
multi output impulse response   100
multi output modeling   67
multi output samples   96
multi output single input   34
multi output step response   102
multiple frequency response   162
multiple root locus   192
multiplexing signal   142

N
natural frequency   32, 105, 164
negative feedback   39
negative feedback modeling   72
Nichol's chart   169
nonperiodic input modeling   138
normalization of frequency response   159
normalization of spectrum   159
Nyquist plot   168

O
optimum damping   111
optimum index   111

output from state space   95
output from transfer function   95
output in time   87
output sample   94

P
P.O.   105
P.O. modeling   145
parabolic signal   89
parallel interconnection   48
parallel system   38
parallel system modeling   71
parallel system pole   163
peak response   105
peak time   105
peak time modeling   145
peak value modeling   144
percent overshoot   105
percent overshoot modeling   145
performance indices   109
performance modeling   144
performance of control system   103
performance on first order   103
performance on second order   105
periodic input   90
periodic signal   90
phase margin   166
phase spectrum   157
plot Nichol's   169
plot Nyquist   168
plot pole   189
plotting bode   160
pole   162
pole on gain range   190
pole plot   189
pole-zero map   162
pole-zero-gain   28
pole-zero-gain MIMO   68
pole-zero-gain SISO   65
pole-zero-gain to polynomial   41
pole-zero-gain to state space   43
polynomial characteristic equation   184
polynomial to pole-zero-gain   40
positive feedback   40
positive feedback modeling   72
prototype second order   105
pulse generator modeling   133

R
ramp input   89

ramp modeling 127
ramp signal 88
random order system 32
real spectrum 157
rectangular wave modeling 133
response in frequency domain 156
response of frequency range 157
response of second order 105
rise time 105
rise time modeling 145
root locus 183
root locus of multiple systems 192
root locus segment 192
roots of C.E. 185
Routh table 186

S

s plane 165
sample code 89
sample of constant 88
sample of input 88
sample of MIMO output 96
samples of impulse response 101
samples of multi output 96
samples of step response 101
second order performance 105
second order response 105
second order system 32,105
segment of root locus 192
series interconnection 47
series system 37
series system modeling 70
series system pole 163
settling time 105
settling time modeling 146
signal code 89
signal combining 142
signal constant 89
signal damped sine 89
signal exponential 89
signal modeling 123
signal multiplexing 142
signal parabolic 89
signal periodic 90
signal ramp 88
signal rectified sine wave 90,91
signal sinc 89
signal sine 89
signal square wave 91
signal step 88
signal triangular wave 93
simulate input-output 94

sinc signal 89
sine signal 89
sine wave 90
sine wave modeling 129
single index modeling 147
single input multi output 33
single input single output 28
SISO 28
SISO interconnection 45
SISO modeling 63,141
spectrum graphing 159
spectrum imaginary 157
spectrum magnitude 157
spectrum normalization 159
spectrum phase 157
spectrum real 157
square wave 91
square wave modeling 134
stability 183
state space form 28
state space MIMO 67
state space model 65
state space SISO 65
state space to pole-zero-gain 43
state space to transfer function 42
steady state 104
steady state modeling 144
steady state value 104
step input modeling 124
step response 100
step response for multi output 102
step response modeling 143
step response of second order 105
step signal 88
stopband 172
system 28
system band reject 172
system bandstop 172
system bandwidth 171
system bode plot 160
system conversion 40
system DC gain 171
system frequency 164
system frequency response 156
system gain margin 166
system highpass 172
system in frequency domain 155
system lowpass 172
system modeling 63
system performance 105
system phase margin 166
system pole-zero 162

system root locus   183
system simulation   95
system spectrum   157
system spectrum graph   159
system stopband   172

T

table Routh   186
time constant   104
time domain input-output   87
time domain response   87
transfer function   28
transfer function MIMO   69
transfer function modeling   64
transfer function of a model   76
transfer function SISO   64
transfer function to state space   42
triangular wave   93
triangular wave modeling   135
triggered wave modeling   138
two frequency wave modeling   131

U

user-defined root locus   188

Z

zero   163
zero-pole-gain modeling   65

Mohammad Nuruzzaman

www.ingramcontent.com/pod-product-compliance
Lightning Source LLC
Chambersburg PA
CBHW031823170526
45157CB00001B/160